5th Asia Conference on Mechanical and Materials Engineering (ACMME 2017)

IOP Conference Series: Materials Science and Engineering
Volume 241

Tokyo, Japan
9 - 11 June 2017

Editor:

Omar S. Es-Said

ISBN: 978-1-5108-5198-6
ISSN: 1757-8981

Printed from e-media with permission by:

Curran Associates, Inc.
57 Morehouse Lane
Red Hook, NY 12571

Some format issues inherent in the e-media version may also appear in this print version.

Printed by Curran Associates, Inc. (2017)

For permission requests, please contact the Institute of Physics
at the address below.

Institute of Physics
Dirac House, Temple Back
Bristol BS1 6BE UK

Phone: 44 1 17 929 7481
Fax: 44 1 17 920 0979

techtracking@iop.org

Additional copies of this publication are available from:

Curran Associates, Inc.
57 Morehouse Lane
Red Hook, NY 12571 USA
Phone: 845-758-0400
Fax: 845-758-2633
Email: curran@proceedings.com
Web: www.proceedings.com

Table of contents

Volume 241

5th Asia Conference on Mechanical and Materials Engineering (ACMME 2017)
9–11 June 2017, Tokyo, Japan

Accepted papers received: 3 October 2017
Published online: 1 November 2017

Preface

Material performance analysis and evaluation

Mechanics and calculation of materials

Mechanical design and automation

Modern manufacturing system and Mechatronics

Thermal theory and engineering application

Energy Engineering

IOP Conf. Series: Materials Science and Engineering **241** (2017) 011001 doi:10.1088/1757-899X/241/1/011001

PREFACE

It is our great pleasure to introduce you the proceedings of 2017 5th Asia Conference on Mechanical and Materials Engineering (ACMME 2017) held in Tokyo, Japan from June 9-11, 2017. ACMME 2017 is dedicated to issues related to mechanical and materials engineering. One of the objectives of the conference is to establish platforms for collaborative research projects in this field, and to find potential opportunities for international cooperation.

The conference program included keynote, oral, and poster presentations from scholars working in the areas of materials science and engineering. It covered recent trends and progress made in the field of mechanical and materials engineering. Professors from USA, Malaysia and Taiwan were invited to deliver keynote speeches regarding the latest information in their respective areas of expertise.

These proceedings present a selection from papers submitted to the conference by universities, research institutes, and industries. All the papers were subject to peer-review by conference committee members and international reviewers. The papers were selected based on their quality and their relevance to the conference. The volume presents recent advances in the field of mechanical and materials engineering as well as various related areas, including Materials Science, Biomaterials, Manufacturing Processes, and Mechanical Engineering, among others.

We would like to express our gratitude to all the members of the conference committee. We would also like to thank the reviewers, who spared their valuable time, for their advice. It has certainly helped improve the quality, accuracy, and relevance of each paper selected for the conference program and for publication. We also wish to thank all the authors who have contributed to this conference, as well as the organizing committee, reviewers, speakers, chairpersons, sponsors, and all the conference participants for their support for ACMME 2017.

Prof. Omar S. Es-Said, Loyola Marymount University, USA
September 5, 2017

IOP Conf. Series: Materials Science and Engineering **241** (2017) 011001 doi:10.1088/1757-899X/241/1/011001

STATEMENT OF PEER REVIEW

All papers published in this volume of *IOP Conference Series: Materials Science and Engineering* have been peer reviewed through processes administered by the proceedings Editors. Reviews were conducted by expert referees to the professional and scientific standards expected of a proceedings journal published by IOP Publishing.

IOP Conf. Series: Materials Science and Engineering **241** (2017) 011002 doi:10.1088/1757-899X/241/1/011002

CONFERENCE COMMITTEE CHAIRS

Prof. Omar S. Es-Said, Loyola Marymount University, USA
Prof. Dr. Mohd Hamdi Bin Abd Shukor, University of Malaya, Malaysia
Prof. C. A. Huang, Chang Gung University, Taiwan

PROGRAM COMMITTEE CHAIRS

Prof. Hsin-Chih Lin, National Taiwan University, Taiwan
Prof. Jiyoung Kim, The University of Texas at Dallas, USA
Prof. Osman Adiguzel, Firat University, Department of Physics, Turkey

INTERNATIONAL TECHNICAL COMMITTEES

Prof. Meng-Kao Yeh, National Tsing Hua University, Taiwan
Assoc. Prof. Kazunori Asano, Kindai University, Japan
Assoc. Prof. Zheng Lu, Tongji Universiy, China
Assoc. Prof. Jit Kai Chin, University of Nottingham Malaysia Campus, Malaysia
Assoc. Prof. Debdulal Das, Indian Institute of Engineering Science and Technology, India
Assoc. Prof. Nouredine Fenineche,UTBM University, France
Assoc. Prof. ŚLUSARCZYK Beata, Czestochowa University of Technology, Poland
Assoc. Prof. KOT Sebastian, Czestochowa University of Technology, Poland
Prof. Achutha Kini,Manipal university, India
Chong Fai Kait, Universiti Teknologi PETRONAS, Malaysia
Saiful Amri Mazlan, Universiti Teknologi Malaysia
Prof. Jagannath Korody, Manipal university, India
Prof. Sathyashankara Sharama, Manipal university, India
Prof. Uma Maheshwera Reddy Paturi, CVR College of Engineering, India
Dr. Ernie Suzana Ali, Universiti Sains Islam Malaysia
Dr. Suman Chatterjee, National Institute of Technology Rourkela, Odisha, India
Dr. Peck Loo Kiew, UCSI University, Malaysia
Dr. Jully Tan, UCSI University, Malaysia
Dr. Hendriko Hendriko, Politeknik Caltex Riau, Indonesia
Dr. Deepa Prabhu,Manipal Institute of Technology, India
Dr. Karthik Silaipillayarputhur,King Faisal University, Saudi Arabia
Dr. Mark Ovinis,Universiti Teknologi Petronas, Malaysia
Dr. Olivier Lavigne, The University of Adelaide, Australia
Gobinda Gopal Khan,Tripura University (A Central University), India
Chumnan Boonyaputthipong, Khon Kaen University, Thailand
Prof. Ita B.I., UNIVERSITY OF CALABAR, Nigeria
Dr. Nanik Indayaningsih, Indonesian Institute of Science-Research Center for Physics,Indonesia
Prof. Noorhana Yahya,Universiti Teknologi PETRONAS, Malaysia

IOP Conf. Series: Materials Science and Engineering **241** (2017) 011003 doi:10.1088/1757-899X/241/1/011003

Peer review statement

All papers published in this volume of *IOP Conference Series: Materials Science and Engineering* have been peer reviewed through processes administered by the proceedings Editors. Reviews were conducted by expert referees to the professional and scientific standards expected of a proceedings journal published by IOP Publishing.

Tunable Properties of Exfoliated Polyvinylalcohol Nanocomposites by In Situ Coprecipitation of Layered Double Hydroxides

Jiajia Liu [1,a], **Richard K.K. Yuen** [2,b] **and Yuan Hu** [1,c]

[1]State Key Laboratory of Fire Science, University of Science and Technology of China, 96 Jinzhai Road, Hefei, Anhui 230026, P.R. China.
[2]Department of Architecture and Civil Engineering, City University of Hong Kong, Tat Chee Avenue Kowloon, Hong Kong.

E-mail: niull@mail.ustc.edu.cn, brichard.yuen@cityu.edu.hk, yuanhu@ustc.edu.cn

Abstract. Poly(vinyl alcohol) (PVA) nanocomposites were prepared by a "one step" method based on the coprecipitation of layered double hydroxide (LDH) nanosheets in the polymer aqueous solution. The morphology, fire resistance properties, mechanical and optical properties of the PVA/LDH nanocomposites were studied. The LDH nanosheets were homogeneously dispersed in the PVA matrix as indicated by X-ray diffraction (XRD) pattern and transmission electron microscopy (TEM) characterization. Meanwhile, the peak of heat release rate (pHRR) and total heat release (THR) were decreased by 58% and 28%, respectively. Storage modulus at 30 oC was increased, and the transmittance of more than 90% at the visible region was obtained upon addition of 5 wt% LDH.

1. Introduction

Poly(vinyl alcohol) PVA is a water soluble polymer with biocompatibility and biodegradability that allows penetration into various fields such as proton-exchange membranes and polymer electrolyte fuel cells [1], separation apparatus [2], permeability membranes [3], drug delivery [4], and membrane manufacturing [5]. However, PVA is highly flammable. Thus, the investigations on halogen-free flame retardant PVA have been one of the most substantial research topics for the PVA modification.

Recently, nanocomposite technology has brought a revolutionary advancement in the area of fire retardation. In the past few decades, the addition of nanofillers such as montmorillonite, layered double hydroxides (LDH), carbon nanotube, graphene/graphite oxide, etc, exhibits favorable effect on improving the flame retardancy, thermal stability and mechanical properties of polymer materials simultaneously at low loading [6, 7]. As the important layered inorganic materials, LDHs are widely applied in catalysis, optical, medical and flame retardant fields [8, 9] Many kinds of nanocomposites based on LDHs and polymeric materials could be prepared with improved properties. As is well-known, the dispersion of nanoadditives in a polymer matrix is a key factor for property enhancements [10]. However, LDH usually stacks together to form layered structure with thickness of tens nanometers due to the strong interlayer electrostatic interactions. Pristine LDH is not suitable for the penetration of giant polymer chains or chain segments into their gallery space. Based on this idea, we prepared exfoliated polymer nanocomposites by a "one step" method based on the coprecipitation of

IOP Conf. Series: Materials Science and Engineering **241** (2017) 012001 doi:10.1088/1757-899X/241/1/012001

LDH nanosheets in the PVA aqueous solution. The PVA solution acts as a dispersant medium that prevents the stacking of the generated sheets, which leads to the exfoliated morphology.
In this work, we reported the synthesis of PVA/LDH nanocomposites using a ''one step'' method, which led to a well-dispersed nanosheets with an exfoliated structure. The incorporation of LDH nanosheets not only enhanced the modulus and flame-retardant properties, but also maintained the optical transparency of PVA.

2. Experimental

9.0 g PVA was dissolved in a 100 mL solution of distilled water under stirring 80 ^{o}C. When the solution turned transparent, the solution was left to cool down at room temperature. A solution of $Zn(NO_3)_2 \cdot 6H_2O$ (0.1672 g) and $Al(NO_3)_3 \cdot 9H_2O$ (0.1052 g) was added dropwise with vigorous stirring under nitrogen. The colloid solution was left for 3h with constant stirring until using the addition of NaOH (0.5 M) to maintain a constant pH of 9. At last the colloid solution was cast onto glass plates and dried at at room temperature for 1 day and further heated at 60 ^{o}C for 3 h to remove residual water. The nanocomposite films were washed by Soxhlet extraction in methanol for 24 h to remove $NaNO_3$ and residual unreacted salts. The total amount of metallic salts was chosen to yield PVA/LDH nanocomposites with 1 wt% filler loading if complete coprecipitation occurred. PVA nanocomposites with different LDH concentrations (3.0 and 5.0 wt%) were also prepared. Table 1 lists the formulations of the PVA/ZnAl-LDH nanocomposites.

3. Results and discussion

The formation of a layered ZnAl-LDH structure in the PVA solution by coprecipitation was analyzed by TEM. Fig. 1e displays the TEM image of the PVA3 obtained after the in situ coprecipitation of ZnAl-LDH in an aqueous solution of PVA. The TEM image reveals that the presence of single layers of inorganic materials in the PVA matrix, which correspond to exfoliated and disorderly LDH nanocomposites. XRD analyses were required to complement the data from the TEM experiments and to further confirm that the platelets were formed by the "one-step" nanocomposite preparation method corresponding to the LDH exfoliated sheets. The diffraction peak of pure PVA (Fig. 1a) is located at $2\theta = 19.5^{o}$. The XRD patterns of the PVA nanocomposites (Figs. 1b and c) after coprecipitation only show the diffraction peak from PVA and no noticeable reflections related to the stacked ZnAl-LDH, which indicate that the disordered or exfoliated layers structure in the PVA matrix. In XRD pattern of PVA3 with higher inorganic content, there are two new weak diffraction peaks corresponding to the (003) and (006) reflection of the ZnAl-LDH phase (JCPSD 22-0700). This peaks may be due to the formation of intercalated/exfoliated nanocomposite, which indicates that the ZnAl-LDH has been partially intercalated and exfoliated in the PVA matrix [11]. In the case of hydrothermal treatment, the (00l) reflections are characterized by high intensities combined with narrow peak widths, indicating that the LDH phase presents a relatively high crystallinity.

Table 1. Formulation and microcombustion calorimeter (MCC) data of PVA and PVA/LDH Nanocomposites

Sample	Component	PHRR (W/g)	THR (kJ/g)
PVA0	PVA	431	28.6
PVA1	PVA/1 wt% ZnAl-LDH	355	24.9
PVA2	PVA/3 wt% ZnAl-LDH	258	23.2
PVA3	PVA/5 wt% ZnAl-LDH	181	20.6

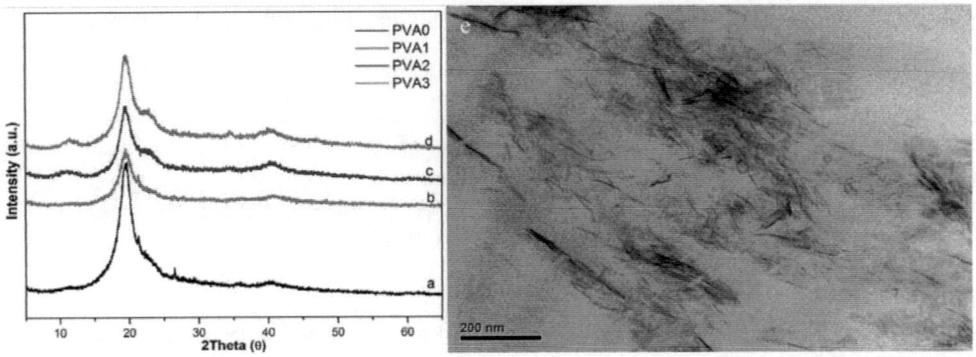

Figure 1. (a) - (d) XRD patterns of PVA and PVA nanocomposites; (e) TEM image of PVA2 ultrathin section.

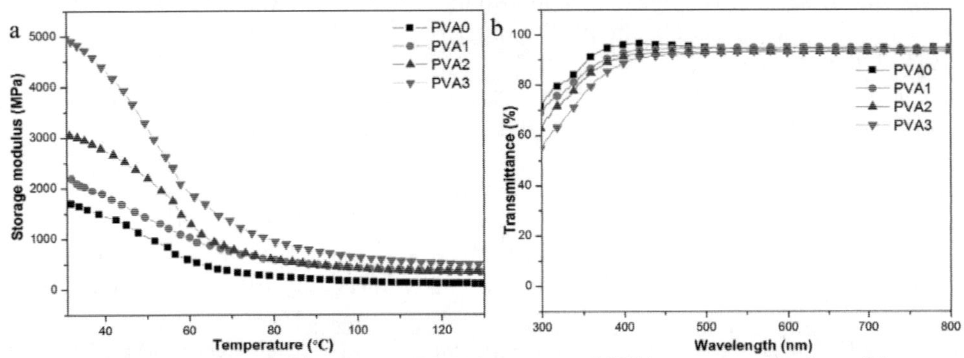

Figure 2. (a) Dynamic mechanical analysis (DMA) curves of PVA nanocomposites; (b) transmittance properties of PVA nanocomposites in the wavelength range of 300-900 nm.

DMA was carried out to evaluate the mechanical properties. The storage modulus and loss tangent curves are shown in Fig. 2a. The storage modulus is a measure of the stiff ness, and it increased in all of the PVA/LDH nanocomposites. PVA3 had the largest increase compared with PVA0 at 30 °C. The strong hydrogen bonds between LDH nanosheets and PVA matrix and the homogeneous dispersion of the LDH nanosheets explain the observed increase.

MCC is an effective bench scale methods for investigating the combustion properties of polymer materials. Several parameters can be obtained from MCC, such as heat release rate (HRR) and total heat release (THR), the corresponding combustion data are presented in Table 1. Compared to the neat PVA, the PVA nanocomposites acquire significant improvements in flame resistance. With the increase of LDH, PHRR decreased from 431 to 181 w/g. Meanwhile, THR decreased from 28.6 to 20.6 kJ/g, which indicated that LDH is an effective material to improve the flame retardancy.

As mentioned above, PVA is widely used in packing and coating materials, which require high transparence in practical applications. Thus, a transparence measurement based on UV-vis transmittance analysis was performed. Fig. 2b shows UV visible transmission spectra of all nanocomposites produced. It can be seen that the optical transparency of PVA/LDH nanocomposites films are very high, even for the case of 5 wt% LDH content. The average transmittance of the PVA/LDH composite films is about 90%, which is almost equal to the neat PVA film. This can be attributed to the UV absorbance of the nanoscale dimensions of the nanosheets in the PVA matrix, which has been reported in many previous literatures. The good compatibility and strong interfacial adhesion between the well dispersed LDH and PVA matrix also play a role in the enhancement of optical transparency.

4. Summary

In this work, LDH nanosheets were incorporated into PVA by a direct coprecipitation method of LDH nanoplatelets within the polymer solution. The LDH nanosheets were homogeneously dispersed in the PVA matrix. By incorporating the LDH nanosheets, mechanical properties and fire resistance properties were significantly enhanced. Meanwhile, the nanocomposites are highly transparent in the visible region.

Acknowledgments
The work was financially supported by the National Basic Research Program of China (973 Program) (2012CB922002), the national natural science foundation of china (No. 21374111), Fundamental Research Funds for the Central Universities (WK2320000032) and the Research Grants Council of the Hong Kong Special Administrative Region (Project No. CityU 122612).

References
[1] M. Helen, B. Viswanathan, S.S. Murthy. Poly (vinyl alcohol)–polyacrylamide blends with cesium salts of heteropolyacid as a polymer electrolyte for direct methanol fuel cell applications. *Journal of applied polymer science*, 116:3437-3447, 2010.
[2] B.V.K. Naidu, M. Sairam, K. Raju. Pervaporation separation of water+ isopropanol mixtures using novel nanocomposite membranes of poly (vinyl alcohol) and polyaniline. *Journal of membrane science*, 260:142-155, 2005.
[3] R. Saito, D. Nakagawa. Synthesis of water vapor barrier membrane by coating of poly (vinyl alcohol) substrate with poly (tert-butyl acrylate)-silica nanocomposite. *Polymers for Advanced Technologies*, 117:285-290, 2009.
[4] A.G. Sullad, L.S. Manjeshwar, T.M. Aminabhavi. Polymeric blend microspheres for controlled release of theophylline. *Journal of applied polymer science*, 117:1361-1370, 2010.
[5] D. Banerjee, A. Jha, K.K. Chattopadhyay. Synthesis and characterization of water soluble functionalized amorphous carbon nanotube-poly (vinyl alcohol) composite. *Macromolecular Research*, 20:1-8, 2012.
[6] P. Kiliaris, C.D. Papaspyrides. Polymer/layered silicate (clay) nanocomposites: an overview of flame retardancy. *Progress in Polymer Science*, 35:902-958, 2010.
[7] J.R. Potts, D.R. Dreyer, C.W. Bielawski. Graphene-based polymer nanocomposites. *Polymer*, 52:5-25, 2011.
[8] L.Z. Qiu, W. Chen, B.J. Qu. Structural characterisation and thermal properties of exfoliated polystyrene/ZnAl layered double hydroxide nanocomposites prepared via solution intercalation. *Polymer Degradation and Stability*, 87:433-440, 2005.
[9] C.M. Becker, A.D. Gabbardo, F. Wypych, S.C. Amic. Mechanical and flame-retardant properties of epoxy/Mg–Al LDH composites. *Composites Part A*, 42:196-202, 2011.
[10] D.Y. Cai, S. Mo. Recent advance in functionalized graphene/polymer nanocomposites. *Journal of Materials Chemistry*, 20:7906-7915, 2010.
[11] W. Chen, B.J. Qu. Structural characteristics and thermal properties of PE-g-MA/MgAl-LDH exfoliation nanocomposites synthesized by solution intercalation. *Chemistry of Materials*, 15: 3208-3213, 2003.

IOP Conf. Series: Materials Science and Engineering **241** (2017) 012002 doi:10.1088/1757-899X/241/1/012002

Parametric Instability of Static Shafts-Disk System Using Finite Element Method

Wahab A.M., Rasid Z.A. and Abu A

Intelligent Dynamic System i-Khoza, Malaysia-Japan International Institute of Technology (MJIIT), Universiti Teknologi Malaysia Kuala Lumpur, Jalan Sultan Yahya Petra 54100 Kuala Lumpur, Malaysia

abdulmalekabdulwahab@gmail.com

Abstract. Parametric instability condition is an important consideration in design process as it can cause failure in machine elements. In this study, parametric instability behaviour was studied for a simple shaft and disk system that was subjected to axial load under pinned-pinned boundary condition. The shaft was modelled based on the Nelson's beam model, which considered translational and rotary inertias, transverse shear deformation and torsional effect. The Floquet's method was used to estimate the solution for Mathieu equation. Finite element codes were developed using MATLAB to establish the instability chart. The effect of additional disk mass on the stability chart was investigated for pinned-pinned boundary conditions. Numerical results and illustrative examples are given. It is found that the additional disk mass decreases the instability region during static condition. The location of the disk as well has significant effect on the instability region of the shaft.

1. Introduction

Work by Bolotin [1] has well presented the general theory of parametric instability and nowadays the subject of parametric instability is going through a period of intensive development. Accurate predictions of parametric instability behaviour become important in design process of high-speed turbo machines, drive train system and any types of rotating machinery as failure to do so may result in fatigue failure of rotating components earlier than expected. In engineering applications, machines may be subjected to periodically changes of stiffness, which can become the source of the parametric excitation.

In order to model and simplified the structure of the rotating machine, beam theory has been used with different geometries and boundary conditions according to their applications. This model can accurately estimate the characteristic and behaviour of the machine. Thus, Chen and Ku [2] has modelled a shaft using Timoshenko beam theory to investigate the effect of gyroscopic towards parametric instability. It was found that rotary inertia and transverse shear deformation contribute to dynamic instability of rotating shaft. Chen and Peng [3] suggested that unstable condition was more at higher mode compare to lower mode. Yim and Yim [4] stated that for shaft and disk system, the size of instability region depended on location of disk. Pei [5] explained that the increase of instability region was due to the width of disk centre orbit of the system growth as axial force fluctuating.

Numerical and analytical methods have been used to estimate the instability condition that could occur in shaft system due to parametric excitation. Finite element method (FEM) is one of numerical method that has been used to estimate the instability region [2], [3], [6], [7]. However, only few studies have

IOP Conf. Series: Materials Science and Engineering **241** (2017) 012002 doi:10.1088/1757-899X/241/1/012002

been conducted to show the effect of disk location towards dynamic instability of the shaft system in torsional motion condition.
Thus, the main idea of current work is to estimate and compare the instability region of shaft-disk system under axial load for different locations of disk. In order to model the shaft in torsional motion, Nelson's beam theory has been used. The FEM model is used to establish the Mathieu's equation. Floquet's theory has been applied to estimate the dynamic instability region. Then, the effect of disk location towards dynamic instability region can be investigated.

2. Material and Method

The parametric instability of shaft using FEM is developed based on theory that stiffness changes as load is applied onto the system. Fig. 1 shows a uniform shaft-disk of length L that is exposed to an axial load P(t).

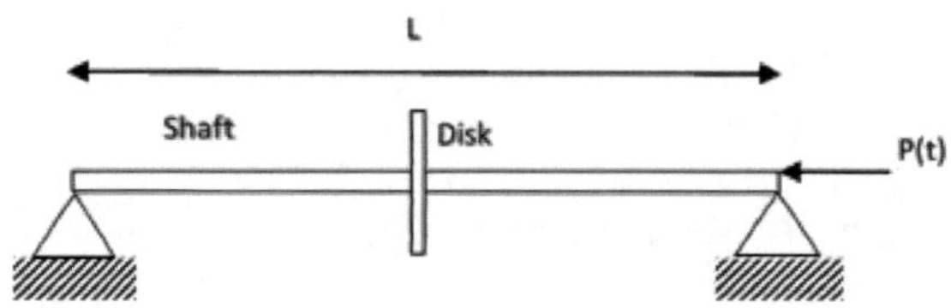

Fig.1. Shaft-disk system subjected to axial load.

Table 1. Geometric of shaft-disk system.

Shaft	Disk	Steel Properties
Length = 300 mm	Diameter, d = 100	Young Modulus, E = 20x10^{10}
Diameter, d = 10 mm	Thickness, t = 3 mm	Density, ρ= 7833 kg/m^3

In this study, the FEM model of shaft is based on the Nelson's model [8] that consists of five degrees of freedom (DOF) per node i.e. two translations, two rotations about y-axis and z-axis and another torsional rotation about x-axis. The kinetic energy for system in Fig. 1 under bending and torsional mode is shown as[10][9];

$$T = \frac{1}{2}\int_{-a}^{+a} \rho A\left(\left(\frac{\partial v}{\partial t}\right)^2 + \left(\frac{\partial w}{\partial t}\right)^2\right) d\hat{x} + \frac{1}{2}\int_{-a}^{+a} \rho I_D\left(\left(\frac{\partial \theta_y}{\partial t}\right)^2 + \left(\frac{\partial \theta_z}{\partial t}\right)^2\right) d\hat{x} - \frac{1}{2}\int_{-a}^{+a} I_P\left(\Omega + \frac{\partial \theta_x}{\partial t}\right)\left(\theta_y \frac{\partial \theta_z}{\partial t} - \theta_z \frac{\partial \theta_y}{\partial t}\right) d\hat{x} + \int_{-a}^{+a} I_p\left(\Omega + \frac{\partial \theta_x^2}{\partial t}\right) d\hat{x} \tag{1}$$

I_D and I_P are the diametric and polar mass moments of inertia per unit length of the shaft element respectively. The potential energy including the shear deformation, bending moment, torsional deflection and axial compression load are represented as;

$$U = \frac{1}{2}\int_{-a}^{+a} EI(\theta_z'^2 + \theta_y'^2) d\hat{x} + \frac{1}{2}\int_{-a}^{+a} k'GA[(v' - \theta_z)^2 + (w' + \theta_y)^2] d\hat{x} + \frac{1}{2}\int_{-a}^{+a} GI_P \theta_x'^2 d\hat{x} - \frac{1}{2}\int_{-a}^{+a} P[v'^2 + w'^2] d\hat{x} \tag{2}$$

where symbol ' denotes the differentiation with respect to axial length, k' is the shear correction factor, A is the cross-sectional area, G is the shear modulus, E is the elastic modulus and I is the cross sectional second moment of inertia. Eq.1 and Eq.2 were transformed into FEM equation. Then Lagrangian theory was applied to establish the equation of motion in FEM that includes the buckling load P;

$$[M]\{\ddot{q}\} - \Omega[G]\{\dot{q}\} + ([K] - P(t)[K_g])\{q\} = 0 \tag{3}$$

The periodic axial compression load is expressed as fundamental static buckling load P*;

$$P(t) = \alpha P^* + \beta P^* \cos\phi t \tag{4}$$

where ϕ, α and β are the axial compress frequency, static and dynamic load factors respectively. Mathieu type is established by inserting Eq. 4 into Eq. 3.

$$[M]\{\ddot{q}\} - \Omega[G]\{\dot{q}\} + ([K] - \alpha P^*[K_g] - \beta P^* \cos\phi t[K_g])\{q\} = 0 \tag{5}$$

Eq. 6 is used to estimate the solution of Eq. 5 which lead to non-trivial solutions that show by linear homogenous equations as Eq. 7.

$$\{q\} = \{a\}\sin(\phi t / 2) + \{b\}\cos(\phi t / 2) \tag{6}$$

$$\begin{vmatrix} [K] - \left(\alpha + \frac{\beta}{2}\right)P^*[K_g] - \frac{\phi^2}{4}[M] & \Omega\frac{\phi}{2}[G] \\ -\Omega\frac{\phi}{2}[G] & [K] - \left(\alpha - \frac{\beta}{2}\right)P^*[K_g] - \frac{\phi^2}{4}[M] \end{vmatrix} = 0 \tag{7}$$

In order to estimate the stability boundaries, the eigenvalue for the determinant equation in Eq. 7 is obtained. Reference [11] explained the step for solving the eigenvalue for quadratic matric equation.

3. Results and Discussion

In order to validate current FEM work, the first four natural frequency of the disk-shaft system were compared with Hili et.al [12] as shown in Table. 2. The comparison was for simply supported steel shaft-disk system in static condition as Fig. 1. Geometry and material properties of the system are shown in Table. 1. MATLAB coding was established to obtain the numerical result of the model. Fig. 2 shows relative error between current work and reference [12]. For mode 1, mode 2 and mode 4 the relative error is below 10%, which is reasonable for the current model. However, at mode 3, the relative error is around 22 % that is still in reasonable condition. The models start to converge at 30th elements as number of elements increased. Therefore, the result at 30 elements has been used for FEM analysis in this paper.

Table 2. Comparison of natural frequency (rad/s) between reference [12] and current work (at 30 number of elements)

Mode	Current work	Reference [12]
1	951.15	948.76
2	3728.46	3845.30
3	5731.10	7376.45
4	12428.60	13552

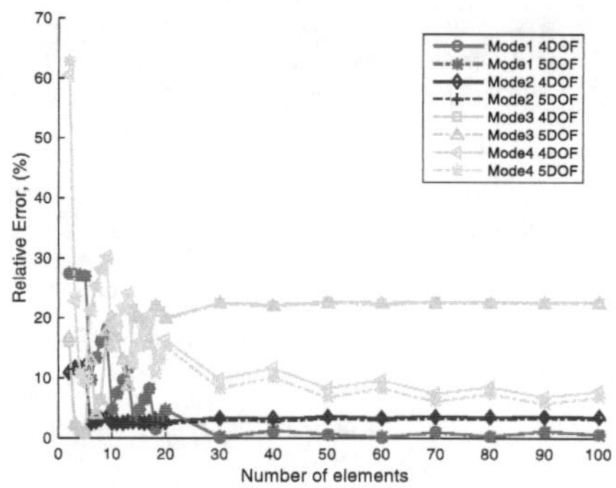

Fig. 2. Convergence of relative error with reference [12]

On the other hand, there are slightly different readings of relative error between Timoshenko beam represented by 4 DOF and Nelson's beam element represented by 5 DOF. It indicated that torsional motion has significant effect on natural frequency of the shaft.

Fig. 3 shows the effect of additional rigid disk towards parametric instability region of shaft system under axial load. The dynamic instability region shifted inwards as additional disk was applied. This condition was expected since in general natural frequency decreased as mass increased and instability region location depends on natural frequency of the system. Chen and Peng [3] stated that parametric instability region occur in between of natural frequency for non-rotating shaft. On the other hand, the instability region slightly narrow especially at higher mode as disk was attached. This showed that shaft-disk system was more stable compared to system with shaft only.

In general, the instability region shifted as location of disk changed. Fig. 3 shows that, at 50% of shaft length, the instability region shifted outwards compared to 75%. Furthermore, the instability region narrowed especially at higher mode as the location of disk is getting nearer to end of the shaft. This showed that shaft system was more stable as disk is located nearer to the end of the system. This results is in line with Yim and Yim [4]. In addition, there are very small different between 4 DOF and 5 DOF model in term of instability region. This condition may due to torsional motion has slightly effect especially on simply supported boundary condition.

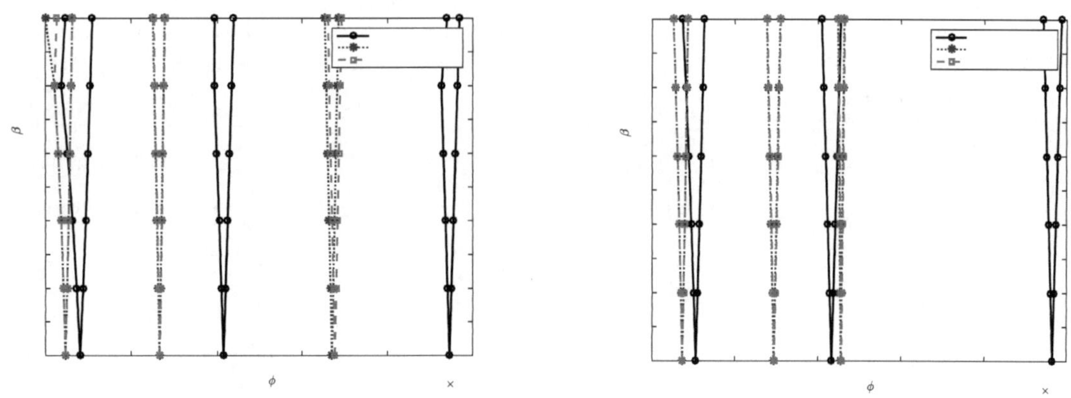

50%. 75%.

Fig.3. Parametric instability region comparison for different disk location from end.

4. Conclusion

The parametric instability region of shaft-disk system with different location of disk has been estimated using FEM and Floquet's theory. This study shows that the location of the disk has significant effect on the parametric instability region of shaft-disk system. The next study will consider the system in rotating condition.

References

[1] V. V. V. Bolotin, "THE DYNAMIC STABILITY OF ELASTIC SYSTEMS. VOLUME 2," DTIC Document, 1962.

[2] L. Chen and D. Ku, "Dynamic stability analysis of a rotating shaft by the finite element method," *J. Sound Vib.*, vol. 143, pp. 143–151, 1990.

[3] L. Chen and W. Peng, "Stability analyses of a Timoshenko shaft with dissimilar lateral moments of inertia," *J. Sound Vib.*, vol. 207, pp. 33–46, 1997.

[4] K. B. Yim and J. T. Yim, "Dynamic stability of a rotor with shear-flexible shaft under axial loads," *J. Mech. Sci. Technol.*, vol. 27, no. 2, pp. 359–366, Feb. 2013.

[5] Y.-C. Pei, H. Lu, and C. Chatwin, "Dynamics of a rotating shaft—disc under a periodic axial force," *Proc. Inst. Mech. Eng. Part K J. Multi-body Dyn.*, vol. 224, no. 2, pp. 211–219, 2010.

[6] M. Sabuncu and K. Evran, "Dynamic stability of a rotating asymmetric cross-section Timoshenko beam subjected to an axial periodic force," *Finite Elem. Anal. Des.*, vol. 41, no. 11–12, pp. 1011–1026, Jun. 2005.

[7] S. Bose, A. Nandi, and S. Neogy, "Flexible spinning and precessing rotor-stability analysis based on different analytical and finite element models," *J. Vib. Control*, vol. 21, no. 12, pp. 2394–2406, Sep. 2015.

[8] H. D. Nelson, "A Finite Rotating Shaft Element Using Timoshenko Beam Theory," *J. Mech. Des.*, vol. 102, no. 4, p. 793, 1980.

[9] J. S. Rao, "Finite Element Methods for Rotor Dynamics," in *History of Rotating Machinery Dynamics SE - 16*, vol. 20, Springer Netherlands, 2011, pp. 269–297.

[10] M. Petyt, *Introduction to Finite Element Vibration Analysis*. Cambridge University Press, 2010.

[11] W. M. Miranda and M. T. C. Faria, "Finite Element Method Applied to the Eigenvalue Analysis of Flexible Rotors Supported by Journal Bearings," *Engineering*, vol. March, pp. 127–137, 2014.

[12] M. A. Hili, T. Fakhfakh, and M. Haddar, "Vibration analysis of a rotating flexible shaft–disk system," *J. Eng. Math.*, vol. 57, no. 4, pp. 351–363, Mar. 2007.

IOP Conf. Series: Materials Science and Engineering **241** (2017) 012003 doi:10.1088/1757-899X/241/1/012003

A comparative study on stress and compliance based structural topology optimization

Hailu Shimels G[1], Dereje Engida W*[2] and Fakhruldin Mohd H[3]

[1, 2, 3] Mechanical Engineering Department, Universiti Teknologi Petronas, Bandar Seri Iskandar, Tronoh, Perak, 32610, Malaysia

*dereje.woldemichael@utp.edu.my

Keywords: Topology optimization, Compliance, Stress constraints

Abstract. Most of structural topology optimization problems have been formulated and solved to either minimize compliance or weight of a structure under volume or stress constraints, respectively. Even if, a lot of researches are conducted on these two formulation techniques separately, there is no clear comparative study between the two approaches. This paper intends to compare these formulation techniques, so that an end user or designer can choose the best one based on the problems they have. Benchmark problems under the same boundary and loading conditions are defined, solved and results are compared based on these formulations. Simulation results shows that the two formulation techniques are dependent on the type of loading and boundary conditions defined. Maximum stress induced in the design domain is higher when the design domains are formulated using compliance based formulations. Optimal layouts from compliance minimization formulation has complex layout than stress based ones which may lead the manufacturing of the optimal layouts to be challenging. Optimal layouts from compliance based formulations are dependent on the material to be distributed. On the other hand, optimal layouts from stress based formulation are dependent on the type of material used to define the design domain. High computational time for stress based topology optimization is still a challenge because of the definition of stress constraints at element level. Results also shows that adjustment of convergence criterions can be an alternative solution to minimize the maximum stress developed in optimal layouts. Therefore, a designer or end user should choose a method of formulation based on the design domain defined and boundary conditions considered.

1. Introduction

Structural optimization, which deals about assemblage of material to sustain loads in the best way, is one type of engineering optimization tools. It has been used in different areas of applications[1-3]. Problems of structural optimization have been modeled and solved using size, shape or topology optimization. In size and shape optimization, some predefined size and shape of solid elements are included in addition to the loading and boundary conditions before the optimization starts[4, 5]. Unlike size and shape optimization, in topology optimization there is no predefined size and shape of elements except the definition of boundary and loading conditions, which makes optimal layout to be independent of the knowledge and experience of the end user. Topology optimization is a mathematical approach which seeks optimal material layout within a given design domain for a given set of boundary and loading conditions. It includes determination of connectivity, geometries of cavities and location of voids in the design domain[6].

Problems in structural topology optimization have been, formulated and solved to minimize compliance of the structure under volume constraint[7, 8]. This type of formulation, which is the most common way of formulating topology optimization problems, has some drawbacks such as variation of results with the amount of material distributed and unable to consider stress and displacement[6, 9]. These limitations may prevent optimal layouts to be applicable in the real-world scenario unless some post processing action such as sizing and shape optimization, is involved to accommodate stress concentrations and major deflections. In this type of formulation, the optimal material distribution is highly affected by the amount of materials to be distributed, which will let the optimal layout be different for different values of the constraint.

For those designs areas and applications where stress is main design variable, it will be difficult to directly take optimal layouts from classical compliance minimization formulations. Researches have been conducting to include stress constraints in topology optimization process[10, 11]. In this type of formulation and solution of topology optimization problems, which considers stress constraints, the optimal lay out always favors the failure criterion other than the amount of material to be distributed. This will make the optimal material layout to be independent of amount of material distributed, safe and secured from failures associated with the loading conditions. Though, this approach is accepted and more realistic from engineering point of view[9], especially for those designs where stress is main concern, it has three challenges associated with the design variables namely singularity phenomenon, local nature of stress constraint and nonlinear behavior of stress constraint.

So far, structural topology optimization problems have been formulated based on the need and experience of the designer and end users. This paper aims to show how the formulation technique used can alter the optimal layout of a given design domain and propose a guideline how to choose an appropriate formulation technique. All topology optimization problems defined in Section 2 are formulated using SIMP method and tested using bench mark problems.

2. Problem formulation

In this paper, weight minimization under stress constraints and classical compliance minimization under volume constraint are considered.

2.1. *Weight Minimization Problem.* Eq.1 shows a stress based topology optimization (STO) problem formulated to minimize weight of a given design domain under stress constraints. Stress constraints are defined based von Mises stress failure theory, which states, “A material will fail when the von Mises stress induced in the material exceeds yield strength”.

$$\begin{aligned}
&\underset{x}{Min}\quad V = \sum_{e=1}^{N} (x_e)^P v_e \\
&\text{Subjected to: } g(x_e) = \frac{\sigma^{vm}}{\sigma_{yield}} < 1 \\
&KU = F \\
&0 < x_{\min} \le x_e \le 1
\end{aligned} \qquad (1)$$

Where, V is the volume (objective function), N total number of elements which defines the design domain, e elements within the design domain, v_e is volume of each element in the design domain, σ_{vms} Von mises stresσ_{yield} is maximum (yield stress).K is global stiffness matrix, U is global displacement vector, F global force vector, x_e is relative density/design variable, x_{min} is the minimum relative density to control the singularity phenomenon associated with the design variable.

2.2. *Classical Compliance Minimization.* Eq.2 shows a classical topology optimization problem to minimize compliance of a given structure under volume constraint[12].

$$\begin{aligned}
&\underset{x}{Min}\quad C(x) = U^T KU = \sum_{e=1}^{N} (x_e)^P u_e^T k_o u_e \\
&\text{Subjected to : } \frac{V(x)}{V} volfrac \\
&KU = F \\
&0 < x_{\min} \le x_e \le 1
\end{aligned} \qquad (2)$$

Where, C is compliance, U and K are global displacement vector and global stiffness matrix, respectively, u_e and k_0 elemental displacement vector and elemental stiffness matrix, respectively, p is the penalization factor, $V(x)$ and V material and design domain volume, respectively, $volfrac$ is volume fraction, F global force vector,x_{min} is the minimum value for the design variable to avoid singularity phenomenon associated with

IOP Conf. Series: Materials Science and Engineering **241** (2017) 012003 doi:10.1088/1757-899X/241/1/012003

removal of material from the design domain and x_e is a vector of design variable.

3. Methodology

All the optimization problems formulated are solved using an optimality criteria method, by which updating scheme for design variables can be expressed as shown in Eq.3[13].

$$\begin{aligned}
&\text{if} \quad x_e\beta_e^{\eta} \le \max(x_{\min}, x_e - m) \\
&\qquad x_e^{new} = \max(x_{\min}, x_e - m) \\
&\text{if} \quad \max(x_{\min}, x_e - m) < x_e\beta_e^{\eta} < \min(1, x_e + m) \\
&\qquad x_e^{new} = x_e\beta_e^{\eta} \\
&\text{if} \quad \min(1, x_e + m) < x_e\beta_e^{\eta} \\
&\qquad x_e^{new} = \min(1, x_e + m)
\end{aligned} \tag{3}$$

Where m is a positive limit, which usually takes a value of 0.2, η is a numerical damping coefficient with a value of 0.5[13] , β which will be dependent on the type of problems defined in Eq. 1 and Eq.2 as shown in Eq. 4 and Eq.5,respectively.

$$\beta_e = \frac{\frac{\partial v}{\partial x}}{\lambda \frac{\partial g}{\partial x}} \tag{4}$$

$$\beta_e = \frac{\frac{\partial c}{\partial c}}{\lambda \frac{\partial v}{\partial x}} \tag{5}$$

Where λ is a Lagrangian multiplier, where the value can be found using bisection method. The sensitivity analysis for problems defined can be found as shown in Eq.6 for stress based topology optimization and Eq.7 for classical compliance minimization problem

$$\frac{\frac{\partial v}{\partial x}}{\frac{\partial g}{\partial x}} = \frac{p x^{p-1} v_e}{(p-q)x^{p-q-1} D_0 B_e u_e - x^{p-q} D_0 B_e \frac{\partial k}{\partial x}\frac{u}{k}} \tag{6}$$

$$\frac{\frac{\partial C}{\partial x}}{\frac{\partial V}{\partial x}} = p x^{p-1} U_e^T K_0 U_e \tag{7}$$

4. Numerical Result

To validate formulated problems simply supported beam and cantilever beams under different loading and boundary conditions were taken[13, 14]. All the design domains are discretized by square rectangular finite elements. The material considered for all cases described in this paper has a Young's modulus of $E = 1.0$ MPa, a Poison's ratio of $v = 0.3$ and a von Mises stress of 1.2Mpa subjected to a unit load[15].

The maximum stresses induced for compliance based formulations is calculated for each iteration and compared to the maximum stress of design domains under stress based formulation. Compliance of the design domains under stress based formulation is calculated for each case study and compared with the respective compliance based formulations.

The design domains and loading conditions for different case studies used in this study are defined followed by their respective optimal material distribution layout, stress distribution plots and variation of maximum stress and compliance. To have a full control on the stress measure in each element all the stress constraints are defined at element level. A Matlab simulation code is used for analysis on Dell –i5- 4GB RAM (3.41)-3.2GHz computer.

4.1. Simply Supported Beam**.** The first case corresponds to a simply supported beam, as shown in Fig.1. The design domain is discretized into 120 X 48 elements. Fig.2a and Fig.2b shows optimal material distribution under stress constrained topology optimization problem and classical compliance minimization problem, respectively.

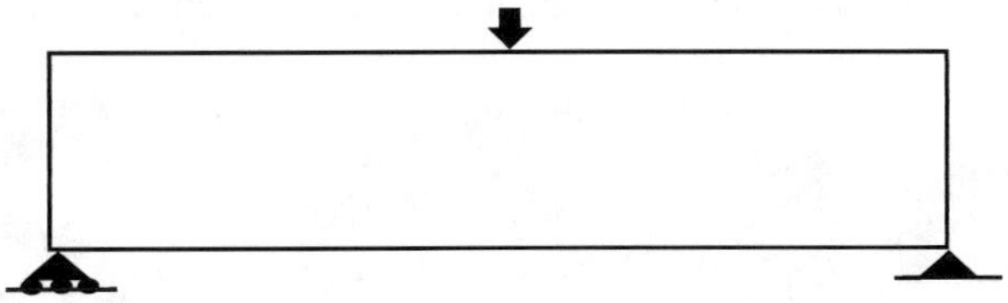

Figure 1 Design domain definition

As it can be seen from the optimal layout plots, the optimal layout from classical compliance minimization problem is more complex than from that of stress based optimization problem. This complex layout will make the manufacturing process to be somewhat difficult and the need for post processing to be higher as shown in Fig.2b.

(a) Material distribution for stress based formualtion

(b) Material distribution for complaince based formualtion

Figure 2 Material distribution for a) stress based formualtion b)Compliance based formualtion

Fig. 3a shows a plot of stress vs number of iterations during the optimization process. From the plot it can be seen that maximum stress induced in classical compliance minimization problem is greater than that of stress based topology optimization problem.

Fig. 3b shows a plot of compliance vs number of iterations during the optimization process. From the plot, the compliance of a classical compliance minimization problem is less than that of stress based topology optimization problem. As the problems gets closer to convergence criterion the difference in the value of the compliance of defined design domain decreases.

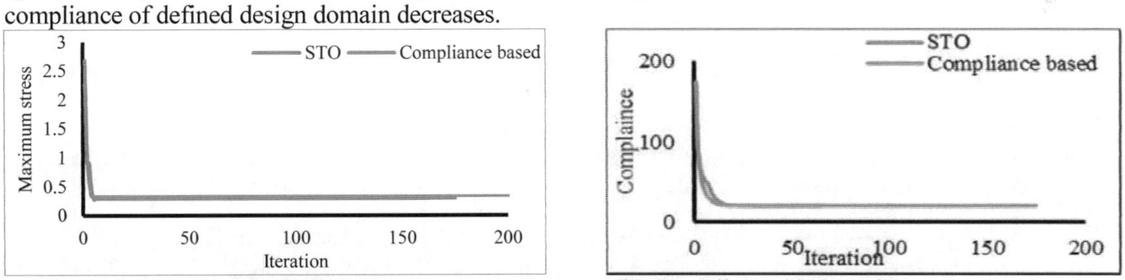

(a) Maximum stress Vs Iteration

(b) Compliance Vs Iteration

Figure 3 variation of stress and compliance with iteration

4.2. Cantilever Beam: Under this case study different loading and boundary conditions of a cantilever beam are considered as shown in Fig.4. All the design domains are defined and discretized into 120X60.

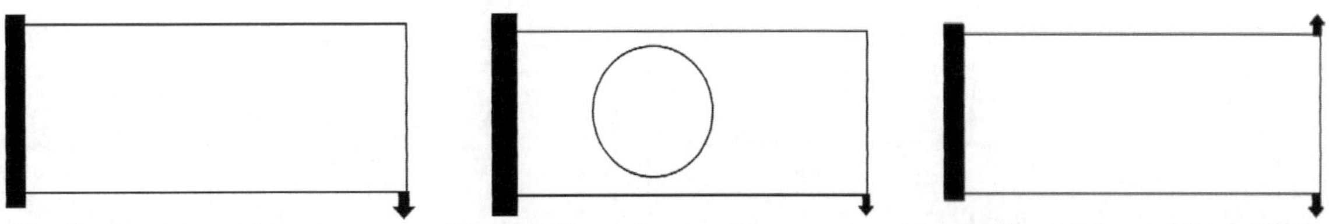

(a) Classical cantilever beam

(b) Cantilever beam with pre-defined shape

(c) Cantilever beam with multiple loading

Figure 4 Design dmain defnition

IOP Conf. Series: Materials Science and Engineering **241** (2017) 012003 doi:10.1088/1757-899X/241/1/012003

Fig.5a, Fig.5b and Fig.5c shows optimal materials distribution of a cantilever beam using stress based formulation under the loading and boundary conditions defined in their respective design domains. Fig.5d, Fig.5e and Fig.5f shows optimal materials distribution of a cantilever beam under compliance for the respective design domains. From the optimal layouts of the cantilever beam under stress and compliance based formulations, the optimal layouts from stress based formulation are simple and clearer than that of optimal layouts from compliance based formulation. This makes the need for post processing for layouts form stress based topology optimization to be less than that of layouts from compliance based formulations.

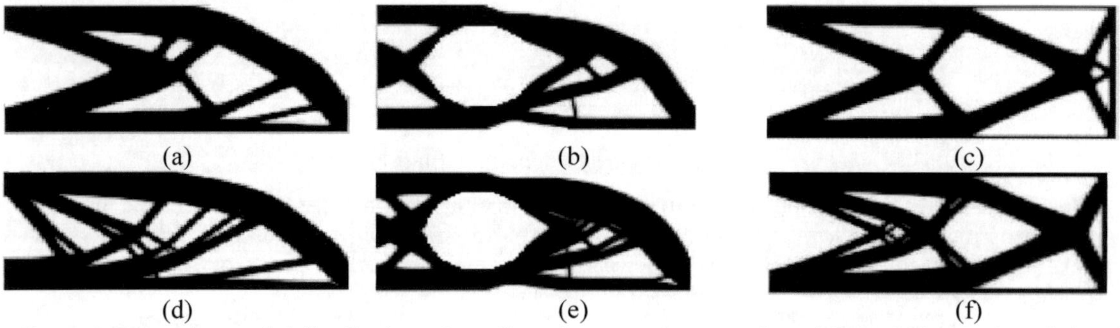

Figure 5 Opitmal material distribution of cantilever beam under stress based formulation (a, b and c) and compliance based formulation (d, e, and f)

The maximum stress induced in the design domains defined is higher when the problems are solved using compliance based formulations except the multiple loading case as shown in Fig.6a, Fig.6c and Fig.6e. This shows, the solutions and formulations of a problem should be based on the type of design domain defined, boundary and loading conditions.

Fig.6b, Fig.6d and Fig.6f shows variation of compliance for a cantilever beam under stress and compliance based formulations considering the loading and boundary conditions defined in Fig 4a, Fig4b and Fig.4c, respectively. For the first two case studies considered in this section, the design domains have higher compliance when the problems are formulated using stress based formulation. But for the third case study considered the change in the compliance of the design domain under the two types of formulation decreases at the problem converges to its optimal solution. From whcih we can conclude that the slection problem formualtion techniques must be selected carefully ahead of generating optimal solutions.

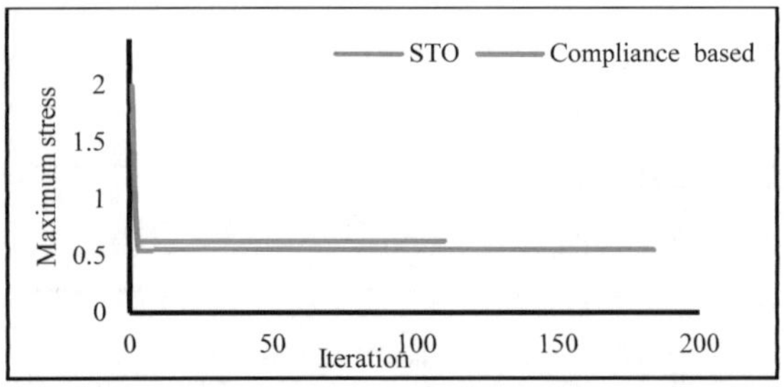

(a) Maximum stress vs iteration for classical Cantilever beam

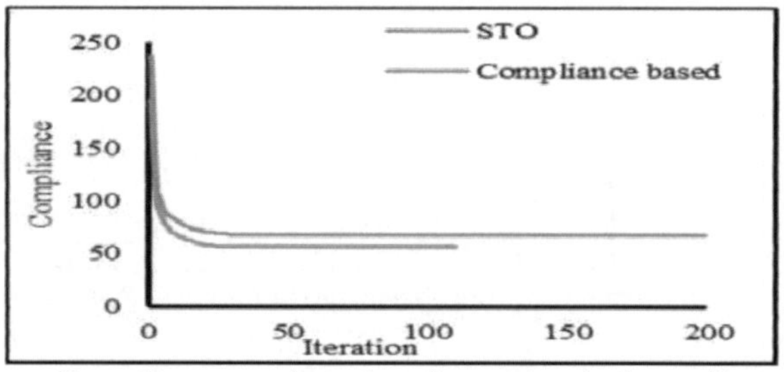

(b) Compliance vs iteration for a classical cantilever beam

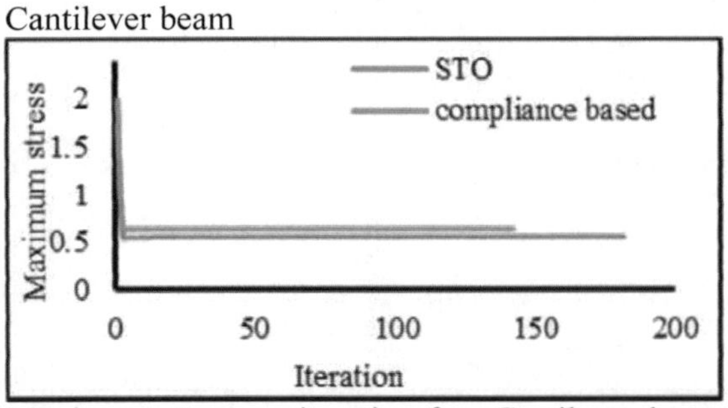

(c) Maximum stress vs iteration for Cantilever beam with pre-defined shape

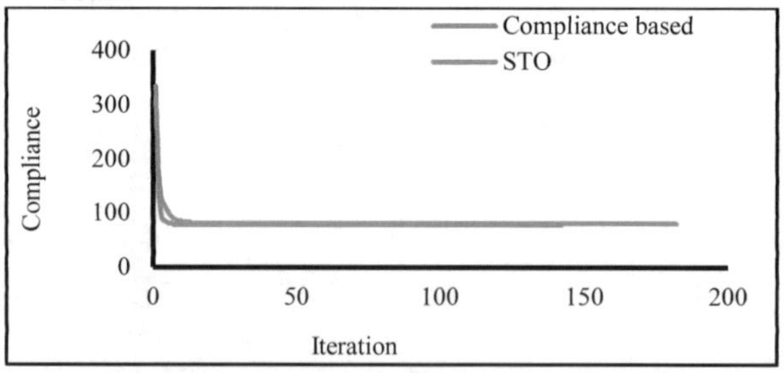

(d) Compliance vs iteration for a cantilever beam with predefined shape

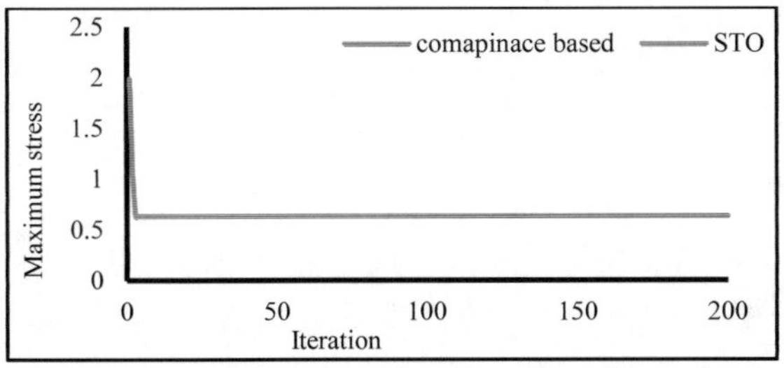

(e) Maximum stress vs iteration for Cantilever beam under multiple loading

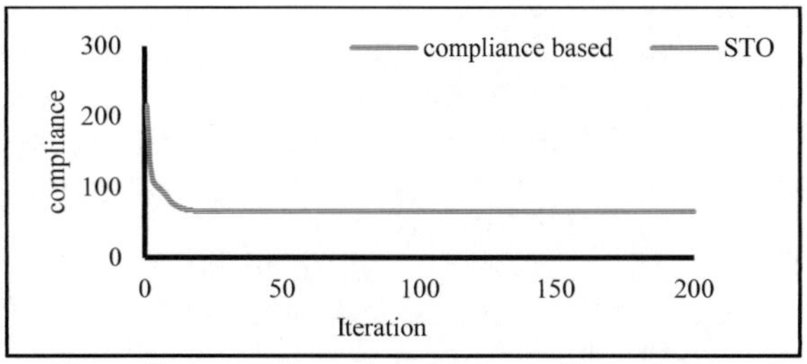

(f) Compliance vs iteration for a cantilever beam with multiple loading

Figure 6 Variation of maximum stress vs iteration (a, c and e) and variation of compliance vs iteration (b, d and f)

From the simulation results of different cases considered in this paper, optimal material distribution of design domains from compliance based formulation have complex optimal layout than stress based formulations. The variation of compliance and maximum stress induced in the design domain are dependent on the type of design domain and loading conditions defined. For designs where stress is main design constraint, most of design using ductile material,ptimal layouts from weight minimization results will be preferable.

5. Conclusion

Based on the benchmark case studies considered and simulation results, both the formulations can generate safe optimal material distribution. For all case studies considered, the maximum stress induced and compliance are dependent on type of problem considered. In stress based topology optimization, material within the design domain is distributed in such a way that it can sustain the applied loads under defined boundary and loading conditions in a best way. In classical compliance minimization problem, the optimal volume fraction, which is defined by the end user, is one of the key input variables. This makes the optimal material distribution and induced stress within the material to be highly dependent on the knowledge and experience of the end-user.

Adjustment of convergence criterions can be an alternative solution to minimize the maximum stress developed in optimal layouts. The effect of convergence criterion on other physical and mechanical properties and advantage of adjusting the convergence criterion over post processing can be studied as a future work. In addition, the effect of algorithms on the optimal layouts, stress distribution, compliance of design domain and maximum stress induced in the design domain, and alternatively a multi objective optimization which includes both objective functions under their respective design constraints can be studied further.

Acknowledgement

This work was supported by Universiti Teknologi PETRONAS and Ministry of Higher Education (MOHE) Malaysia under FRGS grant no FRGS/1/2014/TK01/UTP/02/1.

Reference

[1] Zhu JH,Gu XJ,Zhang WH and Beckers P, *Structural design of aircraft skin stretch-forming die using topology optimization.* Journal of Comput and App Math, 2013. **246**: p. 278-88.

[2] Takezawa A,Yoon GH,Jeong SH,Kobashi M and Kitamura M, *Structural topology optimization with strength and heat conduction constraints.* Comput Methods in App Mechanics and Eng, 2014. **276**: p. 341-61.

[3] Asger NC,Andreas B,Morten NJ,Niels A and Ole S, *Combined shape and topology optimization of 3D structures.* Computers & Graphics, 2015. **46**: p. 25-35.

[4] Bendsøe MP, *Optimal shape design as a material distribution problem.* Struct Optim, 1989. **1**: p. 193-202.

[5] Rozvany GIN,Zhou M and Birker T, *Generalized shape optimization without homogenization.* Struct and Multidisci Optim, 1992. **4**(3): p. 250-52.

[6] Deaton JD and Grandhi RV, *A survey of structural and multidisciplinary continuum topology optimization: post 2000.* Struct and Multidisci Optim, 2014. **49**(1): p. 1-38.

[7] Cai K,Gao Z and Shi J, *Compliance optimization of a continuum with bimodulus material under multiple load cases.* Computer-Aided Design, 2013. **45**(2): p. 195-203.
[8] Tovar A and Khandelwal K, *Topology optimization for minimum compliance using a control strategy.* Eng Struct, 2013. **48**(0): p. 674-82.
[9] Duysinx P,Miegroet LV,Lemaire E ,Brüls O and Bruyneel M, *Topology and generalized shape optimization: Why stress constraints are so important?* Int. J. Simul. Multidisci. Des. Optim. , 2008. **2**(4): p. 253-58.
[10] Amstutz S and AA, *Topological optimization of structures subject to Von Mises stress constraints.* Struct and Multidisci Optim, 2010. **41**(3): p. 407-20.
[11] Cai S and Zhang W, *Stress constrained topology optimization with free-form design domains.* Comput Methods in App Mechanics and Eng, 2015. **289**(0): p. 267-90.
[12] Bendsøe MP, *Optimization of structural topology, shape, and material.* Vol. 414. 1995: Springer.
[13] Erik A,Anders C,Mattias S,Boyan SL and Ole S, *Efficient topology optimization in MATLAB using 88 lines of code.* Struct and Multidisci Optim, 2011. **43**: p. 1-16.
[14] Holmberg E,Torstenfelt B and Klarbring A, *Stress constrained topology optimization.* Struct and Multidisci Optim, 2013. **48**(1): p. 33-47.
[15] Le C,Norato J,Bruns T,Ha C and Tortorelli D, *Stress-based topology optimization for continua.* Struct and Multidisci Optim, 2010. **41**(4): p. 605-20.

IOP Conf. Series: Materials Science and Engineering **241** (2017) 012004 doi:10.1088/1757-899X/241/1/012004

Numerical Assessment of Ultra-high Performance Concrete Material

YIN Hor[1,a], SHIRAI Kazutaka[2,b] and TEO Wee

[1]Graduate School of Engineering, Hokkaido University, Sapporo 060-8628, Japan
[2]Faculty of Engineering, Hokkaido University, Sapporo 060-8628, Japan
[3]Civil and Environmental Engineering Department, Universiti Teknologi PETRONAS, Malaysia

yin_hor@yahoo.com

Abstract. This paper presents numerical assessment of ultra-high performance concrete (UHPC) material using finite element (FE) method. Since UHPC is a relatively new material with superior strength and energy absorption, the concrete models on UHPC in FE analysis are not fully capturing behavior of UHPC. In this study, the calibration of concrete models for UHPC is exploited. The calibrated models show their capability to capture the response of UHPC material quite well compared with experiment. More details of results are discussed.

1. Introduction

Ultra-high performance concrete (UHPC) is a new advancement material in the construction industry. UHPC offers superior properties, such as ultra-high strength of 150–200 MPa, extremely less voids, low permeability, durability enhancement and energy absorption [1-3]. Because of its exceptional properties, UHPC is suitable for the protective structures under severe loading such as earthquakes or blasts [4] and its use has recently been extended into the rehabilitation of concrete structures [5]. An overview on UHPC from recent research studies has been provided in this paper. To understand the behavior of such structures in finite element (FE) method, prior study of UHPC material may be necessary because of its unique properties and until now there is no general solutions in FE calculation. Therefore, more research on calibrated models for UHPC is needed. This paper is intended to report the results from numerical analysis of UHPC material in FE model calibration. More information about the calibrated models are further discussed.

2. Overview on UHPC

2.1. UHPC as Rehabilitation Material for RC Structures

Rehabilitation and strengthening of deteriorated concrete structures is a relatively new technique. There are several studies on mechanical properties of UHPC and its strengthening applications [5-8]. Brühwiler and Denarie [5] carried out four applications of UHPC for the protection of crash barrier wall, strengthening of highway bridge, bridge piers, and industrial slabs. Later, the studies on UHPC strengthening RC members have been conducted extensively [6-8]. The results showed that UHPC enhanced overall response of structures.

2.2. Structural Response of UHPC.

Many of the structural tests on UHPC were under static loading condition [9]. Beyond affordability for full-scale tests, only a handful of them were conducted under impact or blast loads [4, 10, 11]. In general, the resistance of structural UHPC elements have demonstrated to be superior in terms of strength, durability and ductility compared with normal and high strength concrete.

2.3. *Vibration Control of UHPC Structures.*

A tuned mass damper (TMD) is often used to reduce vibration affecting the structures. Due to its superior properties, UHPC structures can significantly minimize the height of the bridge decks which could be more vulnerable to the vibrations. Chin et al. [12] have introduced TMD on the parapet of the UHPC pedestrian cable stayed bridge. Based on their record, TMD reduced the acceleration about 30% when two pedestrians were crossing the bridge [12]. This implies that combinations of UHPC material and vibration control techniques, such as tuned mass dampers (TMDs), make it possible to construct thin and long span bridge structures. In addition, high strength fiber reinforced concrete members are expected to be used as a hysteretic damper for energy dissipation against strong earthquakes owing to its high ductility [13]. The application of vibration control technology and use as a vibration damping device, both sides of the development can be expected to UHPC structures.

3. FE Model of UHPC Material

In this study, finite element (FE) method was employed by using commercial computing software, LS-DYNA [14]. LS-DYNA is often used to simulate severe loading such as impact or blasts in explicit FE calculation numerically [4]. However, it can also be used for static loading condition by activating implicit control cards. Owing to the unique properties of UHPC, the calibration model is needed. In this paper, the calibrated model was based on three-point bending tests conducted in laboratory in Universiti Teknologi PETRONAS, Malaysia, to assess its performance. This specimen has a standard size with cross-section of 100 by 100 mm and a clear span of 400 mm.

Several concrete material models are available in LS-DYNA. In this study, only a concrete damage model, also known as Karagozian & Case (K&C) model [15] (MAT_73R3 in LS-DYNA) was used. Two techniques, an automatic parameter generation and equation of state (EOS) for correcting UHPC stiffness, are deployed to simulate UHPC material under static loads using implicit control.

3.1. *Automatic Generation of Parameters*

The advantage of K&C model is that it requires only the compressive strength of concrete to input the material cards. All remaining parameters may be calculated as functions of the concrete strength. However, some parameters are changeable to represent the actual material test results such as tensile strength of concrete, FT, and tensile softening parameter, b2. The default parameters generated from K&C model are based on the concrete compressive strength of 45.4 MPa, but in this study the compressive strength of UHPC of 138.8 MPa is highly different. Therefore, a study on model calibration in this context may be needed. One possibility is to study, for instance, the effect of b2. The tensile strength of UHPC is set to 11 MPa.

3.2. *Equation of State (EOS)*

EOS tabulated compaction is required for the K&C model. This EOS is written in form of pressure-volumetric strain relationship as follows:

$$P = C(\varepsilon_v) + \gamma T(\varepsilon_v)E \tag{1}$$

where ε_v is the volumetric strain given by the logarithm of the relative concrete volume. Bulk modulus is a function of volumetric strain. Because Young's modulus may be calculated in corresponding to bulk modulus, thus, the technique of using EOS model reflects the Young's modulus resulting in development of the correction of the UHPC stiffness as seen in Fig. 1(b). Detailed description of notations in Eq. (1) above could be found in [16].

IOP Conf. Series: Materials Science and Engineering **241** (2017) 012004 doi:10.1088/1757-899X/241/1/012004

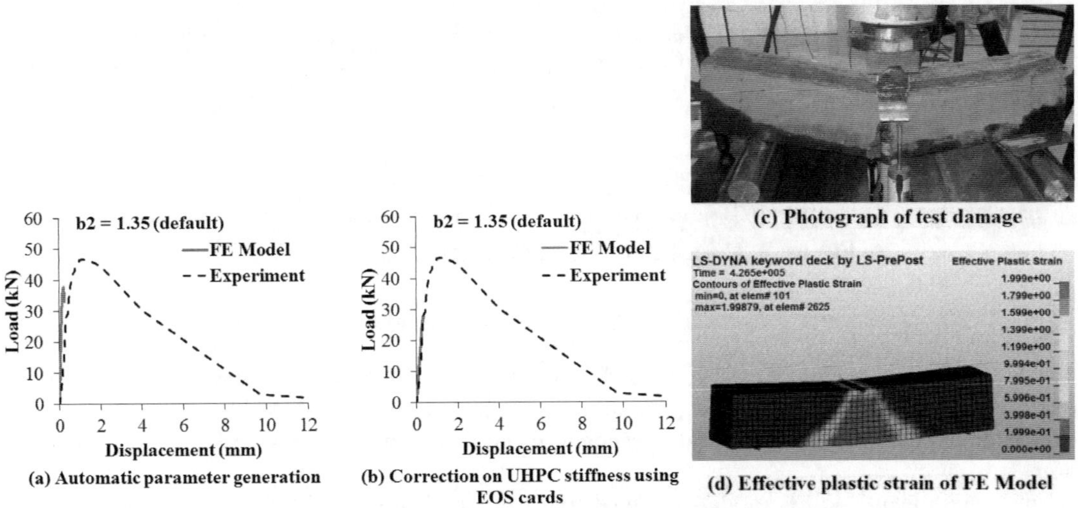

Figure 1. Comparison between numerical and experimental results.

4. Analytical Results and Discussion

Fig. 1 shows the analytical results of load-displacement relationships obtained from numerical FE model compared with laboratory experiment. A test damage and a graphic presentation of effective plastic strains from FE model are shown in Fig. 1(c) and (d), respectively.

From Fig. 1(a) and (b), the results obtained from FE model based on the default value of parameter, b2, (b2 = 1.35) are illustrated in different responses. As seen in Fig. 1(a), use of automatic parameter generation gave a stiffer stiffness compared to experiment; meanwhile in Fig. 1(b), use of EOS shifted the model curve very closely to the actual experimental result. However, both cases showed significantly lower peak loads than those obtained from the experiment.

Several cases of modified b2 were further investigated in this study. Results of the effect of b2 obtained from the automatic parameter generation and from the use of EOS are shown in Fig. 2 and 3, respectively. From Fig. 2 among all the cases, b2 = -25 gave a reasonable result compared to the experiment. When the load increased from zero up to about 25 kN, the stiffness was similar to other cases of b2; but above 25 kN, the analytical result of b2 = -25 has shifted to the experimental position and agreed very well to the peak load as seen in Fig. 2(d). It can be also noted that negative values of b2 influenced very much on the peak load result.

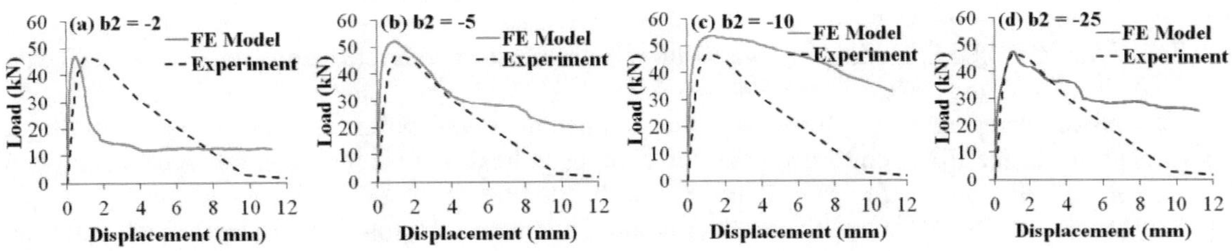

Figure 2. Automatic parameter generation

On the other hand, Fig. 3 shows the results of another simulation method by using EOS to adjust the stiffness of element. From this analysis, results demonstrate a very good agreement on the UHPC stiffness for all studied values of b2, except b2 = -25. In Fig. 3, the best fitting curve goes to Fig. 3(b) which b2 is equal to -5. Similarly, to the automatic parameter generation, the peak load was influenced very much by b2 compared to the default value of b2 (b2 = 1.35).

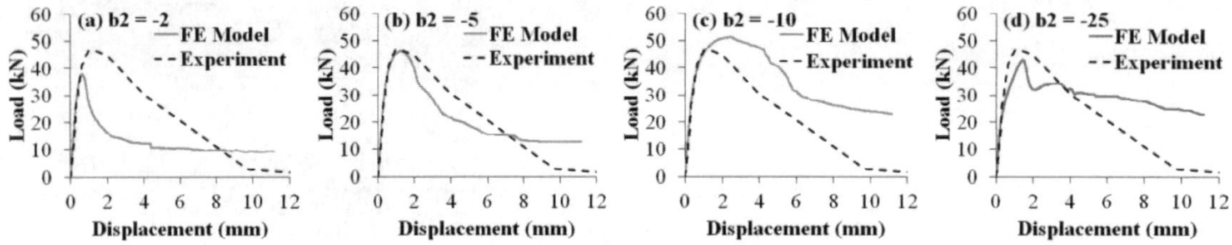

Figure 3. Correction of UHPC stiffness using EOS

5. Conclusions

UHPC is a novel material in construction industry and it has potential for use in protection and strengthening of structures. In this paper, FE model calibration using LS-DYNA for UHPC material based on the three-point bending tests was presented. From this study, the following conclusions could be made:

(1) The results from the FE models showed that the K&C model with the automatic parameter generation was capable of simulating the peak load of UHPC material with a good agreement to experimental results by setting of appropriate b2 parameter.

(2) The analysis based on equation of state (EOS) tabulated compaction in the K&C model showed very good results for both the peak load and the stiffness of UHPC material.

6. References

[1] P. Richard and M. Cheyrezy, "Composition of reactive powder concretes," *Cem. Conc. Res.*, vol. 25, pp. 1501-1511, 1995.

[2] B. A. Tayeh, B. A. Bakar, M. M. Johari, and Y. L. Voo, "Mechanical and permeability properties of the interface between normal concrete substrate and ultra high performance fiber concrete overlay," *Constr. Build. Mater.*, vol. 36, pp. 538-548, 2012.

[3] S.-T. Kang, Y. Lee, Y.-D. Park, and J.-K. Kim, "Tensile fracture properties of an Ultra High Performance Fiber Reinforced Concrete (UHPFRC) with steel fiber," *Comp. Struct.*, vol. 92, pp. 61-71, 2010.

[4] L. Mao, S. J. Barnett, A. Tyas, J. Warren, G. Schleyer, and S. Zaini, "Response of small scale ultra high performance fibre reinforced concrete slabs to blast loading," *Constr. Build. Mater.*, vol. 93, pp. 822-830, 2015.

[5] E. Brühwiler and E. Denarie, "Rehabilitation of concrete structures using ultra-high performance fibre reinforced concrete," in *UHPC-2008: The 2nd Inter. Symp. Ultra High Performance Concrete*, 2008, pp. 05-07.

[6] F. J. Alaee and B. L. Karihaloo, "Retrofitting of reinforced concrete beams with CARDIFRC," *J. Compos. Constr.*, vol. 7, pp. 174-186, 2003.

[7] T. Noshiravani and E. Brühwiler, "Experimental investigation on reinforced ultra-high-performance fiber-reinforced concrete composite beams subjected to combined bending and shear," *ACI Struct. J.*, vol. 110, pp. 251-261, 2013.

[8] A. Lampropoulos, S. A. Paschalis, O. Tsioulou, and S. E. Dritsos, "Strengthening of reinforced concrete beams using ultra high performance fibre reinforced concrete (UHPFRC)," *Eng. Struct.*, vol. 106, pp. 370-384, 2016.

[9] Y. L. Voo, W. K. Poon, and S. J. Foster, "Shear strength of steel fiber-reinforced ultrahigh-performance concrete beams without stirrups," *J. Struct. Eng.*, vol. 136, pp. 1393-1400, 2010.

[10] K. Fujikake, T. Senga, N. Ueda, T. Ohno, and M. Katagiri, "Nonlinear analysis for reactive powder concrete beams under rapid flexural loadings," *J. Advan. Conc. Tech.*, vol. 4, pp. 85-97, 2006.

[11] C. Wu, D. Oehlers, M. Rebentrost, J. Leach, and A. Whittaker, "Blast testing of ultra-high performance fibre and FRP-retrofitted concrete slabs," *Eng. Struct.*, vol. 31, pp. 2060-69, 2009.

[12] W. J. Chin, Y. J. Kim, J.-R. Cho, and J. S. Park, "Dynamic characteristics evaluation of innovative UHPC pedestrian cable stayed bridge," *Eng.,* vol. 4, pp. 869-876, 2012.
[13] H. Fukuyama and H. Suwada, "Experimental response of HPFRCC dampers for structural control," *J. Advan. Conc. Tech.,* vol. 1, pp. 317-326, 2003.
[14] B. Gladman, "LS-DYNA® Keyword User's Manual–vol. I–Version 971," ed: LSTC, 2007.
[15] L. J. Malvar, J. E. Crawford, J. W. Wesevich, and D. Simons, "A plasticity concrete material model for DYNA3D," *Inter. J. Imp. Eng.,* vol. 19, pp. 847-873, 1997.
[16] J. Crawford, Y. Wu, J. Magallanes, H. Choi, and S. Lan, "Use and validation of the release II K&C concrete material model in LS-DYNA," *Karagozian & Case, Glendale,* 2012.

Characterizing the Effects of Micro Electrical Discharge Machining Parameters on Material Removal Rate during Micro EDM Drilling of Tungsten Carbide (WC-Co)

Mehdi Hourmand[1], Ahmed A. D. Sarhan[1], Mohd Sayuti[1]

Department of Mechanical Engineering. Faculty of Engineering, University of Malaya (UM), Kuala Lumpur, Malaysia

m.hourmand@gmail.com

Abstract. Micro-dies, molds and miniaturized products can be manufactured using micro EDM process. In this research, EDM machine and on-machine fabricated CuW micro-electrode were utilized to produce the micro holes in WC-16%Co. The effects of voltage, current, pulse ON time, pulse OFF time, capacitor and rotating speed on Material removal rate (MRR) during micro EDM drilling of WC-16% Co was analyzed using fractional factorial design method. ANOVA analysis shows that increasing current, rotating speed, capacitor and decreasing voltage and pulse ON time lead to the amplify in MRR. It was found that out of all the factors, current and capacitor had the most significant effect on MRR, while the effect of capacitor was more than current. Eventually, it can be concluded that micro holes can be produced using EDM machine.

1. Introduction

Tungsten carbide (WC) and its composite (WC-Co) have high strength, hardness, wear and corrosion resistance over large scale of temperatures [1]. Thus, they are suitable materials for manufacturing dies, cutting tools and other special tools and components [1]. EDM process can remove the conductive materials with various strengths, toughness and hardness by take placing spark between the electrode and workpiece submerged in the dielectric [2]. Adding aluminum powder to the dielectric in EDM process helps to augment material removal rate (MRR) and decrease surface roughness [3]. The characteristics of micro EDM and EDM are approximately similar [1]. But in micro EDM, the electrode size, discharge energy and axis movement resolution are at micron range [1]. There are no tool deformation, chatter, mechanical stress and vibration errors in EDM and micro EDM process because of no direct contact and negligible force between electrode and workpiece [1]. Thus, micro EDM is the most effective process for machining of WC [1].

Moreover, Jahan et al. [4] reported that the highest MRR was achieved in micro EDM of WC-10%Co by using CuW micro-electrode among AgW, CuW and W micro-electrodes. In order to achieve higher MRR, lower electrode wear and better surface roughness in micro-EDM of WC, the negative polarity of micro-electrode was suggested [5]. Jahan et al. [6] compared the effect of transistor and RC-type pulse generators during micro-EDM of WC. Their results showed that micro EDM performance relies on discharge energy. It has been stated that controlling the RC-type pulse generator is easier as only voltage and capacitance determine the electrical discharge energy. In RC-type micro EDM, Jahan et al. [4, 6] applied voltage and capacitance (capacitor) as parameters during micro EDM of WC–10%Co and 0.6% others. In addition, in transistor-type micro EDM, they used voltage, resistance, pulse ON time, pulse

OFF time [6, 7] and electrode rotational speed [7] as parameters during micro EDM of WC–10%Co and 0.6% others [6, 7].

Due to the low cost and easy availability of EDM machines as compared to micro EDM and other hybrid machines, it has been used in this research. CuW micro-electrode is used to produce the micro holes in WC-16%Co. There has been less research done on the effects of micro EDM parameters such as voltage, current, pulse On time, pulse OFF time, rotating speed and capacitor on micro EDM drilling of WC-16%Co using any typical EDM. Thus the purpose of this research work is to examine how Material removal rate (MRR) is effected by the micro EDM parameters during micro EDM drilling of WC-Co by the usage of an EDM machine.

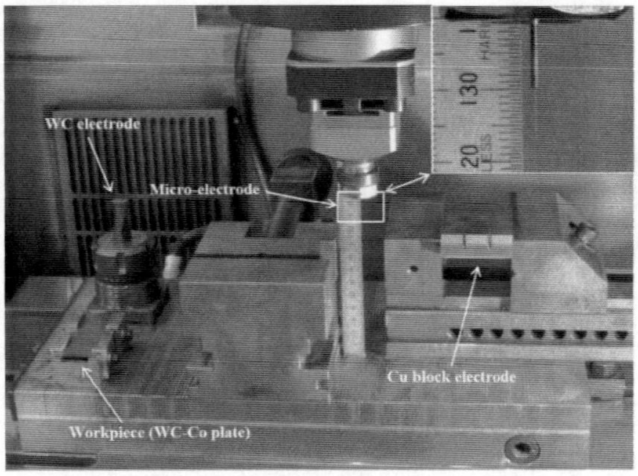

Figure 1. Experimental setup for fabricating micro-electrodes and micro holes EDM drilling

2. Experimental details

In this research, an AG40L Sodick electrical discharge machine was used to replace the micro EDM machine. Fig. 1 demonstrates the experimental setup for fabricating micro-electrodes and micro holes EDM drilling. CuW material (30%Cu + 70%W) with 1 mm diameter was selected as the raw material for the micro-electrode and WC-16%Co with 0.331 mm thickness was chosen as the workpiece. Copper was selected for the block electrode to fabricate the micro-electrode (Fig. 1) and it was aligned with a dial test indicator with 2 μm accuracy [8]. A WC electrode was used to dress the tapered micro-electrode after micro holes EDM drilling in order to produce a straight electrode [5]. The polarity of the micro-electrode during fabrication and electrode dressing was positive, while negative polarity helped produce the micro holes [5]. Positive polarity was applied on the micro-electrode during micro-electrode fabrication and dressing while negative polarity of micro-electrode assisted in machining micro holes. Oil-based dielectric fluid mixed with aluminum powder at 1.5 g/lit was also used.

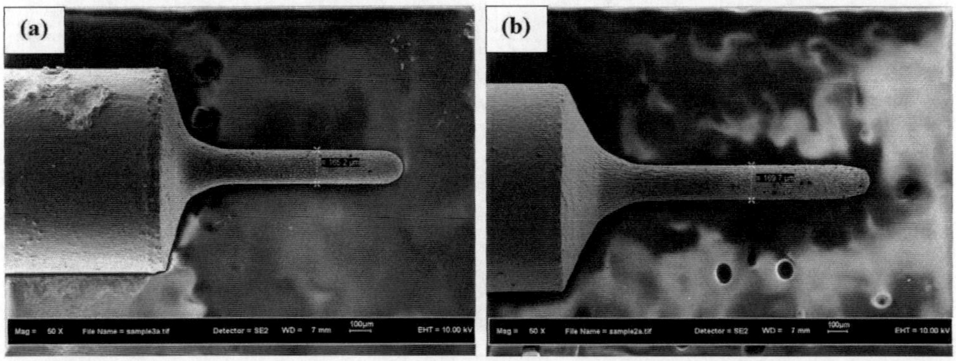

Figure 2. (a) Micro-electrodes with 165.2 μm diameter (b) Micro-electrodes with 169.7 μm diameter

Table 1. The levels of machining parameters

Symbol	Parameters	Unit	Level		
			Low	*Center*	*High*
A	Voltage (V)	V	50	70	90
B	Current (Ip)	A	0.4	2.7	5
C	Pulse on time (t_{ON})	µs	1	3	5
D	Pulse OFF time (t_{OFF})	µs	1	5	9
E	Rotating speed (R)	rpm	0	10	20
F	Capacitor (C)	µF	0	0.05	0.1

The through micro holes were produced with two micro-electrodes with 165.2 and 169.7 µm diameter and around 6.8 mm initial length (Fig. 2). These two electrodes were produced on-machine by horizontal moving block electrical discharge grinding (horizontal moving BEDG [HM-BEDG]) [8] based on the experimental setup shown in Figure 1. The micro-electrodes and hole diameters were measured with FESEM. Voltage, current, pulse ON time, pulse OFF time, rotating speed and capacitor were adopted as the parameters to analyze MRR and overcut during micro-EDM drilling of WC-Co using EDM machine. Table 1 shows the machining parameter levels and symbols. MRR was calculated with the following equations [5].

$$\text{MRR} = \{\frac{\pi}{3}[r_{Top}^2 + r_{Top}\, r_{Bottom} + r_{Bottom}^2] \times h\} \div t \quad (1)$$

where r_{Top} (mm) is the radius at the entrance side, r_{Bottom} (mm) is the radius at exit side, h is the workpiece thickness, t (min) is the machining time to make a hole at a particular setting.

The experiments were designed with 2_{IV}^{6-2} fractional factorial design and 3 center points consisting of 19 runs including 16 (=2^{6-2}) and 3 center points. These (resolution IV designs) are designs in which no main effect is aliased with any other main effect or with any two-factor interactions, but two-factor interactions are aliased with each other. The factorial effects defining contrast are I = ABEF = ACDF = BCDE. The alias relationships are as follows:

A = BEF = CDF, B = AEF = CDE, C = ADF = BDE, D = ACF = BCE, E = ABF = BCD, F = ABE = ACD, AB = EF, AC = DF, AD = CF, AE = BF, AF = BE = CD, BC = DE, BD = CE, ABC = ADE = BDF = CEF, ABD = ACE = BCF = DEF

The three-, four-, five- and six-factor interactions are considered errors. Therefore, the aliased of two-factor interactions are important in this case.

3. Results, Analysis and Discussion

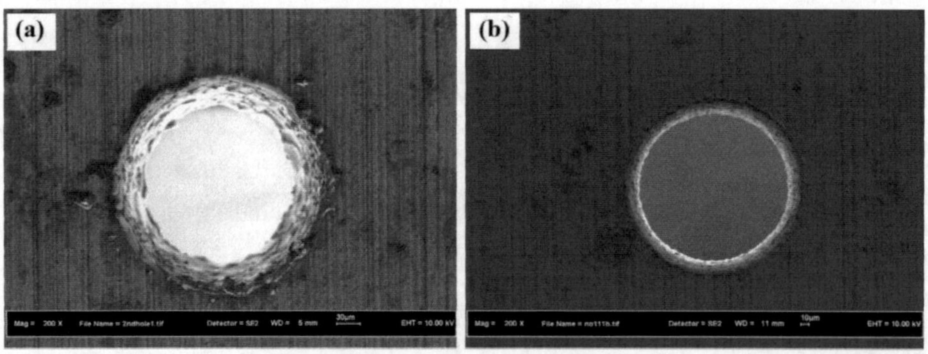

Figure 3. (a) worst micro hole (90 V, 5 A, 5 µs, 9 µs, 20 rpm, 0.1 µF) (b) one of the best micro hole (50 V, 0.40 A, 1µs, 1 µs, 0 rpm, 0 µF)

IOP Conf. Series: Materials Science and Engineering **241** (2017) 012005 doi:10.1088/1757-899X/241/1/012005

Table 2. ANOVA table for MRR

Source	F-Value	Prob > F	Status
Model	152.44	< 0.0001	significant
A-Voltage	18.72	0.0015	
B-Current	101.71	< 0.0001	
C-Pulse ON time	11.40	0.0070	
E-Rotating speed	15.87	0.0026	
F-Capacitor	878.59	< 0.0001	
AB	24.22	0.0006	
AF	16.60	0.0022	
Curvature	8.52	0.0153	significant
Residual			
Lack of Fit	15.84	0.0607	not significant
R-Squared = 0.9907		Adj R-Squared = 0.9842	
Pred R-Squared = 0.9634		Adeq Precision = 31.929	

Figure 3 shows the worst micro holes and one of the best micro-holes were produced as a function of different parameters. The experimental outcomes for MRR were analyzed using Design Expert software. The significance of the models, individual model terms and lack of fit was tested using ANOVA (Analysis of variance). The MRR response for ANOVA validation underwent a Square root transformation [2]. Table 2 demonstrates the ANOVA tables for MRR. If the "Prob > F" value is less than 0.05, it is significant [2]. In this research, A, B, C, E, F, AB and AF were significant model terms for MRR response. Moreover, B and F were the most significant factors. But AB, EF, AF, BE and CD were possibly significant due to aliase.

Insignificant lack of fit is preferable for MRR responses. The results for MRR reveals that an R-Squared value approaching 1 is preferable. Moreover, there were minor differences between Adj. R-Squared and Pred. R-Squared showing that the models had an admissible transaction between input and output parameters. Adeq. Precision greater than 4 is desirable, as it measures the signal-to-noise ratio [2]. Also, residuals had constant variance and normal distribution after using the square root transformation to analyze and model MRR response.

From Figure 4 which captures the response graph for MRR it can be seen that MRR increased with amplifying current, rotating speed and capacitor and decreasing voltage and pulse ON time. The current and capacitor were the most significant factors; however, the effect of the capacitor was greater than the current. Out of all the factors, it was observed that current and capacitor were the most contributing factors with capacitor showing more effect than the current. The voltage controls the discharge gap between the electrode and workpiece, therefore, increasing the voltage augments the discharge gap as well [2]. Appropriate voltage according to current, pulse ON time, pulse OFF time, capacitor and rotating speed results in improved MRR. Figure 4(a) indicates that with voltage greater than 50 V, the amount of MRR decreased as the discharge gap was larger than the suitable range [2] for low amounts of energy in micro EDM [current (0.4 - 5 A), pulse ON time (1 - 5 μs) and capacitor (0 - 0.1 μF) are low]. According to Figure 4(b), MRR augmented with increasing current because of rising amount of spark energy and heat transmitted to the workpiece for melting and vaporization. Figure 4(c) depicts the effect of pulse ON time on MRR. Sparks take place during pulse ON time. The amount of spark energy augments by increasing the pulse ON time [9].

During long pulse ON time, sparks initially get generated between the microelectrode and workpiece, after which some of the sparks generate between the micro-electrode and chip and some between the micro-electrode and workpiece due to the difficulty in removing chips between the micro-electrode and workpiece. The size of chip is decreased by occurring sparks between the micro-electrode and chips. Hence, machining efficiency is deteriorated [9]. The occurring sparks between the micro-electrode and chips cause the chip size to decrease. Thus, machining efficiency is reduced [9]. One more factor that contributes to the inefficiency of machining is the collection of chips inside the micro hole leading to unstable servo operation. As a result, MRR deteriorated with increasing pulse ON time (Figure 4(c)). Figure 4(d) illustrates that MRR increased by increasing the rotating speed. This is due to the fact that higher rotating speed helped in removing the chips from the distance between workpiece and micro-electrode. According to Figure 4(e), MRR increased sharply by amplifying capacitor because of increasing spark energy transmitted to the workpiece. This reveals that capacitor was the most significant factor for MRR.

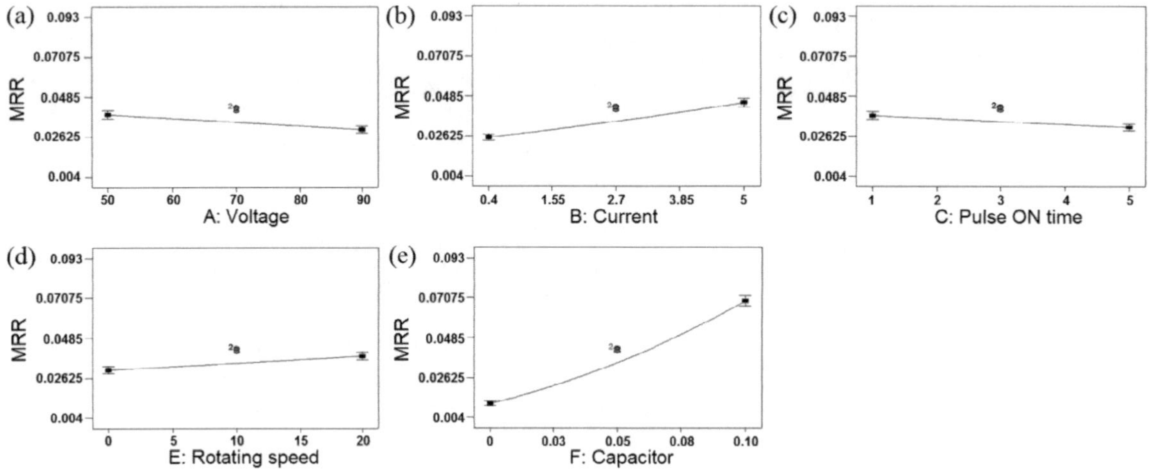

Figure 4. Response graphs for MRR (a) Voltage. (b) Current. (c) Pulse ON time. (d) Rotating speed. (e) Capacitor

4. Conclusion

This research has been conducted to investigate the effects of micro EDM drilling of WC-16%Co using a CuW microelectrode and EDM machine. ANOVA analysis shows that increasing current, rotating speed, capacitor and decreasing voltage and pulse ON time lead to the amplify in MRR. It has been observed that current and capacitor contributed the most effect in the increase of MRR. Amongst the two parameters, the effect of capacitor on MRR was greater than that of the current. Thus it can be established that the capacitor contributes the most in increasing the MRR. In the end, it can be concluded that for producing micro holes, EDM is efficient and can be used.

5. References

[1] M.P. Jahan, M. Rahman, Y.S. Wong: Int. J. Mach. Tool. Manuf. Vol. 51 (2011), p. 837-858.
[2] M. Hourmand, S. Farahany, A.A. Sarhan, M.Y. Noordin: Int. J. Adv. Manuf. Technol. Vol. 77 (2015), p. 831-838.
[4] S. Assarzadeh, M. Ghoreishi: Int. J. Adv. Manuf. Technol. Vol. 64 (2013), p. 1459-1477.
[5] M.P. Jahan, Y.S. Wong, M. Rahman: J. Mater. Process. Technol. Vol. 209 (2009), p. 3956-3967.
[6] M.P. Jahan, Y. San Wong, M. Rahman: Int. J. Adv. Manuf. Technol. Vol. 46 (2010), p. 1145-1160.
[7] M. Jahan, Y. Wong, M. Rahman: J. Mater. Process. Technol. Vol. 209 (2009), p. 1706-1716.
[8] M.P. Jahan, Y.S. Wong, M. Rahman: Mach. Sci. Technol. Vol. 16 (2012), p. 131-156.
[9] M. Hourmand, M. Noordin, Adv. Mat. Res. Vol. 845 (2014), p. 980-984.
[10] E.C. Jameson: Electrical Discharge Machining (Society of Manufacturing Engineers, Dearbern, Michigan 2001).

Evaluation of the Properties Magnesium Phosphate Cement with Emulsified Asphalt

Jia-Chong Du[1, a], Ruei-Siang Shen[1] and Yu-Zhun Zhou[1]

[1]Department of Construction and Spatial Design, Tungnan University, Department of Construction and Spatial Design, Shen Ken, New Taipei City, Taiwan

[a]cctu@mail.tnu.edu.tw

Abstract. Three type mixtures of magnesium phosphate cement with emulsified asphalt for evaluation their properties. The mixtures of the samples were fabricated and allowed them 2 hours, seven and twenty eight days curing before tested by compressive strength, Marshall stability and indirect tensile strength to probe into their engineering properties. The test results show that all tests have the greatest values at the 28 days curing and too much asphalt emulsion may cause too soft as result of low stability. The compressive strength of Type-III mixture has the greatest value, no matter what curing time is. The Marshall stability test and indirect tensile strength of the Type-III mixture are qualified by the specification required for fast maintenance. The more asphalt emulsion added, the less compressive strength has.

1. Introduction

A variety of pavement patching materials have been developed due to material engineer effort. The use of either hot mix asphalt or hydraulic cement for pavement patching materials has some deficiencies which are mentioned in my researches [1].

For rapid asphalt patching material series, cold patch asphalt repair materials are common used for emergency pothole repair. However, these materials are typically used to make temporary repairs until proper maintenance can be done. In addition, their qualities are often criticized by the government agencies. Because of too soft or too much binder, the mixture can face instability problems in summer [2]. The pavement distress such as pushing or shoving, raveling, and dishing is the most frequently encountered in-service failure when cold patch asphalt repair materials are used.

Dishing occurred during compaction under traffic that leaves a depression in the repaired surface; it is a result of improper compaction or installation techniques. Other failure mechanisms may include freeze-thaw deterioration, poor skid resistance, and lack of adhesion to the side or bottom of the pothole [3].

In the early 1980s, hydraulic cement series had been developed for fast hardness of early strength concrete such as "one-day cement" from Japan, "Swiftcrete" from British,"Draifach" from Germany, "Supercement" from Italian, "Pyrament" and "phosphate cement" from American, and aluminate cement [4]. However, magnesium phosphate cements (MPC) formed through an acidbase reaction between MgO and a soluble acid phosphate (typically an ammonium or potassium phosphate) were developed considerably greater compressive and tensile strengths than hydraulic cement, and given they could take less energy to produce. It is a wonder why they are not more commonly used these days because the promotion and proliferation of hydraulic cement occurred when energy was cheap and health concerns of the public were simply not an issue [5. 6].

If a cheap source of potassium hydrogen phosphate could be found, then MPCs could make a valuable contribution to reducing global warming and improving building biology from the point of view of occupant health [6].

Asphalt series, is a mainstream of pavement, but the properties of patching materials are still insufficient which is mentioned above. On the contrary, the hydraulic cement used has the advantage of long service life, high strength and low price for maintenance. However, its rigid characteristic results in brittle failure, especially, damage occurred in the angle corner and on the edge, and the noise is also loud when driving a vehicle.

Based on the energy conservation and carbon reduction, pavement patching materials should admit to the requirement of green environmental protection, and fast maintenance is necessity under the restriction of the construction time in conformity with the traffic environment demanded in Taiwan. Thus, the objects of this study are to evaluate the properties of nonpoisonous and harmless MPC with emulsified asphalt (MPC-EA) mixture.

2. Materials and Methods

MPC typically comprises an alkaline earth phosphate salt, magnesium oxide and type F fly ash. MPC tend to cure rapidly, and exhibit high chemical stability, high compressive strength, low porosity and permeability to water. Due to the rapid cure at room temperature, MPC is considered for fast maintenance.

Cationic rapid setting (CRS-1) asphalt emulsion used for this study was met the specification requirement of the ASTM D2172. The properties of river sand used as fine aggregate and CRS-1 were tested listed in Table 1 and Table 2, respectively. The mixtures of the samples were fabricated by the ratios as shown in Table 3 for test and were allowed them curing 2 hours, seven and twenty eight days before tested by compressive strength, Marshall stable test, and indirect tensile strength for their engineering properties evaluated.

Table 1 The properties of fine aggregate

Materials	Specific gravity	Water absorption	FM
Fine Aggregate	2.66	0.81%	2.18

Table 2 CRS-1test results

Properties	Specification	CRS-1
Specific gravity	-	1.03
Viscosity, saybolt furol at 50°C	20-100	92
Particle charge test	Positive	Positive
Sieve test, %	0.1 max	0.08
Residue by distillation, %	60 min	65.8
Sieve content, %	0.3 max	0.25
Penetration at 25°C (100g, 5s)	100-250	120
Stability (1 day)	1% max	0.43
Storage stability (5 days)	5% max	1.06
Ductility at 25°C, cm	40 min	49.4

Table 3 Mixture ratios of mortar

Materials	Type-I	Type-II	Type-III
MPC	450	450	450
Fine Aggregate	1000	1000	1000
Fly Ash	45	45	45
Asphalt Emulsion	27%	7%	7%
water	7%	7%	10%

3. Results and Discussion

3.1. Compressive Strength

Using 2" or 50 mm cube specimens were tested by MTS with a constant loading rate of 5 cm/min, in accordance with ASTM D109.The results of compressive test are shown in Figure 1. As it can be seen, the curing time increases with the compressive strength values increased because MPC is a kind of hydraulic cement which compressive strength developed depends on curing time. The more asphalt emulsion added, the less compressive strength has, as a result, the Type-I have the minimum values. The Type-III mixture has the maximum compressive strength values, regardless of the curing time of 2-hour, 7-day and 28-day. This situation may be explained that enough water added can provide a better chemical reaction and produce a higher compressive strength.

3.2. Marshall Stability Test

In accordance with ASTM D6927, the Marshall stability is the resistance to plastic flow of cylindrical specimens of an asphalt mixture. A load is applied to the specimen at a constant strain rate of 50.8 mm/min until the maximum load is reached. The maximum force (called stability) at that force are read and recorded. The test results as shown in Figure 2 indicate that the Type-III has a very high value above 800 kg at 2-hour curing which means that the stability satisfies the specification required for heavy traffic [7]. Moreover, too much asphalt emulsion may cause too soft as result of low stability like Type-I mixture.

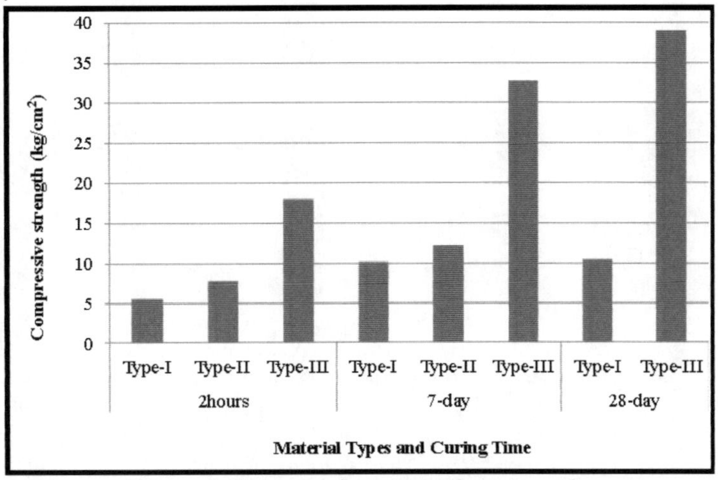

Figure 1 Compressive strength test results

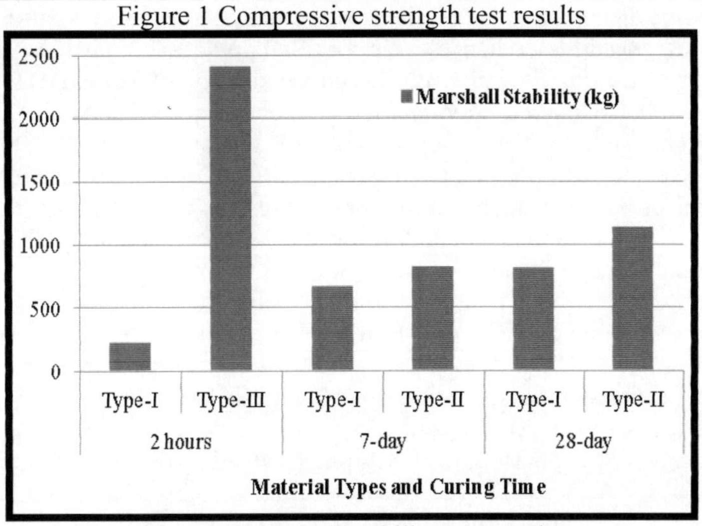

Figure 2 Marshall stability test results

3.3. Indirect Tensile Strength

The test of indirect tensile strength, in accordance with ASTM D6931, may use to evaluate the relative cohesion of the MPC-EA mixtures in conjunction with laboratory mix design testing and to estimate the potential for rutting or cracking. The results can also be used to determine the potential for field pavement moisture damage when results are obtained on both moisture-conditioned and unconditioned specimens. The results of test for Type III mixture, due to its better compressive strength and stability than Type I and II, are shown in Figure 3. The curing time increases with the indirect tensile strength increased. Compared with cold patch asphalt repair materials w, the strength is reasonable [8].

4. Conclusions

Based on the results from the laboratory test using MPC-EA, the following conclusions are made:

- The compressive strength of Type-III mixture has the greatest value, no matter what curing time is. The more asphalt emulsion added, the less compressive strength has.
- For Marshall stability test and indirect tensile strength, the Type-III mixture is qualified by the specification required for fast maintenance.

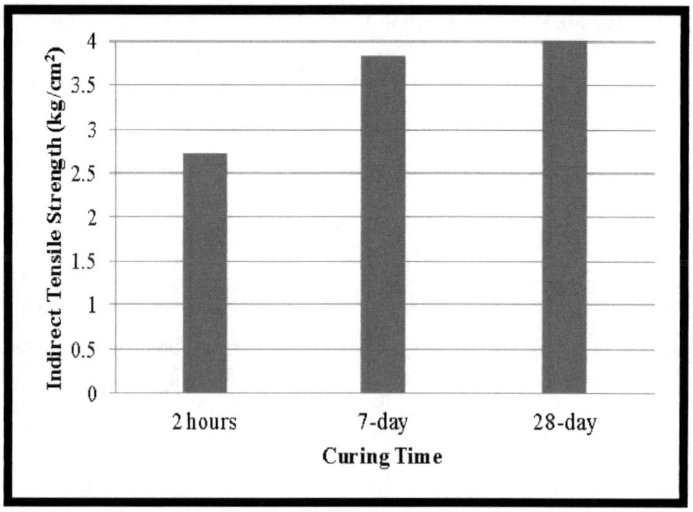

Figure 3 Type-III mixture indirect tensile strength test results

5. References

[1] Jia Chong Du, Ming-Feng Kuo, and J C Yeh: Properties of Cement Asphalt Emulsion Mortar for Pavement, Advanced Materials Research Vol. 723 pp 466-473, (2013).

[2] Estakhri, C. K., L. M. Jimenez, and J. W. Button: Evaluation of Texas DOT Item 334, Hot-Mix, Cold-Laid Asphalt Concrete Paving Mixtures. Publication FHWA/TX-00/1717-1, Federal Highway Administration, Texas Department of Transportation, College Station, (1999).

[3] Samrat Chatterjee, Ronald P. White, Andre Smit, Jolanda Prozzi, and Jorge A. Prozzi: Development of mix design and testing procedures for cold patching mixtures. FHWA/TX-05/0-4872-1, Publication Federal Highway Administration, Texas Department of Transportation, College Station, (2006).

[4] Information on http://www.concretebasics.org/types.html,2012/11/23.

[5] Swanson, George. Building Biology Based New Building Protocol. Magnesium Oxide, Magnesium Chloride, and Phosphate-based Cements. http://www.greenhomebuilding.com/pdf/MgO-GENERAL.pdf

[6] Information on http://www.tececo.com/history.magnesium_cements.php

[7] Asphalt Institute. Mix Design Methods for Asphalt, 7th ed., MS-02. Asphalt Institute. Lexington, KY (2015).

[8] Minegishi, J., Takeda, T., Tatsushita, M., Ohki, H. and Wataya, S: A Study on Cold Mixtures for Pothole Repair in Tokyo. The 11th International Conference on Asphalt Pavements. ISAP, Nagoya (2010).

Acknowledgements

The authors thank Ministry of Science and Technology of Taiwan (MOST 104-2221-E-236-003-) which supported this study.

The Effectiveness of IRI Compared To SDI System for Assessing the Quality And Performance Of Materials Used In Flexible Pavement In Java, Indonesia

Hary Agus Rahardjo[1, a], Suparman[2]

[1,2]Universitas Persada Indonesia Y.A.I, Indonesia

[a]email: haryagus30@yahoo.co.id

Abstract. Technological developments regarding the tools and methods used to determine the level of damage to roads, does not necessarily assure benefits to its users. Need further analysis and testing to determine its effectiveness. In terms of damage to the road using flexible pavement, IRI (International Roughness Index) method needs to be studied and compared with the SDI (Surface Distress Index) method which has been used in Indonesia. Damage to the roads using flexible pavement indicates the quality of materials used in flexible pavement. The method used in this study is a survey directly in the field using both methods. Then the results were compared and analyzed using statistical analysis. The purpose of this study was to determine the effectiveness of IRI compared to SDI. The result reveals that there are no significant differences between IRI and SDI. In other words, IRI can be used singly replace SDI.

1. Introduction

For the past couple of decades, roads in Indonesia are using flexible pavement. Besides being easy to build, have a cheaper price compared to rigid pavement, as well as easier to fix. In addition, they have a great advantage that their surface can be used again for rehabilitation [1].

Related to efforts for the roads maintenance and repair in order to function properly, the initial step necessary is the identification of the damage that requires repair. This identification includes the type, level, area of damage, and the locations. In Indonesia, relating to weaknesses in law enforcement, overloading is also among the most important causes of the deterioration of flexible pavements as indicated by Sadeghi [2].

There are several methods that can be used for that purpose, among others, IRI (International Roughness Index) and SDI (Surface Distress Index). The use of IRI or SDI is not only useful for controlling the quality of flexible pavement but also can be used to determine the performance of the materials used in the construction of flexible pavement. The more damaged areas in flexible pavement, the lower the performance of the material used.

With advances in technology, the IRI method considered more practical. But the question that arises is whether IRI method accurate enough and suitable to be applied in Indonesia. It is given that the flexible pavement in Indonesia, though still function properly, is often uneven or bumpy, unlike in developed countries where they have the smooth and flat surface roads. Finally, using a combination of IRI and SDI methods simultaneously is common in practice [3].

2. Aim and Objectives

The aim of this study is to investigate the effectiveness of IRI compared to SDI. Therefore, the objectives of writing this paper is to prove that the use of IRI effective enough for the road conditions in Indonesia, so that it can be used singly replaces the SDI method. The results of this study then can be a reference for the management of the road in all regions in Indonesia.

3. Methodology

This research began with a review of literature on IRI and SDI method to get an overview of the functions, mechanisms and results of measurements obtained with a focus on its application for road construction. Furthermore, data acquisition in the form of primary data obtained through direct surveys on roads in the West Java region using IRI and SDI method. The 22.5 km of roads divided into segments, each segment length is 100 m. Thus, this case study uses both qualitative and quantitative data [4]. The collected data is then analyzed using statistical analysis method.

4. IRI and SDI to Observe Flexible Pavement

Flexible pavements consist of natural granular material and then covered with one or more waterproof bituminous surface layers, so that they become flexible. Tropical country like Indonesia with a fairly high rainfall, flexible pavement damage often occurs. Moreover, the overloaded vehicles, which are common in developing countries, also accelerated the deterioration of the road [5]. Therefore, it is necessary to arrange the maintenance and repair of roads continuously.

According to government regulations, road maintenance consists of routine maintenance and periodic maintenance. Routine maintenance is the maintenance performed every year and includes a coating of pavement and gravel roads, vegetation control, turf, as well as cleaning and painting of traffic signs. While periodic maintenance normally carried out every four or five years and includes work overlay, closing holes, drainage improvements, road shoulders, and replacement of traffic signs [6].

To determine the level of damage or condition of existing road, then IRI or SDI method is used. IRI can describe a longitudinal profile of a road and used as a standard road surface unevenness [7]. The IRI, which is a profile-based statistic, is shown to be an ideal predictor (or independent variable) since it has the advantage of being repeatable, reproducible, and stable with time [8]. But must be careful at the time of measurement, because of the variation of speed during the measurement process has a large influence on the IRI measurement [9]. SDI is an assessment system of pavement condition conducted by visual observation directly in the field and can be used as a reference in maintenance effort. SDI value that indicates in good condition does not necessarily indicate similar conditions to the value of IRI, because SDI values obtained from the road surface conditions while the value of IRI is more directed to the road surface roughness [10].

5. Result and Discussion

Length of roads surveyed is 22,500 kilometers long starting from Ciawi Bogor City up to Puncak, Bogor Regency, West Java Province. Road section consists of two 2-way lanes, and pavement width is 7-10 meters with 3.5-5 meters per lane. Damage to roads is observed directly every 1.00 meter based on the method of SDI, while the IRI system by using the tool HAWKEYE. Having recapitulated, frequency distribution data of the road conditions, both for the value of IRI and SDI values are shown in Table 1 and Table 2 and Figure 1 and Figure 2 below.

IOP Conf. Series: Materials Science and Engineering **241** (2017) 012007 doi:10.1088/1757-899X/241/1/012007

Table.1. IRI Frequency Distribution

Category	IRI	Frequency	%
Good	<4	14,600	64.89%
Fair	4 - 8	7,700	34.22%
Poor	8 - 12	200	0.89%
Bad	>12	-	0.00%

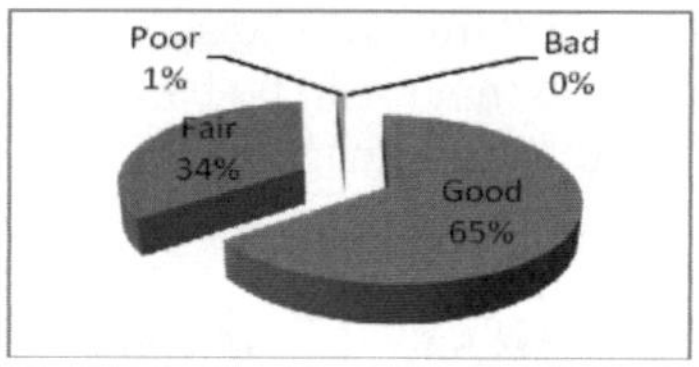

Fig.1. IRI Frequency Distribution

Table.2. SDI Frequency Distribution

Category	SDI	Frequency	%
Good	<50	14,000	62.22%
Fair	50 - 100	8,200	36.44%
Poor	100 - 150	300	1.33%
Bad	>150	-	0.00%

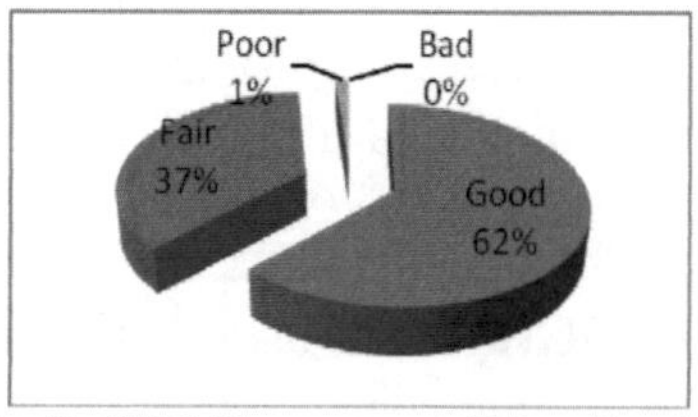

Fig.2. SDI Frequency Distribution

The table shows that there is no significant difference between the results of IRI and SDI. For the category of good conditions, IRI states that 64.89% roads are in Good condition, whereas 62.22% expressed by SDI. The IRI results appear higher compared with the results of SDI. But for Fair and Poor condition, results of IRI by 34.22% and 0.89% is lower than the results of SDI which constituted 36.44% to fair condition and 1:33% for Poor condition. There is an inconsistency between the resulting IRI and SDI, sometimes higher, sometimes lower. This differs from the findings of Sinurat, who is stating that the results of IRI relatively lower than the results of SDI [11].

But to know the effectiveness of IRI compared to SDI or to know is there any similarity between the two methods, it is then performed on the comparative sample Hypothesis independent non-parametric statistics. The test results of two mean between IRI and SDI t - Test: Paired Two Sample for Mean is shown in Table 3 below.

Table 3. t – Test: Paired Two Sample for Mean

	IRI	*SDI*
Mean	4.368387097	55.57983871
Variance	1.7760367	795.4870705
Observations	62	62
Pearson Correlation	0.715083787	
Hypothesized Mean Difference	0	
db	61	
t Stat	-14.78837517	
P(T<=t) two-tail	7.72045E-22	

In this hypothesis test, define Ho: there is no significant difference between the mean test based on the value of the road surface roughness (IRI) with the road surface distress index (SDI), with alpha = 5%. Statistical calculation at above table shows that the p value (7.72%)> 5% and t table (1.999)> t stat. Thus, Ho is accepted, it means that there is no difference between IRI and SDI significantly. In other words IRI can be used as effectively as SDI. Thus, the combination use of IRI and SDI simultaneously [3] is no longer needed for the conditions in Indonesia. As stated by David N. Ammons and T. Dwane Brinson, it could be added here, however, that comparisons of cities across the street condition ratings are viable when each city uses the same rating system and when the ratings are considered approximate rather than absolute [12].

IOP Conf. Series: Materials Science and Engineering **241** (2017) 012007 doi:10.1088/1757-899X/241/1/012007

6. Conclusion

From the discussion above it can be concluded that there is no significant difference between the value generated by IRI or SDI. In other words, IRI can be used singly replace SDI. Despite fluctuating results or inconsistent result between the two methods, but the difference is not significant. For further research, it is recommended to consider the level of vehicle speed that is used to perform measurements using the IRI method. It is intended to get more accurate results.

7. References

[1] Saurabh Jain et al. *Design of Rigid and Flexible Pavements by Various Methods & Their Cost Analysis of Each Method.* International Journal of Engineering Research and Applications. Vol. 3, Issue 5, Sep-Oct 2013, pp.119-123.

[2] J. M. Sadeghi. And M. Fathali. *Deterioration Analysis of Flexible Pavements under Overweight Vehicles*. Journal of Transportation Engineering © ASCE. Vol.133. No. 11. November 2007. Pp.625-633.

[3] M. Yoga Mandala Putra. *Functional and Structural Evaluation Method Using AASHTO 1993 in Flexible Pavement Management. Case Study: Medan – Lubukpakam segment.* Civil Engineering Journal. ISSN 0853-2982. Vol. 20 No. 3 Desember 2013. Pp. 245-254.

[4] Yin, Robert K. *Case Study Research: Design and Methods*. 5th edition. Thousand Oaks, California: SAGE Publications, Inc. 2014.

[5] J. M. Sadeghi. And M. Fathali. *Deterioration Analysis of Flexible Pavements under Overweight Vehicles*. Journal of Transportation Engineering © ASCE / Vol.133. No. 11. November 2007. Pp.625-633.

[6] Regulation of the Minister of Public Works of the Republic of Indonesia Number 13 / PRT / M / 2011 on Procedures for Road Maintenance and Surveillance.

[7] Park, Kyungwon. et al. *Applicability of the International Roughness Indexas a Predictor of Asphalt Pavement Condition*. Journal of Transportation Engineering © ASCE / DECEMBER 2007. pp.706-709.

[8] Williams, Lakeasha N. et.al. *Predicting Pavement Condition Index Using International Roughness Index in a Dense Urban Area.*Journal of Civil Engineering Research. 2015, 5(1): pp. 10-17.

[9] Du, Yuchuan. et al. *Measurement of International Roughness Index byUsing Z-Axis Accelerometers and GP.* Research Article. Hindawi Publishing Corporation. Volume 2014, Article ID 928980, 10 pages.

[10] Minarti, Eka. Et.al. *Observation of Road Damage Based on Surface Distress Index (SDI) and International Roughness Index (IRI).* Civil Engineering Journal. Syah Kuala University. ISSN 2302-0253. Volume 3, No. 3, Agustus 2014. pp. 163- 173.

[11] Doan Sinurat. *Comparative Study on Determining The Road, Base On Visual Observation and Parvid Equipment*. Civil Engineering Journal, University of North Sumatera. Vol.3 No.1. 2014.

[12] David N. Ammons and T. Dwane Brinson. *Condition Ratings: Their Use among North Carolina Cities.* PUBLIC MANAGEMENT BULLETIN. School of Government. The University of North Carolina at Chapel Hill Street . NO. 5. January 2011. Pp.1-12.

IOP Conf. Series: Materials Science and Engineering **241** (2017) 012008 doi:10.1088/1757-899X/241/1/012008

Corrosion Behavior of Active Screen Plasma Nitrided 38CrMoAl Steel under Marine Environment

Yang Li[1, 2], He Yongyong[1], Mao JunYuan[1] and Zhang Lei[2]

[1]State Key Laboratory of Tribology, Tsinghua University, Beijing, 100084, PR China
[2]Department of Materials Science and Engineering, Yantai University, Yantai, 264005, PR China

metalytu@163.com

Abstract. The 38CrMoAl steels were nitrided at different temperatures for 7 h using active screen plasma discharge. The analysis showed that the thick compound layer composed of ε-$Fe_{2-3}N$ and γ′-Fe_4N was formed on the surface. The corrosion behavior was evaluated by measuring the anodic polarization curves in natural sea water (similar 3.5% NaCl solution), and observation of corroded surface were conducted. The electromechanical measurements indicated that the corrosion potential of the nitrided specimens shifted to a nobler value compared to that of untreated specimens. Passive regions were also observed in the polarization curves for all the nitrided specimens. These results indicate that active screen plasma nitriding can enhance the corrosion resistance of the 38CrMoAl steel under marine environment.

1. Introduction

Plasma nitriding is an effective surface modification technique to improve the surface hardness and corrosion resistance of various steels. By the nitriding process, the α-Fe solid solution becomes enriched with nitrogen and, if the chemical potential of nitrogen in the nitriding atmosphere is sufficiently high, different iron nitrides may form. Many researches have been made to investigate the effects of processing parameters on the microstructures and properties of modified layers formed on various steels by nitriding over past several decades [1-3]. Nitrided layers can been subdivided into a diffusion layer and a compound layer which consist predominantly of iron nitrides ε-$Fe_{2-3}N$ and γ′-Fe_4N and a diffusion layer. The diffusion layer determines the strength of the nitrided layer, as well as its fatigue strength, while the compound layer determines the tribological characteristics and corrosion resistance.

Recent years, many results have been published about the influence of various nitriding techniques on the corrosion behaviors of stainless steels [4-6]. However, very little work about corrosion behavior of nitrided low alloy steels has been done on these aspects [7]. For instance, A. Basu et al. [8] studied the effect of plasma nitriding on the corrosion resistance of high carbon low alloy steel. The evident improvement of corrosion resistance was due to the maximum coverage of nitrides after nitrided at 560 oC for 5 h. Basso et al. [9] reported the plasma nitrocarburizing of AISI H13 tool steel with different processing time from 2 to 20 h and the influence of microstructure of nitrided layer on its corrosion behavior.

Corrosion behavior of nitrided low alloy steels depends on the type of the nitride formed in the compound layer. Nitrided steels are more resistant to general corrosion in comparison with untreated ones, but are subjected to pitting corrosion. The objective of this paper is to investigate the effect of active screen plasma nitriding on the corrosion resistance of 38CrMoAl low alloy steel in 3.5% NaCl solution.

2. Experiments

The chemical composition of the 38CrMoAl steel used for plasma nitriding is (wt%): 0.37~0.42 C, 1.30~1.60 Cr, 0.70~1.10 Al, 0.50~0.80 Mn, 0.30~0.46 Si and Fe balance. The diameter of 25 mm steel bars were cut into 6 mm thick discs. The surfaces of specimens were mechanically polished to a mirror finish before being placed into the nitriding chamber.

The nitriding was performed in active screen discharge plasma nitriding furnace at temperatures ranging from 500 to 530°C for 7 h. The N_2 and H_2 gas was used as nitriding agent to provide active nitrogen atoms for diffusion into the substrate. The working pressure during nitriding was 500 Pa.

The microstructure of the nitrided specimens was investigated using optical microscopy and scanning electron microscopy (SEM). X-ray diffraction (XRD) analysis was carried out to determine the phase composition in the nitrided layer produced by nitriding. The electrochemical experiments were performed with the CHI660e electrochemical working stations in solution of 3.5% NaCl at room temperature. All potentials recorded throughout the test are with respect to a saturated calomel electrode (SCE).

3. Results and Discussion

The cross sections of the 38CrMoAl steels nitrided at 500°C and 530°C for 7 h are presented in Fig.1. The nitrided layer are consisted of the outer compound zone, a thin and dense white layer, and the inner diffusion zone in accordance with usually nitrided low alloy steels. Metallographic observation shows that the thickness of the compound layer is changed from 7 to 11 μm with the nitriding temperature increased from 500°C and 530°C . It can be seen that the layer thickness increases with the increasing operating time and treatment temperature [10].

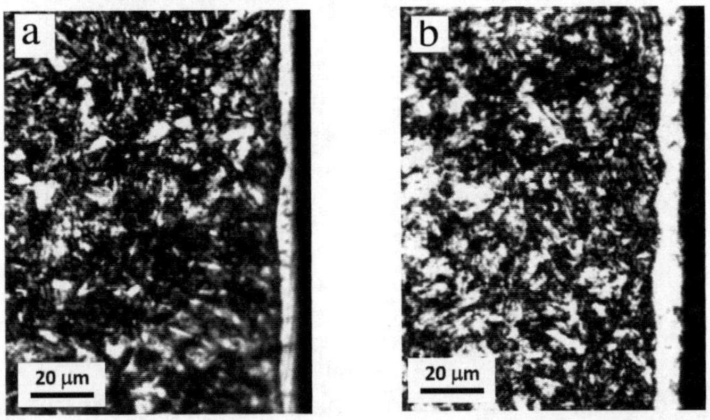

Figure 1. Cross-sections images of nitrided 38CrMoAl steel.

X-ray diffraction patterns obtained from the specimens nitrided at different temperatures are presented in Fig.2. The X-ray analysis confirmed the presence of ε-$Fe_{2-3}N$ and γ′-Fe_4N phases in the compound layer formed at the surface region for all nitrided specimens. The details of the surface properties are given elsewhere [7, 10].

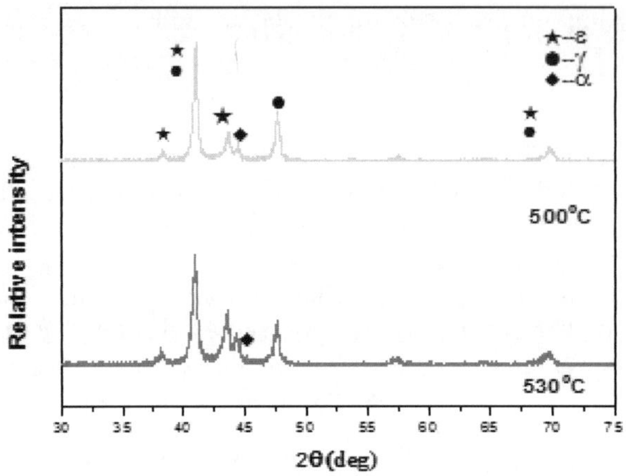

Figure 2. XRD patterns of the specimens nitrided at 500 °C and 530 °C for 7 h.

Fig.3 shows the polarization curves for the untreated and nitrided specimens at temperatures of 500°C and 530°C measured in 3.5 % NaCl solution. The 38CrMoAl substrate has a corrosion potential of -540 mV and exhibits a fast corrosion current increase as the potentia. The corrosion potential for the nitrided specimens at 500°C and 530°C temperatures is -310 mV, -200 mV respectively and is higher than that of 38CrMoAl steel. This result showed that the compound layer formed during nitriding treatment caused the corrosion potential of the nitrided 38CrMoAl steel shift in the noble direction evidently.

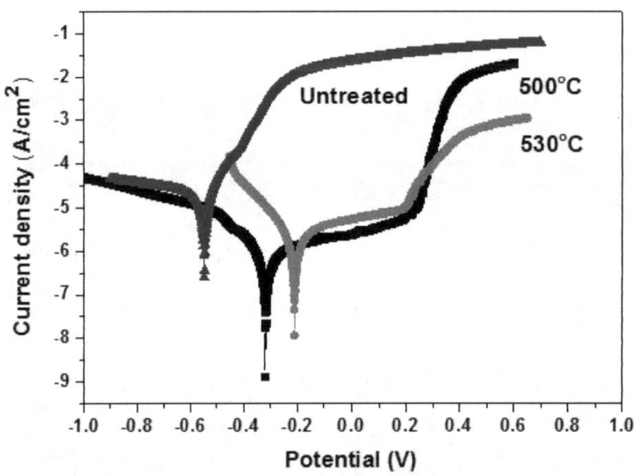

Figure 3. Potentiodynamic curves of anodic polarization of the different specimens in marine solution.

In contrast to the untreated 38CrMoAl steel, all the nitrided specimens show a passivation behavior in the potential range from -200 mV to350 mV the resulted in a very low anodic current in the passive regions. Clearly, the existence and thickness of compound layer plays an important role in determining the corrosion resistance of the nitrided specimens. These results are in agreement with the observations provided by Wen [11]. Ensinger et al. [12] also concluded that a dense nitrided layer irrespective of the type of the nitrides would always enhance corrosion resistance of steel.

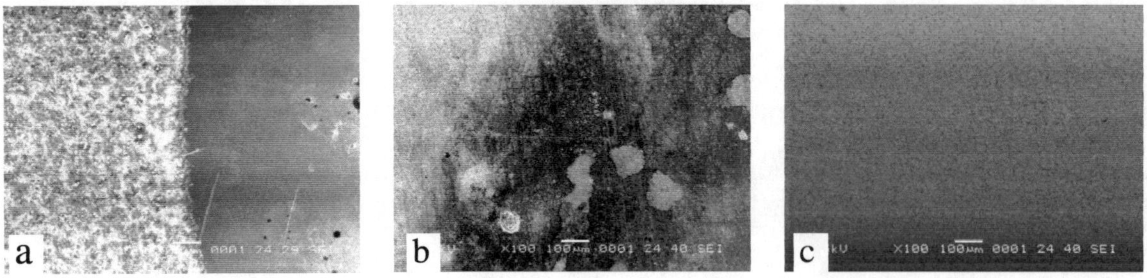

Figure 4. SEM images of the different specimens after polarization tests.

Figs.4 showed the corroded surface morphologies of the unnitirded and nitrided specimens, which had been subjected to the anodic polarization measurements up to a potential of 700 mV. From Fig.4a it can be seen that a large amount of patches of corrosion product and pits were left on the surface of the unnitrided specimen and this showed that severely general and pitting corrosion happened simultaneously during polarization test in 3.5% NaCl solution.

The nitrided specimen at 530 °C for 4 h (Fig.4b) which has a lower dissolution current, shows considerably lower amount of pits regarding the other specimens. Once these pits were formed in some weak locations and penetrated through the compound layer they would expanded in the matrix under the compound layer because the compound layer is nobler than substrate. The results obtained in the present investigation were similar to those previously published for the nitrided low alloy steels [7, 10, 11]. A. Alsaran et al. [13] indicated that the presence of many pits distributed on the surface is a typical result of chloride attack.

The enhanced corrosion resistance can be attributed to the presence of a nitride layer rich in ε-nitride on the surface. As nitride is a noble phase, the formation of more nitride helps to cover the surface from corrosion attack and therefore the corrosion resistance improves with the increase in nitriding temperature(Fig.4c). This has been well-documented in literatures and is also demonstrated in this study by corrosion tests in sodium chloride solution [7, 10]. The local conditions are altered such that further pit growth is promoted. The anodic and cathodic electrochemical reactions that comprise corrosion separate spatially during pitting. The common cathodic reactions that must accompany the dissolution occurring in the pit, such as the oxygen reduction reaction ($O_2+2H_2O+4e^- \rightarrow 4OH^-$), result in a local increase in the pH at the cathodic sites. The acidity developed in the pit is not neutralized by the cathodic reaction because of the spacial separation of the anodic and cathodic reactions.

4. Conclusions

The 38CrMoAl substrate exhibits very poor corrosion resistance in the 3.5% NaCl solution with high dissolving rate. It can be clearly seen that nitriding processing led to a higher corrosion potential and an evident passive region with a very low passive current density in the polarization curves compared to those of the un-nitrided specimen. Hence, the higher corrosion potential and low passive current density of the nitrided low alloy steel can be explained by the formation of a dense compound layer on the nitrided substrate surface during nitriding. The results indicate that the plasma nitriding can significantly improve the corrosion resistance of the 38CrMoAl steel.

5. References

[1] E.A. Ochoa, C.A. Figueroa, F. Alvarez, Surf. Coat. Technol. 200 (2005) 2165.
[2] F.Z. Bouanis, C. Jama, M. Traisnel, F. Bentiss, Corros. Sci. 52 (2010) 3180.
[3] L.C. Gontijo, R. Machado, E.J. Miola, L.C. Casteletti, P.A.P. Nascente, Surf. Coat. Technol. 183 (2004) 10.
[4] A. Fossati, F. Borgioli, E. Galvanetto, T. Bacci, Corros. Sci. 48 (2006) 1513.
[5] C.X. Li, T. Bell, Corros. Sci. 46 (2004) 1527.
[6] M. Kuczynska-Wydorska, J. Flis, Corros. Sci. 50 (2008) 523.

[7] H. Dong, Y. Sun, T. Bell, Surf. Coat. Technol. 90 (1997) 91.
[8] A. Basu, J. Dutta Majumdar, J. Alphonsa, S. Mukherjee, I. Manna, Mater. Lett. 62 (2008)3117.
[9] R.L.O. Basso, R.J. Candal, C.A. Figueria, D. Wisnivesky, F. Alvarez, Surf. Coat. Technol. 203 (2009) 1293.
[10] Y. Li, L. Wang, D. Zhang, L. Shen, J. Alloys Compd. 497 (2010) 285-289.
[11] D.C. Wen, Appl. Surf. Sci. 256 (2009) 797-804.
[12] W. Ensinger, A. Schroer, G.K. Wolf, Surf. Coat. Technol. 52 (1992)217–221.
[13] A. Alsaran, H. Altun, M. Karakan, A. Çelik, Surf. Coat. Technol. 176 (2004) 344-348.

Acknowledgement
The project was supported by the National Key Basic Research Program of China (973) (2014CB046404), National Natural Science Foundation of China (51301149), and China Postdoctoral Science Foundation funded project (2015M570090).

IOP Conf. Series: Materials Science and Engineering **241** (2017) 012009 doi:10.1088/1757-899X/241/1/012009

The scattering data analysis of the correlated X-ray scattering for biomacromolecules in solution

Shengjun Liu [1, 2]

[1]Department of electronic and information engineering, Xiamen City University, Xiamen, China 361008
[2]Department of Applied Physics, Stanford University, Stanford, CA 94305

shengjun@xmcu.cn

Abstract. The correlated X-ray scattering (CXS) was introduced and compared with other methods which were used to study the structures of the biomacromolecules in solution, such as nuclear magnetic resonance (NMR), electron microscope (EM) and small-angle x-ray scattering (SAXS). Then the corrected results of the sum, the mean, the median and the variance of the scattering intensity that we collected from xFEL (x-ray Free Electron Laser) experiments in SACLA were calculated. Furthermore, the density scanning method was used to pick out the useful cluster and the other 28 useless clusters were filtered. Finally, our results will lay a good foundation for the application of correlated X-ray scattering (CXS) in computing the angular correlation of the scattering intensity with the one chosen cluster effectively in future.

1. Introduction

Generally, there are three methods to determine the biomacromolecules and its complex: protein crystallography, nuclear magnetic resonance and cryo EM(electron microscopy). Among them, nuclear magnetic can only be used to determine the proteins with light molecule weight which is less than 40,000[1]. However, the cryo EM can be used to determine the proteins and its complex with the heavy molecule weight which is more than 100,000 [2-4]. And it developed so fast that the resolution can reach 3.4 Å[1,5-9]. In addition, there is no restriction about the molecule weight for the protein crystallography. While the only requirement is that the protein have to be crystallised. And it is mainly be determined by the diffraction ability of the protein. We can find that the structure of the protein which is determined by the protein crystallography can reach 88% in the protein data bank. But this method is restricted by the size and quality of the protein crystal. Cultivating the crystal especially the complex is a exploring process which is short of effective theoretical guide[1,6-9].

At present, a great achievements were obtained because of the application of synchrotron radiation in the biology, especially the structural biology based on the protein crystallography[1,6-9]. With the development of the modern laser technology such as the X-ray free electron lasers (xFELs), some new methods such as SAXS and CXS are used to determine the structure of the objects whose substances are difficult to be crystallised such as proteins [10]. These new methods not only can be used to determine the details of the proteins' structure which can be comparable with the protein crystallography, but also it provides a new way to study the structure in its native environment in a time-resolved manner[10,11].

IOP Conf. Series: Materials Science and Engineering **241** (2017) 012009 doi:10.1088/1757-899X/241/1/012009

When the molecule solution is exposed in the X-ray, the photos scattered by the same particle are correlated. If the movement of the molecule is weak during the X-ray exposures, the differences of the momentum transfer between the correlated photos can reveal the structure of the molecule. And the correlation between the photos are ignored in the small angel X-rat scattering (SAXS) [12, 13]. The diffract-before-destroy nature of serial nanocrystallography can be well applied into single-particle and solution scattering [6, 14-18]. Because the CXS can reveal more details information of the molecule's structure in solution as expected by Kam [19, 20]. So the method to analyse the scattering data for correlated X-ray scattering (CXS) was presented in this paper.

2. Experiments and data analysisormatting

To study the distribution of DNA folds in native environments especially in complex environments, angular correlations of scattered x-rays resulting from xFEL (x-ray Free Electron Laser) experiments was introduced. For calculating the angular correlations more accurately, the useless scattering images have to be filtered. We choose the polyethylene glycol (peg) solution as the sample. Many scattering patterns are collected from the femtosecond X-ray scattering experiments at an XFEL light source in SACLA as figure 1. In our experiment, the photo energy is

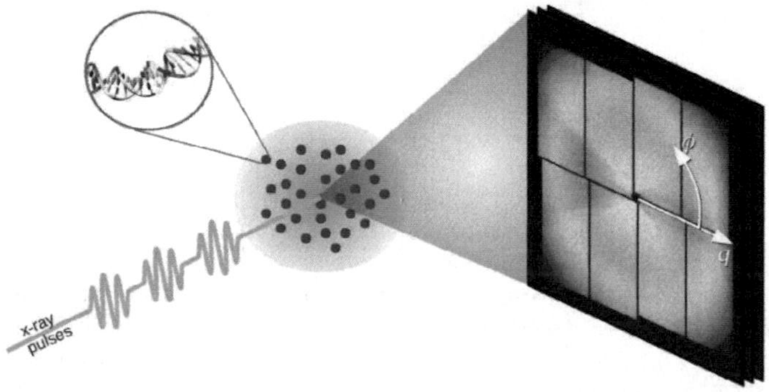

Figure 1. X-ray pulsed exposing a solution of DNA

6.7 kev and the distance between the sample and the detector is 75mm. We tried 20 runs for peg. Among these runs, one run has a fixed number of exposures. In this case, the full run has 5050 exposures (the first 50 of which have the shutter closed). And the data from other beam line monitors exists for all 5050 exposures. So we choose four effective runs from the 20 runs for analysis. And the run number is from 296010 to 296013. On the other hand, we want to focus on the data images themselves. There are so many scattering images during the experiment. So it is important to filter the useless scattering images.

Firstly, we calculated the sum, the mean, the median and the variance of the scattering intensity that we collected from the detector. Then they were divided by the values of photo diode. So we can obtain the corrected results as figure 2. Because there are obvious deviation of the corrected median and variance for run 296010.So we took the median and vari ance of run 296010 as an example for further analysis. Base on the density scanning, seven and twenty-nine clusters were obtained in figure 3. For further study in detail, we choose the twenty-nine clusters and the dark red cluster of it was chosen to be used to calculate the correlation in future and the other twenty eight clusters were filtered.

3. Summary

We choose the polyethylene glycol (peg) solution as the sample in our experiments. Many scattering patterns are collected from the femtosecond X-ray scattering experiments at an XFEL light source in SACLA. Then the sum, the mean, the median and the variance of the scattering intensity that we collected from the detector were calculated. And all these values were divided by the photo diodes which produced the corrected results. A good filtering method was putforwarded and some clusters were showed. One cluster was chosen for calculating the correlation of the scattering intensity and the

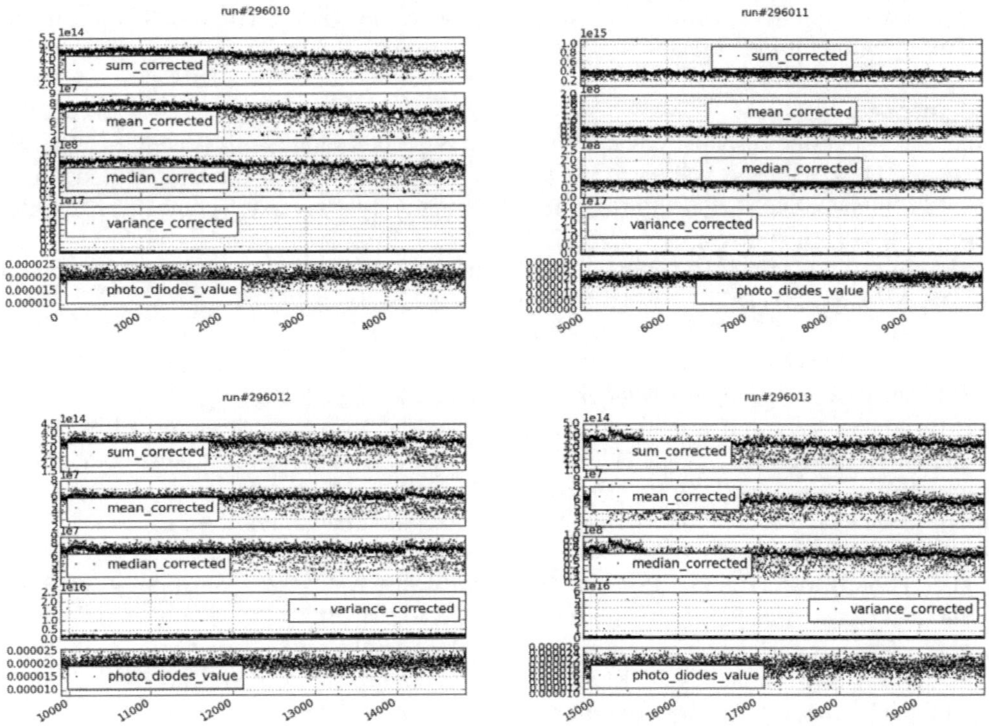

Figure 2. The corrected results of the sum, the mean, the median and the variance of the scattering intensity for four effective runs from 296010 to 296012.

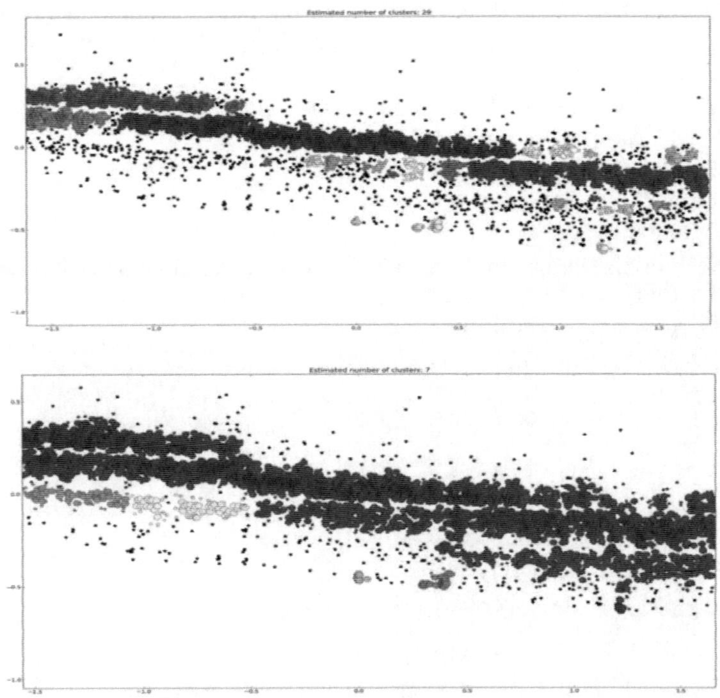

Figure 3. 29 clusters (upper) and 7 clusters (lower)

other clusters were filtered. The density based scanning method will be very usefull for computing the correlation in future. Through the computing the correlation of the scattering intensity, we can obtain the structure information of the biomacromolecules in solution.

Acknowledgments

This research was supported in part by National Nature Science Foundation of China for theoretical physics grant no. 11547238, the visiting scholar research program sponsored by Fujian Provincial Office of Education in 2015, Technical program of Xiamen City University grant no. KYKJ2014-1, National Insti- tutes of Health research grant 251 R01-GM097463, Stanford NIH Biotech- nology Training grant no. 5T32GM008412-20 and US Department of Ener- gy Office of Science under contract no. DE-AC02-05CH11231.

References

[1] Yao D Q and Zhang R G 2014 *Chemistry of Life* **34** 592
[2] Tan H B, Song X Q, Yan H and Xin H X 2017 *Journal of Molecular Structure* **1129** 23
[3] Li A, Ding W L, Wang R Y, He J H, Wang X H, Sun Y X, Gu Y 2017 *Journal of Natural Gas Science and Engineering* **37** 29
[4] Sardar S K and De S K 2016 *Journal of Microscopy and Ultrastructure* **4** 211
[5] Liao M, Cao E, Julius D et al 2013 *Nature* **504** 107
[6] Liu H G, Poon B K, Saldin D K, Spence J C H and Zwart P H 2013 *Acta Cryst.* A 2013 **69** 365
[7] Liu H G, Hexemer A and Zwart P H 2012 *J. Appl. Cryst.* **45** 587
[8] Liu H G, Morris R J, Hexemer A, Grandison S and Zwart P H 2012 *Acta Cryst.* A **68** 278
[9] Liu H G, Poon B K, Janssen A J E M and Zwart P H 2012 *Acta Cryst.* A **68** 561
[10] Emma P et al 2010 *Nature Photonics* **4** 641
[11] Chen G, Modestino M A, Poon B K, Schirotzek A, Marchesini S, Segalman R A, Hexemer A and Zwart P H 2012 *J. Synchrotron Rad.* **19** 695
[12] Derek M., Herschel W, Qiao S L, Kevin S R, Thomas J L, Gundolf S , Garrett N, Ganesh S, Kensuke T, Yasumasa J, Makina Y, Daniel R and Sebastian D 2016 *IUCrJ* **3** 420
[13] Derek M, Thomas J L, Sung J M, Jonas S, Clément L, Herschel W, Aina E C, Michael S, Shirley S, James S, Vijay P, Danielr R and Sebastian D 2014 *Phil. Trans. R. Soc.* B **369** 20130315
[14] Neutze R, Wouts R, Weckert D, and Hajdu J 2000 *Nature* **406** 752
[15] Barty A et al 2012 *Nature Photonics* **6** 35
[16] Seibert M M et al 2011 *Nature* **470** 78
[17] Kuang Q, Xu J C, Liang Y R, Xie F W, Tian F, Zhou S M and Liu X X 2017 *Food Hydrocolloids* **62** 43
[18] Sakai K, Takeshita H, Haraguchi T, Suzuki H, Shashi F, Kume T, Fukuyama A, Nonomura S and Imari T 2017 *Thin Solid Films* **621** 32
[19] Kam Z 1977 *Macromolecules* **10** 927
[20] Kam Z, Koch M and Bordas J 1981 *Proc. Natl Acad. Sci. USA* **78** 3559

IOP Conf. Series: Materials Science and Engineering **241** (2017) 012010 doi:10.1088/1757-899X/241/1/012010

Effect of Secondary Orientation on the Mechanical Behavior of a Unit Cell Model with a Film-cooling Hole in Single Crystal

Gang Cao, Zhixun Wen, Dashun Liu and Zhufeng Yue

School of Mechanics, Civil Engineering and Architecture, Department of Engineering Mechanics, Northwestern Polytechnical University, Xi'an 710072, PR China

caogang@nwpu.edu.cn

Abstract: A 3D unit cell model with a hole has been set up to analyze the stress and strain distribution near the hole of nickel-based single crystal. The emphasis is placed on the influence of secondary orientation on the mechanical behavior of cell model. The crystallographic slip theory was applied to calculate the elasto-plastic deformation. The results show that, the maximum Mises stress and the maximum resolved shear stress are strongly dependent on the secondary orientation. The minimum plastic strain energy density criterion predicts that the optimal orientation of the hole is <100> crystallographic orientation. In addition, the secondary orientation has limited influence on the stress distribution near the hole. But the influence of the secondary orientation on the plastic strain energy density distribution around the hole is obvious.

1. Introduction

At present, most single crystal turbine blades are produced by casting with the airfoil stacking direction under control, and the other two orientations are determined but not controlled, named secondary orientation [1]. Therefore, every single crystal blade has a different secondary orientation [2-4]. Recent reviews indicated that secondary orientation has a significant impact on mechanism behavior because the development of the plastic zone at the crack tip is governed by the crystal anisotropy [5,6]. So understanding the stress and strain response in the cooling holes for single crystal turbine blade is a key feature for aero-jet engines and gas turbines manufacturers.

The present work is an attempt to find out the influence of secondary orientation on the stress and strength of cooled turbine blade. In section two, the FEA model (unit cell model) and the secondary orientation will be defined. Crystallographic constitutive model and parameters are given out in this paper. Finally, we will discuss the simulating results and indicate the optimal orientation of the cooling holes to increase the fracture strength of cooled turbine blade.

2. Crystallographic Theory and Cell Model

The resolved shear stress is the component of the traction along the direction of slip and related to the Cauchy stress through orientation tensor as:

$$\tau^{(\alpha)} = \sigma : \mathrm{P}^{(\alpha)}. \tag{1}$$

Where $P^{(\alpha)}$ is Schmid factor of slip system α. In the rate dependent crystallographic constitutive model, the shear strain rate on the α-th slip system is assumed to be related to the resolved shear stress $\tau^{(\alpha)}$,

$$\dot{\gamma}^{(\alpha)} = \dot{\gamma}_0^{(\alpha)}\left[\frac{\tau^{(\alpha)}}{g^{(\alpha)}}\right]\left[\left|\frac{\tau^{(\alpha)}}{g^{(\alpha)}}\right|\right]^{\frac{1}{m}-1} \tag{2}$$

Where $g^{(\alpha)}$ is the reference shear stress, $\dot{\gamma}_0^{(\alpha)}$ is the reference shear strain rate and m is the strain rate sensitivity exponent. For single crystal, the expression of strain energy density $\mathrm{d}W / \mathrm{d}V = \int_0^{\varepsilon_{ij}} \sigma_{ij}\mathrm{d}\varepsilon_{ij}$ can be written more explicitly as:

$$\left(\frac{\mathrm{d}W}{\mathrm{d}V}\right)_p = \sum_{\alpha=1}^{n}\int_0^{\gamma^{(\alpha)}} \tau^{(\alpha)}\mathrm{d}\gamma^{(\alpha)} \tag{3}$$

In this paper, all the material parameters are as follows, E=918300 MPa，G=972100 MPa，μ=0.386, $\dot{\gamma}_0^{(\alpha)}$=0.003, m=0.02, τ_0=324.9MPa, h_0/τ_0=1.2, τ_s/τ_0=1.17.

A type of cooled turbine blade and the unit cell model are shown in Fig. 1 and Fig. 2, respectively. The cell model is taken from the cooled turbine blade. The dimension of the cell model is 2 mm in length, 2 mm in width and 2 mm in thickness. The radius of the film cooling hole is 0.5 mm. The secondary orientation is defined as the angle of the [0 0 1] orientation relative to the random crystal axis of turbine blade, referred to as α. The definition of α can be seen from Fig. 3.

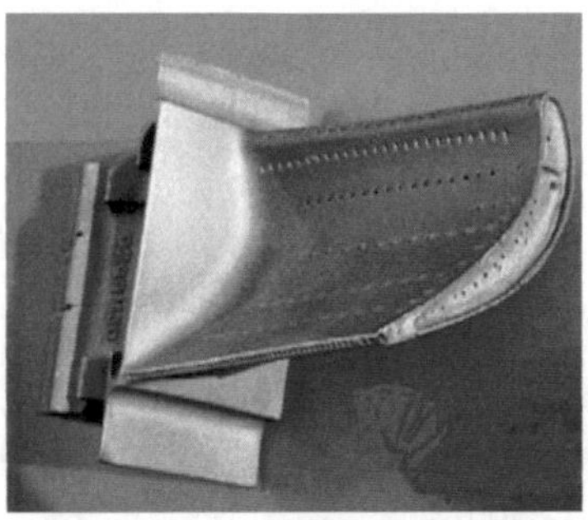

Figure 1. Cooled turbine blade

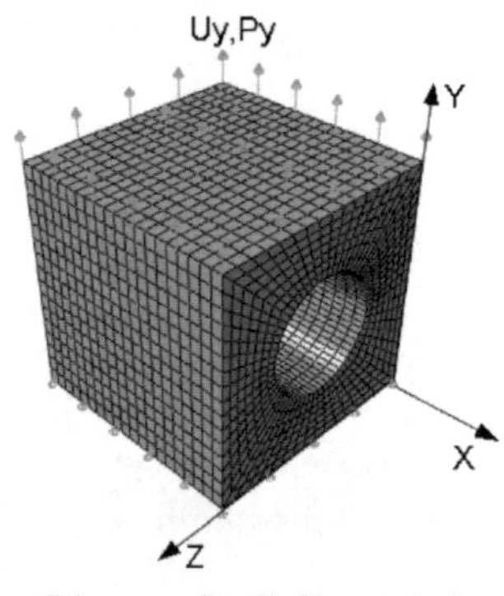

Figure 2. Cell model

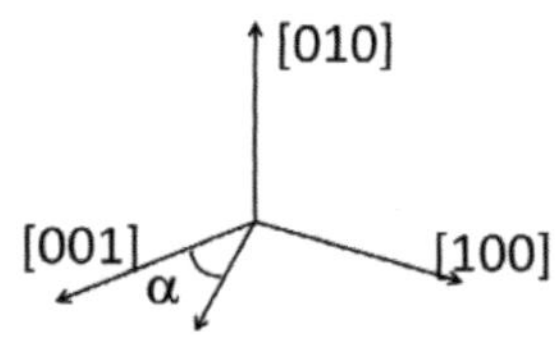

Figure 3. Secondary orientation

3. Mechanical Behavior Near the Hole

Due to the symmetry, the secondary orientation from 0° to 45° is calculated here for simplification. Fig. 4(a)-4(c) show the distribution of Mises stress, maximum resolved shear stress and plastic strain energy density around the hole at the plane x=-2 mm, respectively. In different secondary orientations, both the Mises stress and maximum resolved shear stress around the hole present the similar distribution like the letter 'W', as shown in Fig. 4(a) and Fig. 4(b). The Mises stress along the circle angle of 0° and 180° is the largest, while in the circle with the angle of 0° and 270° is the lowest. It is found that the angle with the maximum resolved shear stress is not the same as that the maximum Mises stress locates. The influence of secondary orientation on the Mises stress and the maximum resolved shear stress along the circle is not obvious. But the secondary orientation has obvious influence on the plastic strain energy density distribution, as shown in Fig. 4(c). Specially, the influence of secondary orientation on the plastic strain energy density at 0° and 180° is remarkable and the change trend is shown in Fig. 4(d). The maximum plastic strain energy density occurs at the circle angles of 0°, 180°±10° and 350°. As expected, the plastic strain energy density criterion would predict crack growth to initiate from these places around the cooling hole.

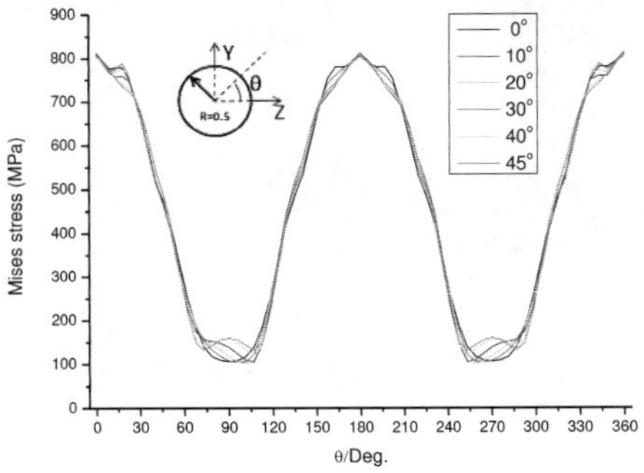

(a) The maximum Mises stress

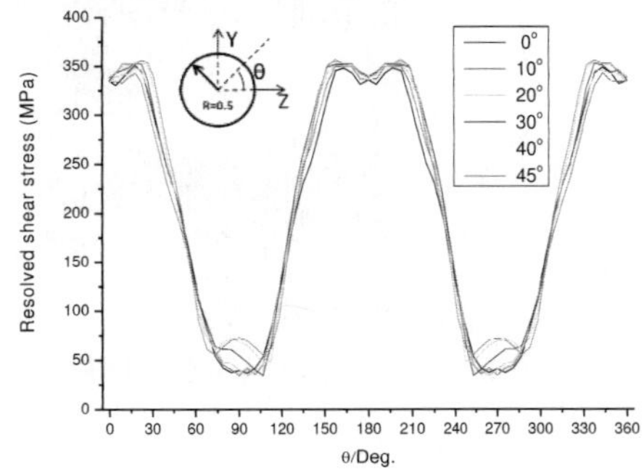

(b) The maximum resolved shear stress

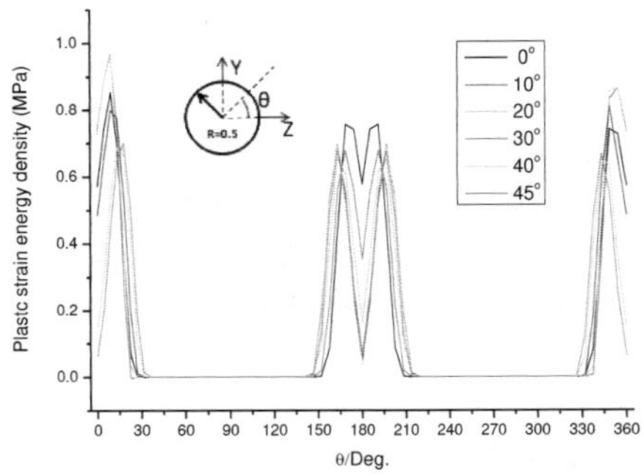

(c) The plastic strain energy density

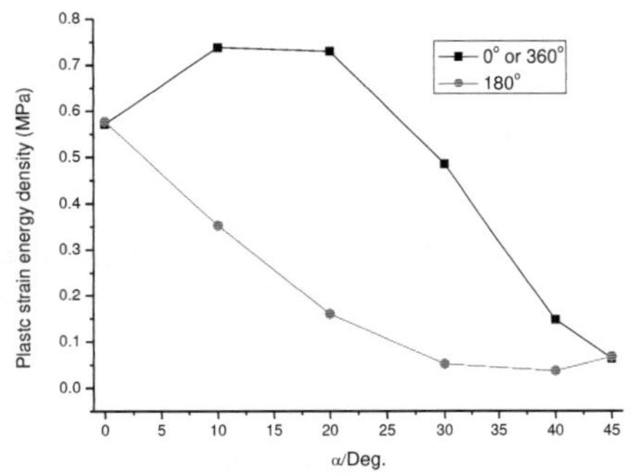

(d) The angles of 0° and 180°

Figure 4. Effect of secondary orientation around the hole

4. Slip Systems Near the Hole

Table 1. Activated slip systems on the surfaces and in the cores of different secondary orientations.

Angle of Secondary orientation	Activated slip systems (strong activation)		Activated slip systems (weak activation)	
	Surface	Core	Surface	Core
0°	2,6,8,10	8,10	3,5,7,12	2,6
10°	2,6,8	8,10	3,5,7,10,12	2,6
20°	2,6	8,10	3,5,7,8,10,12	-
30°	2,3,5,6	8,10	7,8,10,12	-
40°	2,3,5,6	8,10	7,8,10,12	12
45°	2,3,5,6	10,12	7,8,10,12	7,8

The slip system numbered 3, 5, 7 and 12 are not activated at the core, but is weakly active on the surface for secondary orientation α=0°. For different secondary orientation α from 0° to 45°, the slip system numbered 3 and 5 are weakly activated on the surface for 0°, 10° and 20°, but is strongly activated for 30°, 40° and 45°. At the core of unit cell hole, the strong activated slip systems are 8 and 10 except for the secondary orientation 45°, as shown in Table.1. Fig. 5(a) was consistent with the respective slip system number. It can be observed that at the dominant slip field of the core is much more even than that at the surface.

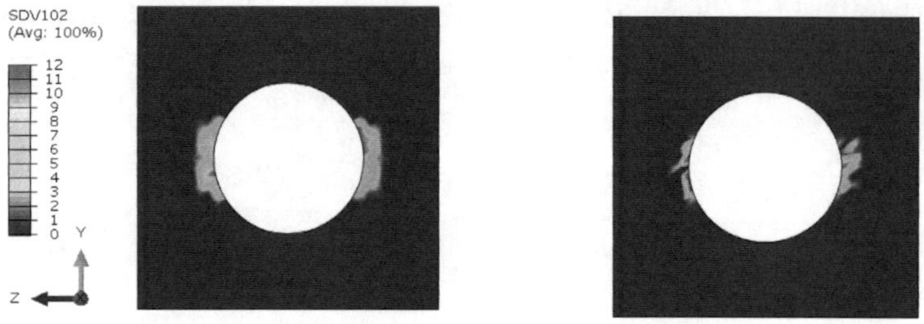

(a) Mid-plane position (b) Surface position

Figure 5. Contours of dominant slip system for secondary orientation

5. Summary

The mechanical response of single crystal with secondary orientation was studied with the cell model by crystallographic FEM. The following summarize are necessary:

1. The distributions of Mises stress and maximum resolved shear stress around the cooling holes are slightly varied with the secondary orientation. But the secondary orientation has obvious influence on the plastic strain energy density distribution around the hole. The initiate cracks occur mainly in the circle with the angle of about 0° and 180°.
2. Comparison between surface and core behaviors, there exist different dominant slip systems and their numbers vary with different secondary orientations.

6. References

[1] Yiyb J. Kim, S.-M. Kim, Int. J. Heat Mass Transf. Vol.47 (2004), p. 245–256.
[2] J.E. Nuñez, G. Glinka., Eng. Fract. Mech. Vol.71 (2004), p.1791–1803.
[3] H.Z. Mao, Z.X. Wen, Z.F. Yue, Mater. Sci. Eng., A Vol.587 (2013), p.79-84.
[4] Q.M. Yu, Z.F. Yue , Z.X. Wen, Mater. Sci. Eng., A Vol.477 (2008), p.319–327.
[5] P.A. Sabnis, M. Mazière, S. Forest, Int. J. Plast., Vol.28 (2012), p. 102–123
[6] D.S. Liu, B.Z. Wang, Z.X. Wen, Adv. Mater. Res. Vol.284 (2011), p.1678-1683.

Acknowledgments

The work was supported by the National Natural Science Foundation of China (51210008, 51375388).

Feasibility Study for Installing Machine in Production Line to Avoid Particle Contamination Based on CFD Simulation

Adisorn Khaokom[a], Jatuporn Thongsri[b]*

Computer Simulation in Engineering Research Group, College of Advanced Manufacturing Innovation, King Mongkut's Institute of Technology Ladkrabang, Bangkok, 10520, Thailand
[a]A.khaokom@gmail.com, [b]Jatuporn.th@kmitl.ac.th

Abstract. Ventilation system inside production line for electronic component production needs to meet the factory standard. Because it can eliminate small particles which may cause of human or machine in production as well as it can distribute the circulating air temperature uniformly. CFD is used in this research in order to study the feasibility and plan for machine layout in production line before actual installation. The simulation shows the airflow in every area inside production line. From simulation with releasing the particles from human and machine is found that this ventilation system generates airflow that makes most particles float out of the machines and no particle downs to the conveyor, it results to contamination. In addition, the simulation also shows the range of 19-26 ^{0}C air temperature that meets the factory standard. The results of this research are the parts of the data to renovate the production line to get more efficiency and proper on the production.

1. Introduction

At present, the electronics industry has developed modern production technology for qualified and good quality products. The component production is essential to work in clean rooms with low particle contamination. Most of particles in clean room are always caused by friction of machine at work or any procedures, as well as humans that including the working suit, hair, skin even breathes itself. The amounts of particles occur depends on each individual human activity [1]. Generally, the sizes of any contaminant, dust and small particles are from 0.001-100 microns. The particles larger than 5 microns are typically found on various surfaces and have the potential to float in the air for short time.

This research is cooperated project with one of electronics components manufacturing factories in Thailand, which plans to improve the production line installing the new machines with new higher efficiency and technology production. Machine layout in production line inside clean room will undoubtedly affect to airflow pattern and temperature distribution [2]. Therefore, if study the feasibility and plan in advance before install machine in production line actually, it makes production quality good and reduces production cost.

From studying, it was found that Computational Fluid Dynamics (CFD) has been applied for simulating in order to solve any problems related to airflow, temperature and contamination in buildings and environment. For example, CFD is applied for simulating the airflow in the clean room of the LCD industry [3] and in the classroom [4] that the simulation results can improve airflow quality using CFD. CFD is applied for studying the cooling by simulating the airflow in large-scale factories by CFD that the simulation results can be used for cooling generated in the factories efficiently [5]. CFD is applied for monitoring the particle transportation in the office with ceiling ventilator attaching [6]. CFD is applied for studying the emergence and the distribution of human particles in the indoor environment [7]. As well as CFD has been used to simulate the airflow in small environments of clean rooms for hard disk drive factories [8] and in automatic machinery for hard disk drives production [9]. That the simulation results can be used to determine the best air conditions and can be really used in the factory to reduce the contamination of particles that may occur in the automatic machinery for hard disk drive production.

So CFD is used in this research to simulate airflows, temperature and particle traces in production line under actual ventilation system of factory in order to assess the air quality that if the machines have been installed whether the efficiency of the airflow will meet the factory standard. The results of this research will be used to improve the production line for further production.

2. Methodology

2.1. Production Line and Machinery Models

The production line including machines to be installed used for the simulation is in the Class 1000. The model must be a simple model in order to simplify the simulation setting any environmental conditions as clean room in the actual factory [8]. The air from air conditioning flows into the FFU and Ulpa Filter on the ceiling and flows out the return shaft, return ceiling and all 3 sides of wall of the specific simulation. The laminar is also installed above the machine that the laminar sucks the air inside the room from the top of the laminar through the HEPA filter, can eliminates the particles floating in the air. Then release the clean air into the machine for driving the particles to outside machine in order to minimize particle contamination from the machine and human.

To solve the problem with CFD, aims to find the direction of airflow occurred, heat transfer in a clean room, and particle behavior caused by machinery and human. The model is created with decreasing the complexity in the irrelevant parts in order to simplify for simulation by measuring the size of the real machine. The machine consists of three main parts: 1) Machinery is function of producing the products; 2) Conveyor is function of transporting the components to the machinery; 3) Human is responsible for bringing in and out of production as shown in Fig. 2.

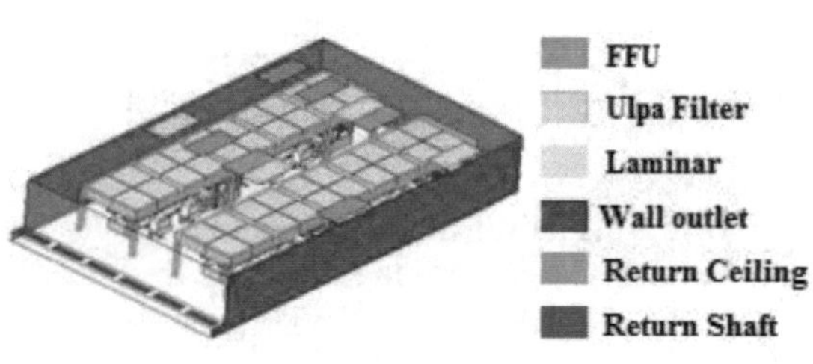

Figure 1. Production line

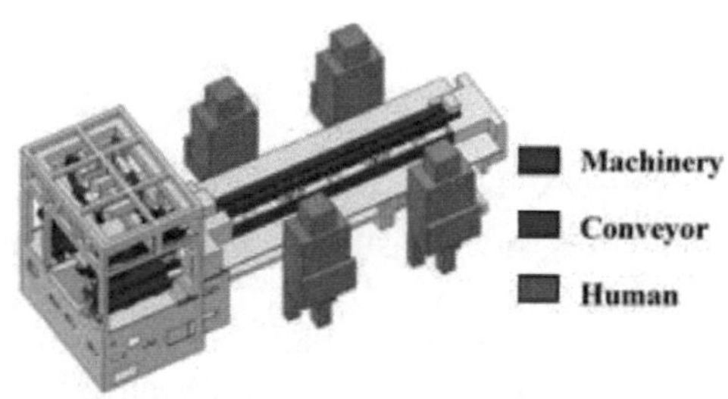

Figure 2. Machinery Model with human inside

2.2. Fluid and Mesh Models

The FLUENT program is used based on the models shown in Figs.1 and 2 to create the fluid and mesh models respectively. Then the mesh model created will be used in further simulation. The fluid and mesh models used in this work are shown in Fig. 3, respectively. The type, size and quality of the mesh affect to the time for calculation and the result in the process of this mesh model creation.

The mesh model created, has 1.41 million nodes, 7.3 millions elements of tetrahedron and 0.93 of the maximum skewness. Based on the mesh analysis, it was found that this mesh model is suitable for both quality of the result and the time for calculation.

IOP Conf. Series: Materials Science and Engineering **241** (2017) 012011 doi:10.1088/1757-899X/241/1/012011

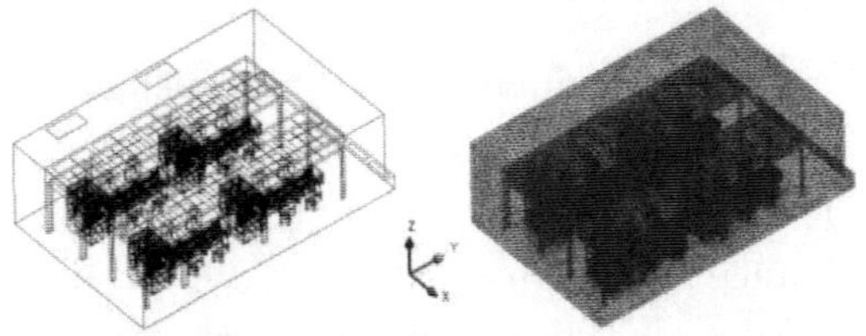

Figure 3. Fluid and mesh models

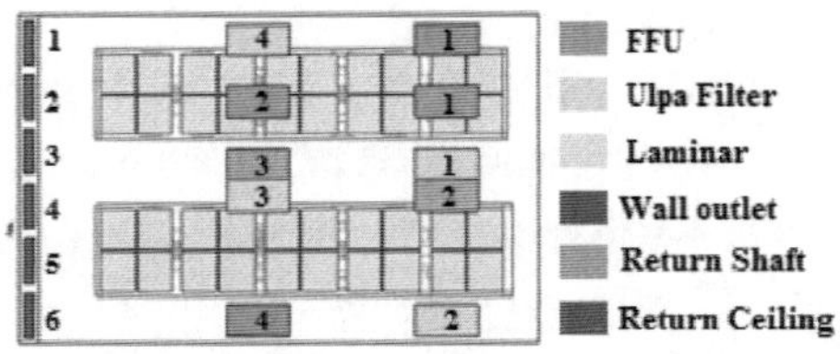

Figure 4. The position of ventilation system

2.3. Mathematical Modeling

The conservation equations consist of mass (1), momentum (2), and energy (3) conservations.

$$\partial\rho / \partial t + \nabla \cdot (\rho\vec{v}) = S_m . \quad (1)$$

$$\partial(\rho\vec{v}) / \partial t + \nabla \cdot (\rho\vec{v}\vec{v}) = -\nabla p + \rho\vec{g} + \vec{F} , \quad (2)$$

$$\partial(\rho E) / \partial t + \nabla \cdot (\vec{v}(\rho E + p)) = \nabla\left(\sum_j h_j J_j\right) + S_h . \quad (3)$$

Where ρ is the static pressure and $\rho\vec{g}$ and $\vec{F}$ are the gravitational body force and external body forces, respectively. S_m and S_h are user-defined source terms.

The Transition Shear Stress Transport (Transition SST) [10] is used for turbulence model in this paper because of its high reliability. Transition SST consists of turbulence kinetic energy (k), specific dissipation rate (ω), transition momentum thickness Reynolds number (Reθt), and intermittency (γ) equations. This paper is limited space, therefore the equation cannot be expressed and described at all. You can study more information in the Reference [11]. The computer must solve 3 conservation equations and 4 turbulence equations of total 7 equations based on the principle of CFD to find the solution in the calculation of turbulence model.

2.4. Fluent Settings

Based on environmental conditions measurements from the real factory that affect to small specific simulations, it was found that the air flows from 2 ways: FFU and Ulpa Filter on the ceiling area with temperature of 20.5^{0}C and 13^{0}C respectively. The conditions of inlet air are shown in Table 1.

When the air flows into his specific environmental simulation, laminar installed on machine will suck the air in the room from the top of the laminar itself through the HEPA filter. Then release the clean air to the machine for driving the particles to outside machine and flow out to this environment in 3 ways: return shaft, return ceiling and 3 sides of wall of the specific simulation. The conditions of out-flow air are shown in Table 2. By the position specified in Figure 4.

In addition, laminar with velocity of 0.45 m/s is installed on machine to blow air into the machine in order to remove the particles out. The shear stress transport of turbulence model with pressure-velocity as coupling, momentum, turbulence kinetic energy and turbulence dissipation values determined as second order upwind, human temperature as 37^{0}C, machinery temperature as 27^{0}C, and environment temperature and pressure as 24.5^{0}C and 101,250 Pascal, respectively. For the other values are set automatically by the program as appropriate. This calculation is a steady state of 1,000 cycles for accurate convergence.

Table 1. Condition of In-flow Air.

FFU (Position)	Mass Flow Rate (kg/s)	Ulpa Filter (Position)	Mass Flow Rate (kg/s)
1	0.449	1	0.384
2	0.568	2	0.398
3	0.449	3	0.36
4	0.449	4	0.36

Table 2. Condition of Out-flow air

Return Shaft (Position)	Outflow (%)	Wall (Position)	Outflow (%)	Return Ceiling (Position)	Outflow (%)
1	7.56	1	1.90	1	6.07
2	7.56	2	0.46	2	5.02
3	12.62	3	2.17		
4	12.32				
5	14.61				
6	12.06				

IOP Conf. Series: Materials Science and Engineering **241** (2017) 012011 doi:10.1088/1757-899X/241/1/012011

3. Results and Discussion

Airflow in every area of ventilation system was found from this simulation. And this condition may result in vortex in some areas as shown in Fig. 5 that it occurs in the areas away from the conveyor and it does not affect to production line. Fig. 6 shows the flow directions that the air flows through the machine then flow out at Return Shaft. The air velocity average at the conveyor line is 0.24 m/s that it's more than the velocity of 0.2 m/s. Therefore, it's considered as the air circulates at the conveyor is effective.

By simulating to release 400 particles within the defined time as shown in Fig. 7 found that the particles caused of human are blew out and eliminated by Return Shaft and laminar, so it's no contamination at conveyor line area. Most particles caused of machine are eliminated by the same way however, there is potential risky for contamination. In the worst case, 3.17% contaminated particles caused of machine is found from testing releasing 4000 particles.

The simulated results show the temperature distribution of the Ventilation system as shown in Fig. 8. The temperature average at the conveyor is 20.71^{0}C; all points are less than 22^{0}C that meet the factory requirement. However, the figure shows that the temperatures coherences are possible to have humidity, so it's interesting for further study.

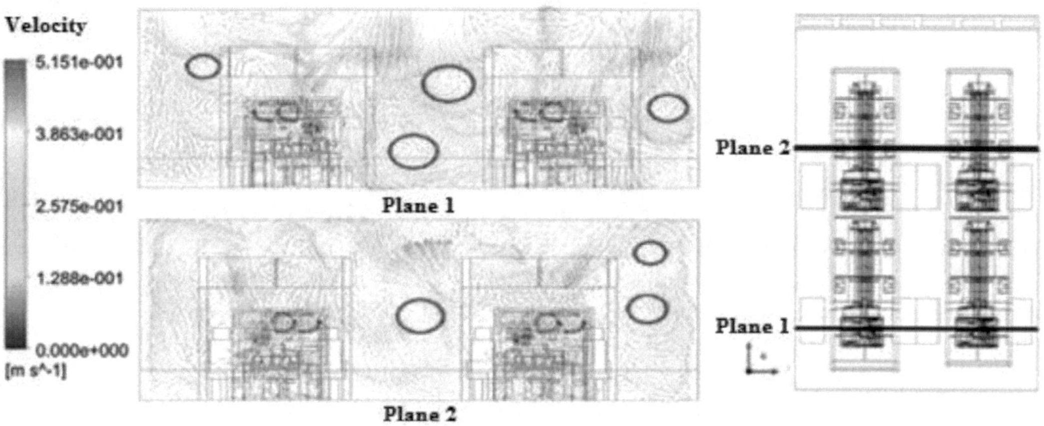

Figure 5. Airflow directions in XZ plane

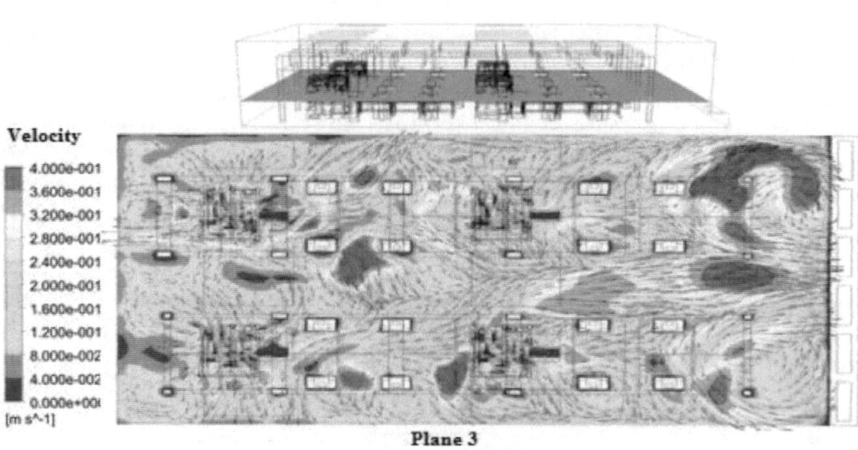

Figure 6. Airflow direction in ZY plane

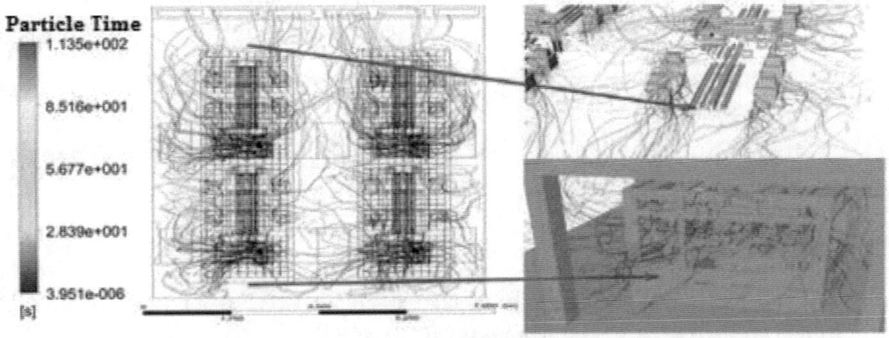

Figure 7. Particle traces caused by machine and human

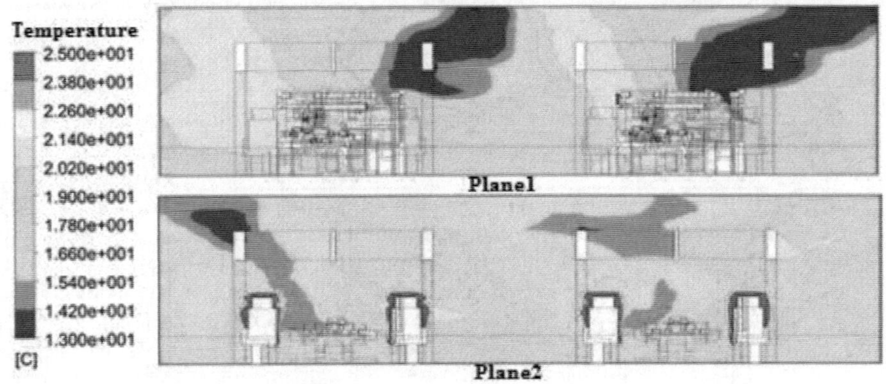

Figure 8. Temperature inside the production line in XY plane

4. Conclusion

In this study, CFD has been used to simulate the airflow, particle traces and temperature distribution in production line with planning to install in clean room of factory. The simulation result with various conditions including actual ventilation system of factory was found that the air flow in production line has efficiency to avoid the contamination caused of machine and human. In addition, the temperature in production line is proper and meets the factory requirement. The information from this research will be used for renovating production line of industrial factory.

Acknowledgment

The authors would like to thank College of Advanced Manufacturing Innovation (AMI) and King Mongkut's Institute of Technology Ladkrabang (KMITL) for supporting this research.

References

[1] McDonagh A and Byrne M 2014 *J Environ Radioact.* **127** 119-26
[2] Thongsri J, Pimsarn M and Tangsopha W 2016 *KKU Eng. J.* **43** 250-53
[3] Noh K C, Kim H S and Oh M D 2010 *Build. Environ.* **45** 825-31
[4] Noh K C and Yook S J 2016 *Build. Environ.* **119** 111-18
[5] Tanasic N, Jankes G and Skistad H 2011 *Build. Environ.* **43** 1200-06
[6] Makhoul A, Ghali K, Ghaddar N and Chakroun W 2013 *Build. Environ.* **63** 97
[7] Tao Y, Inthavong K, and Tu J 2016 *Indoor Built Environ.* **26** 185–98
[8] Montrisavatkul P, Kaewkhaw P, Tangsopa W and Thongsri J 2016 *The 4th Academic Science and Technology Conference 2016* (Thailand: ASTC2016) pp 295-301
[9] Thongsri J and Pimsarn M 2015 *Int. J. Precis. Eng. Manuf.* **16** 509–15
[10] Menter F R 2009 Int. J. Comput. Fluid Dyn. **23** 305-16
[11] Ansys Inc 2013 *Ansys fluent 15 user's guide* (USA: Ansys Inc)

IOP Conf. Series: Materials Science and Engineering **241** (2017) 012012 doi:10.1088/1757-899X/241/1/012012

Study of the viscosity of hydrocarbon mixtures under pressure and temperature: A critical model of the corresponding states to double reference in the modeling domain.

Aziz ETTAHIR[1a], Christian BONED[2], Bernard LAGOURETTE[2], Kamal KETTANI[1], Khaoula AMARRAY[1] and Mohammed GAROUM[1].

[1]Materials, Energy, Acoustics Team (Meat), University Mohammed V of Rabat – Morocco.
[2]Laboratory of Complex Fluids and Their reservoirs (LFCR), University of Pau – French.
[a] corresponding author.

ettahiraziz@hotmail.com, christiaboned@univ-pau.fr, bernardlagourette@univ-pau.fr, kamal.kettani@gmx.com, khaoula.amarray@gmail.com, garoum1@yahoo.fr

Abstract. The studied predictive model of behavior viscosimetric is the model of K.A. Petersen [1]. The dominant idea of this method is to characterize the viscosity of a fluid from two models taken as a reference in passing through a reduced pressure. The method is corresponding state with two references. This study shows that this method is dependent on the choice of reference and for each of the possibilities of C_{10}/C_6H_6 and C_1/C_{10} references . The results were investigated for four different weight ratios. It shows that the introduction of an adjusted coefficient does not improve significantly compared to results without adjustment factor, which appears to be the best choice. Regarding the influence of the choice of references, generally the two couples appear suitable but we noted that the choice is not necessary. In the case of mixtures containing at least one aromatic, the results are correct, especially if one takes the ratio of adjustment and our ratio without adjustment compared to that of K. A. PETERSEN[1]. The experimental results of the viscosity exhibit a good agreement with the calculated values. We can predict that the relative improvement is the finding that the introduction of the second body of reference (C_{10}) from the model states corresponding to a reference (C_1) of the authors.

1. Introduction

A more sophisticated variant of the model states corresponding to a reference [1] does not introduce the form factor and which retains the powers of critical viscosity $\eta c=C\ Tc^{1/6}\ M^{-1/2}\ Pc^{-2/3}$ is the model of corresponding states to double reference. It was initially proposed by A. S. Teja and P. Ricz (1981) [2] and taken up by K. Aasberg-Pedersen (1991) [3]. In order to experiment properly and to test this model and critique their performance on mixtures of hydrocarbons, we used the experimental database (15000 points) that we created [4] on the viscosity and the density, depending on the pressure and temperature, light and heavy hydrocarbons, and their mixtures.

2. Methods

Four our study, we selected viscosity data on pure substances and mixtures of hydrocarbons of known compositions. For the latter, it is important to know precisely the composition of mixtures because the critical viscosity $\eta cm = C\ Tcm^{1/6}\ Mm^{-1/2}\ Pcm^{-2/3}$ involved the terms Mm, Pcm, Tcm. Considering the results of the literature, we chose three rules of mixtures, PEDERSEN [3], ETTAHIR [5] and WONG [6]. Recall here that the rules of PEDERSEN and WONG lead to the same Tcm and Pcm values and differ

IOP Conf. Series: Materials Science and Engineering **241** (2017) 012012 doi:10.1088/1757-899X/241/1/012012

only for ωc. But with the rule PEDERSEN, we adopt the definition of Mm proposed by K. AASBERG-PEDERSEN (1991) [2], [3] explained by the following equation:

$$Mm = 8.67358\ 10^{-3}\ (Mw^{1.56079} - Mn^{1.56079}) + Mn\ ;$$

$$Mw = \sum xi\ Mi^2 / \sum xi\ Mi \ et\ \ Mn = \sum xi\ Mi$$

For rules of WONG-ETTAHIR we used the classical linear law: $Mm = \sum xi\ Mi$.

In addition, we have considered two possibilities of double reference, first methane (C_1) and decane (C_{10}), on the other decane (C_{10}) and benzene (C_6H_6). In the latter case, we check if it is possible to facilitate the return of the viscosity of heavy compounds like alkanes or aromatic, given the references adopted.

Finally, we wanted to test several distinct definitions of the factor K involved in the equation below, rather than limiting itself to the acentric factor (ω) as in the literature. K. AASBERG-PEDERSEN (1991) [2], [3] admit the neighboring relation:

$$\ln \eta r(P,T) = \ln \eta r1(P1,T1) + K \ln (\eta r2(P2,T2)/\eta r1(P1,T1)) \quad avec\ K = (IP-IP1)/(IP2-IP1)$$

Where IP denotes an arbitrary prior parameter which is used to characterize the weighting between two references .Initially this parameter corresponds to IP = ω and the last authors proposed to use IP = M (Molar mass). Since ηr = η / ηc we can finally write:

$$\eta(P,T) = \eta c.\ \eta 1(P_1, T_1)/\ \eta c_1.[\eta_2(P_2,T_2).\ \eta c_1 / \eta_1(P_1,T_1).\ \eta c2]K$$

Where η1 and η2 are evaluated under the conditions corresponding to the temperature and the reduced pressure of the system studied, that is to say: $T_i = T.Tc_i/Tc$ et $P_i = P.Pc_i/Pc$ $(i=1,2)$.
The indices 1 and 2 are relative to the two reference bodies of which the characteristics are known and K is a weighting factor. One of the major disadvantages of the method is clear: it is linked to the fact that its use is a priori limited by the reduced pressure and temperature domain common to the two references: indeed, η, η1 and η2 must be evaluated for Same temperatures and pressures. So, and from our database of references we have limited ourselves to the following areas of use:

$0.476 \leq Tr \leq 2.740$	$Pr \leq 15.0$	Methane
$0.395 \leq Tr \leq 0.773$	$Pr \leq 49.6$	Decane
$0.395 \leq Tr \leq 0.664$	$Pr \leq 20.4$	Benzene

As mentioned above we opted for a comparative study of the factor K according to the following four weights:

- $K\omega = (\omega - \omega 1)/(\omega_2 - \omega_1)$ **(Eq 1)**
- $K_M = (M - M_1)/(M_2 - M_1)$ **(Eq 2)**
- $K_A = (A_\omega K_\omega + A_M K_M + A_P K_P + A_T K_T)/4$ **(Eq 3)**
- $K_{Total} = (K_\omega + K_M + K_P + K_T)/4$ **(Eq 4)**

With $Kp = (Pc - Pc_1)/(Pc_2 - Pc_1)$ et $K_T = (Tc - Tc_1)/(Tc_2 - Tc_1)$

The weighting factors K_{Total} and K_A have been introduced of kinds to better differentiate two neighboring systems, for example two isomers which necessarily admit the same K_M. Indeed, it is intuitive to think that one of the four quantities ω, M, Pc, Tc is in itself sufficiently characteristic of the system studied. As regards the use of K_M, the comparison of the results obtained with this report only reflects the evaluation difference of Mm since for the two rules WONG and ETTAHIR Tcm and Pcm are identical. At the end In K_A the coefficients A_ω, A_M, A_P, A_T are adjustable quantities. For the latter weighting coefficient and for each of the two possibilities of double reference C_1/C_{10} and C_{10}/C_6H_6 we have chosen the same pure bodies to constitute an adjustment basis, but taking into account the overlap constraint the number of points retained By body changes from one reference torque to the other. For the C_1/C_{10} pair, 18 pure bodies and 280 experimental points are involved, whereas for C_{10}/C_6H_6 there are 15 pure bodies and 168 experimental points. The values of the coefficients obtained after numerical adjustment are:
Reference C_1/C_{10} :

$A_\omega = -0.338049087$ $A_M = 2.096563583$ $A_T = 0.144785919$ $A_P = 2.096699584$

Reference C_{10}/C_6H_6 :

$A_\omega = 4.765634686$ $A_M = 10.08678000$ $A_T = -3.056947926$ $A_P = -7.795467124$

In order to asses and compare the performance of various References, we defined the following quantities:

$$Dev(i) = 100 \times (\eta_{exp} - \eta_{cal}) / \eta_{exp} \qquad Dev\ abs(i) = |Dev(i)| \geq 0 \qquad DM = Max(Dev\ abs(i))$$

$$AAD = 1/Nb \times \sum_{i=1}^{Nb} Dev\ abs(i)$$

$$Bias = 1/Nb \times \sum_{i=1}^{Nb} Dev\ abs(i)$$

In which Nb is the number of experimental points, η_{exp} the measured viscosity value and η_{cal} the value calculated using a given reference. The quantity AAD (Average Absolute Deviation) indicates how close the calculate curves are to the experimental curves and the quantity Bias indicates how well the experimental points are distributed to either side of the calculated curves. Finally DM characterizes the maximum error that can be generated using a given reference.

Table 1. Comparison of results obtained based on coefficients K_A adjustment.

Reference		K_M	K_ω	K_A	K_{Total}
C_1/C_{10}	AAD	13.8	9.9	7.4	10.3
	Bias	-11.0	-4.5	-0.1	-5.5
	DM	40.6	43.0	28.5	38.7
C_{10}/C_6H_6	AAD	19.1	20.5	15.7	19.8
	Bias	-16.1	-16.2	-14.0	-16.2
	DM	63.4	63.3	60.5	64.2

There is in both cases an improvement on all tree features of the adjustment relating to K_A and the improvement is very clear in the case of the couple C_1/C_{10}.

3. Discussion of results

3.1. Results obtained for the whole of the base in the field of Recovery:

The application of the method of double reference to the determination of the viscosity of mixtures requires the prior definition of pseudo-critical coordinates. The given results are demonstrated by using the residual viscosity correlation [7] [8] [9] [10]. We restricted to the three rules of mixtures PEDERSEN, WONG and ETTAHIR. The presented results are obtained on a set of mixtures of hydrocarbons derived from our database in the area of recovery.

Initially we used the couple C_1/C_{10} as double reference. The mixtures were in turn separated into two main groups. The first included five binary containing CO_2 and C1 or the second group consists of 16 binary, ternary and quaternary one containing neither C_1 nor CO_2. Table 2 shows the values of mean deviation and we can see that the mixing rule ETTAHIR was less well for the return of mixtures with C_1 and CO_2. For mixtures without C_1 or CO_2, the results are similar for all three rules. It appears that the best relationships are K_A and K_{Total}. These results are in the same way as for pure substances [5]. Table 3 provides detailed results of all the mixes for the average deviation. We noticed that the binary cyclohexane are the least represented. Figure 1 shows the average deviation of mixing, obtained with the rule of Pedersen and weight K_{Total}. We also noted that the maximum deviations are relatively low except for binaries consisting of cyclohexane.

In a second step, we have taken the couple C_1/Benzene as a double reference. Note that no point has been upheld in our database for the mixture C_1+C_4 and ethanol+toluene [11] due to the constraints collection already exposed. In this case the maximum deviation (51.7%) is obtained for the mixture of C_7+C_8 when it was observed for the binary cyclohexane+C_8 with other references. The results of both

groups are provided in table 4. For all data, the report K_{Total} (combined with the rule and Pedersen WONG) leads to the smallest average deviation. With this couple C_1/Benzène, binaries cyclohexane containers are represented better with the couple C_1/C_{10} (see figure 1 and 2).
With the exception of this class of mixtures, the pair definitely needed C_1/C_{10} similarly to the case of pure [5]. This confirms once more that the representativeness of the model is of benefit conditional on the adoption of remote references by their chain length rather than the nature of chemical families (see figure 2). With this pair the viscosity of mixtures of light paraffin in the various possible phases as well as mixtures of heavy bodies is described correctly (see figure I-off). A wider use of the procedure with respect to natural gas or petroleum fractions, system more complex than synthetic blends taken into account in the database, may be feasible if the pseudo-critical parameters are accessible.

Table 2. Average deviation for the reference C_1/C_{10} (Application of all the 2 groups without extrapolation).

Group	Nb	Rule	KM	Kω	KA	KTotal
Group 1 :	520	1	12.1	13.0	10.5	11.1
(Mixture without C1 and CO2)	520	2	11.3	12.9	11.1	11.3
	531	3	12.6	15.9	13.7	13.9
Group 2 :	499	1	10.6	11.1	11.0	10.0
(mixture with C1 and CO2)	499	2	12.9	10.5	13.2	11.1
	345	3	21.5	20.2	21.8	20.7
Group 1 + Group 2	1019	1	11.4	12.1	10.8	10.6
	1019	2	12.1	11.7	12.1	11.2
	876	3	16.1	17.6	16.9	16.6

1: Rule PEDERSEN, 2: Rule WONG, 3: Rule ETTAHIR, Nb: Number of points used in the field of recovery (without extrapolation)

Table 3. Mean deviation for the reference C_1/C_{10}.

Mixture	Reference	Nb	Rule	KM	Kω	KA	KTotal
Toluene+n-Hexane	[12]	32	1	3.9	3.0	4.0	5.0
		32	2	3.9	3.0	4.1	5.0
		32	3	4.3	3.4	3.9	5.5
Toluene+ Pristane	[13]	32	1	11.1	9.3	4.0	3.3
		32	2	3.5	7.8	8.6	4.1
		30	3	13.8	25.1	21.2	19.3
Toluene+Heptamethylenonane	[13]	33	1	11.2	9.1	5.3	7.4
		33	2	6.6	8.6	6.3	7.0
		38	3	8.3	18.5	14.1	12.9
Toluene+1-Methylnaphthalene	[13]	31	1	14.9	5.8	5.4	7.7
		31	2	13.9	5.8	5.8	7.2
		30	3	9.3	6.4	9.0	4.4
n-Butylbenzene+n-Tetradecane	[14]	30	1	15.5	6.0	4.5	7.0
		30	2	14.2	7.3	3.9	6.9
		30	3	11.3	3.4	3.2	4.8
n-Decane+n- Tetradecane	[14]	36	1	4.5	4.0	6.1	5.9
		36	2	4.6	4.0	6.6	6.1
		36	3	5.3	5.1	8.2	8.0
Nonylbenzene+n-Heptane	[14]	32	1	7.9	36.9	5.5	11.7
		32	2	7.8	39.5	5.9	13.9
		34	3	12.1	43.4	10.7	20.8
n-Decane+n-Tetradecane+	[14]	26	1	11.7	4.2	4.4	6.8
nButhylebenzene		26	2	10.8	5.2	3.7	6.7
		28	3	9.3	2.9	3.0	4.8
Cyclohexane+n-Hexane	[15]	28	1	23.4	24.9	26.1	22.0
		28	2	23.4	24.8	26.1	22.0
		28	3	23.1	24.7	25.9	21.8
Cyclohexane+n-Octane	[16]	51	1	22.1	23.8	23.8	20.8
		51	2	22.3	23.3	24.1	20.8
		51	3	23.3	24.9	24.8	21.9
Cyclohexane+n-Dodecane	[16]	25	1	29.3	34.6	32.9	31.1
		25	2	31.2	33.0	34.3	31.5
		26	3	36.2	39.8	38.3	37.2
Cyclohexane+n-Hexadecane	[16]	17	1	33.3	42.1	38.8	38.3
		17	2	37.4	39.4	42.0	39.4
		19	3	50.0	54.7	52.3	52.4
n-Hexane+n-Dodecane	[17]	2	1	8.9	9.7	9.1	8.6
		2	2	11.6	8.6	11.2	9.5
		2	3	30.1	28.2	27.4	27.4
n-Octane+n-Dodecane	[17]	5	1	4.6	4.7	5.0	4.8
		5	2	4.9	4.5	5.2	4.9

IOP Conf. Series: Materials Science and Engineering **241** (2017) 012012 doi:10.1088/1757-899X/241/1/012012

		5	3	6.6	6.2	7.5	7.1
n-Decane+n-Heptane	[18]	4	1	5.9	8.0	7.6	9.4
		4	2	5.5	8.3	7.0	9.1
		4	3	5.3	5.6	5.4	6.1
n-Decane+n-Hexadecane	[18]	14	1	5.7	3.7	4.6	5.2
		14	2	5.1	3.7	5.6	5.7
		16	3	5.0	6.7	9.3	10.2
n-Decane+n-Dodecane+ n-Tetradecane+n-Hexadecane	[18]	32	1	5.0	3.3	5.1	5.4
		32	2	4.3	3.3	5.7	5.6
		32	3	3.5	4.5	7.8	8.2
n-Octane+n-Heptane	[19]	90	1	4.5	6.0	5.3	6.6
		90	2	4.5	6.0	5.3	6.6
		90	3	4.4	5.8	5.2	6.4
Methane+n-Decane	[20] [21]	242	1	15.4	16.9	14.5	14.8
	[22] [23]	242	2	19.5	15.7	18.0	16.6
		128	3	34.6	34.2	32.2	33.6
Methane+Propane	[21] [24]	76	1	9.4	6.7	13.0	8.6
		76	2	10.5	6.6	14.0	9.2
		65	3	20.3	17.6	21.2	18.7
Methane+n-Butane	[21]	16	1	5.3	2.8	6.3	2.9
		16	2	5.8	2.6	6.7	3.0
		0	3	-	-	-	-
Methane+Ethane	[22]	112	1	4.6	6.4	4.8	4.3
		112	2	4.4	6.5	5.1	4.3
		112	3	4.8	4.1	8.2	4.8
n-Decane+CO2	[24]	53	1	4.7	3.4	6.4	4.2
		53	2	6.7	3.3	8.6	4.9
		40	3	28.3	24.7	27.5	27.0

Table 4. Mean deviation for the reference C_{10}/Benzene (Application of all the 2 groups without extrapolation).

Group	Nb	Rule	K_M	K_ω	K_A	K_{Total}
Group 1 :	315	1	15.3	15.9	14.1	15.6
Mixture without C_1 and CO2	315	2	15.1	15.9	14.2	15.5
	351	3	15.3	16.6	15.0	15.9
Group 2 :	215	1	16.7	19.2	36.7	12.6
mixture with C_1 and CO2	215	2	18.3	19.4	37.9	13.9
	194	3	28.5	30.7	56.1	24.5
Group 1 + Group 2	530	1	15.9	17.2	23.3	14.4
	530	2	16.4	17.3	23.8	14.9
	545	3	20.0	21.6	29.6	19.0

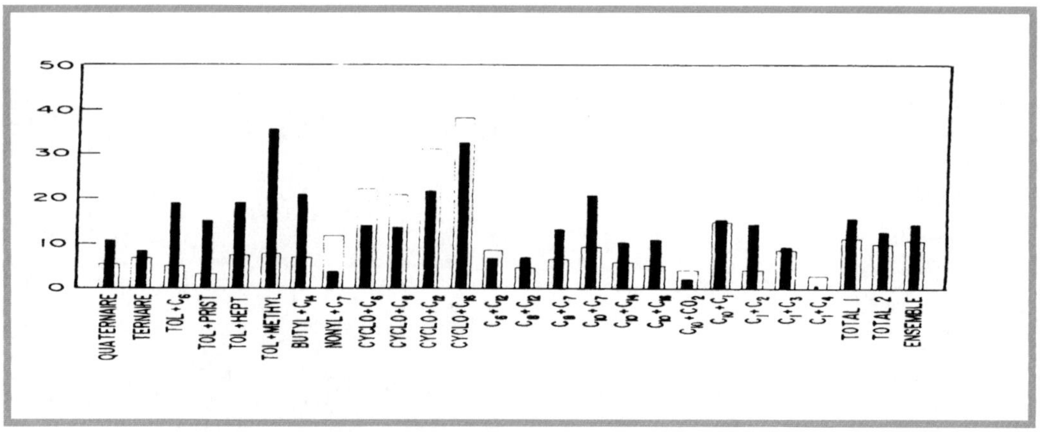

Figure 1. Average deviations obtained for PEDERSEN's rule and report of K_{Total} for mixtures without extrapolation:

■ Reference C_1/C_{10} □ Reference C_{10}/Benzene

IOP Conf. Series: Materials Science and Engineering **241** (2017) 012012 doi:10.1088/1757-899X/241/1/012012

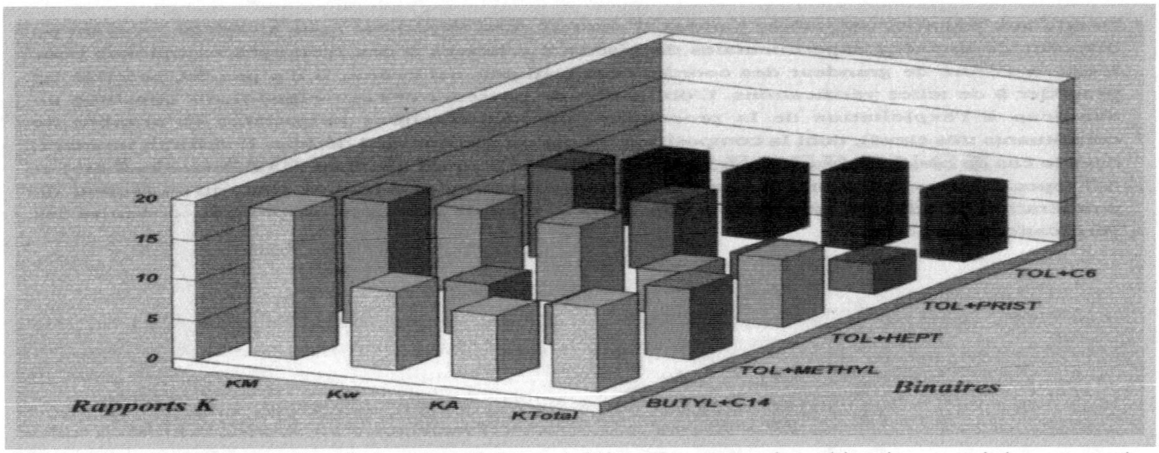

Figure 2. Comparison of performance of the weighting K reports about binaries containing aromatic (Average deviation and reference C_1/C_{10}).

3.2. Results obtained on the entire base after extrapolation outside the area of recovery:

As before we used the couple of references C_1/C_{10} by extrapolating from the base to field $0476 \leq Tr \leq Pr \leq 2.74$ and 100, so with 4712 pairs P, T instead of 1007 without extrapolation. The results are provided in Table 5 do not show an overall deterioration in respect to the average deviation. Therefore, for the areas studied Pr and Tr, stress recovery can be lifted without difficulty. These good results by using the parameters KA and K_{Total} deserve to be highlighted. Figure 3 is typical of the average deviation found with the rule of Pedersen and the weight K_{Total} and warrants that the results obtained with an extrapolation compare favorably to those in the absence of extrapolation.

Table 5: Average deviation for the reference C_{10}/Benzene (Application of the 2 groups together with extrapolation).

	Nb	Rule	K_M	K_ω	K_A	K_{Total}
Group 1 :	1328	1	14.4	15.1	12.2	12.6
Mixture without C1 and CO2	1328	2	13.6	15.2	12.6	12.7
	1379	3	14.5	17.4	14.5	15.0
Group 2 :	3384	1	9.7	11.5	8.2	10.0
Mixture with C1 and CO2	3384	2	9.5	11.3	8.3	9.8
	3271	3	12.5	12.8	12.1	12.6
Group 1 + Group 2	4712	1	11.0	12.5	9.3	10.7
	4712	2	10.7	12.4	9.5	10.6
	4650	3	13.1	14.2	12.8	13.3

1: Rule PEDERSEN, 2: Rule WONG, 3: Rule ETTAHIR, Nb: Number of points used in the field of recovery (without extrapolation)

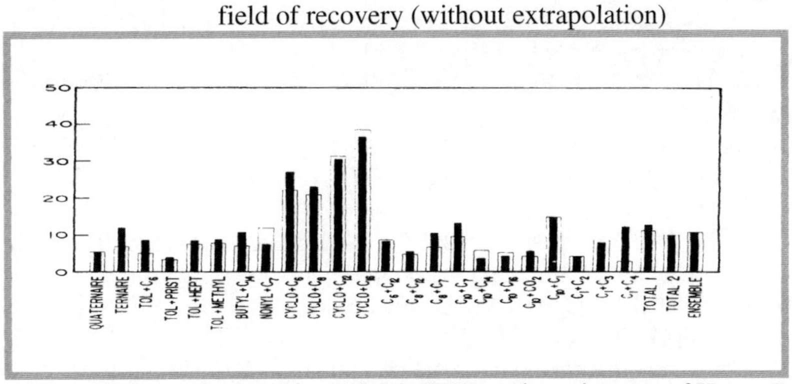

Figure 3. Average deviations obtained for PEDERSEN's rule and report of K_{Total}: Case of mixture (with and without extrapolation) and reference C_1/C_{10}:

☐ without extrapolation ■ with extrapolation

4. Conclusion

We examined the performance of the model states corresponding to double references. The results show that that the introduction of the adjusted coefficients in the ratio of K_A does not improve significantly compared with K_{Total} and it seems to be the best choice. The Reports ω, K_M and K, were easy but they presented not good results compared to K_A and K_{Total}. This finding is confirmed for the pure body in our work presented in ETTAHIR thesis. About the influence of the choice of references, we found that the two sets of references appear proper, but we note that the choice is not indifferent by using the C_{10}/Benzene and occurs cyclohexane and minimizes the maximum deviation from the couple C_1/C_{10}. This generates the highest average deviation for this group of mixtures, regardless of gear choice K. However, only the two lends itself to an extension beyond the field of recovery, outlook favorable for lifting the restriction of the constraint method. In addition, analysis of results indicates that the rules are those WONG PEDERSEN and to be given priority, the rule is less effective ETTAHIR although its average deviation (13.3%) on the field compared with the extrapolated K_{Total} is very respectable. But the other two rules are better, approaching 10%.
Note that the performance of this predictive model vis-à-vis other elements that ring, are much more convincing because the deviations observed are considerably reduced compared to those induced by other models tested, such as the model of residual viscosity and the model states corresponding to a reference, especially in the case of liquid mixtures and heavy bodies. Small deviations were also found for the mixture CO_2+C_{10}, despite a high concentration of CO_2 (greater than 50%), promising results on the ability of the model to simulate the influence of CO_2 on the viscosity of fluids in reservoirs.
The viscosities calculated for the quaternary C_{10}+C_{12}+C_{14}+C_{16}+C_{10} butylbenzene the ternary C_{14} and other heavy components in mixtures, are very close to experimental results, unlike the case of model states corresponding to a reference. The marked improvement of the representation is undoubtedly the introduction of the second reference (C_{10}), which promotes long carbon chains. With the single reference model (C_1), only lights paraffin were well represented. These performances suggest therefore predict potential capabilities of the model to be extended to heavy oil.

References

[1] PEDERSEN K.S., FREDENSLUND A., Chemical Engineering Science, Vol42, N°1, (1987),182-186.
[2] EJA A.S., RICE P., Industrial Engineering Chimical Fundamental. Vol20,(1981),77-83.
[3] AASBERG-PEDERSEN K.,These, Licentiatafhanding, Dnmarks Tekniske Hojskole. (1991).
[4] ETTAHIR A.Thèse de doctorat d'Université - Université de Pau et des Pays de l'Adour (France). (1993) .
[5] AASBERG-PEDERSEN K., KNUDSEN K., FREDENSLUND, A.Fluid Equilibria,Vol 70,(1991),293-308.
[6] A. ETTAHIR, BONED C., LAGOURETTE B., P. XANS. Colloque CIMTEC2010, Faenza Italy, 13- 18 Juin 2010.
[7] ALLIEZ J., BONED C., LAGOURETTE B., ETTAHIR A., High Pressure Food Science, Bioscience an chemistry, N°222, (1998), 75-80.
[8] WONG D.S.H. SANDLER S.I, TEJA A.S,Industrial Engineering Chimical Fundamental.Vol23,(1984),38-44.
[9] JOSSI J.A., STIEL L.I., THODOS G., A.I.Ch.E. Journal, Vol8, (1962),59-63.
[10] DYMOND J.H., ROBERTSON J., ISDALE J.D.,International Journal of Thermophysics, Vol12,N°2,(1991), 275-287.
[11] A. ETTAHIR, BONED C., LAGOURETTE B., P. XANS, K. KETTANI, M. REGRAGUI, 7ème congress of mechanic (19-22 Avril 2005)-Rabat-Morocco.
[12] ETTAHIR A., BONED C., LAGOURETTE B., XANSE P., International Journal of Thermophysics, Vol16,N°6,(1995), 1309-1334).
[13] KANTI M., Thèse de doctorat - Université de Pau et des Pays de l'Adour (France). (1988) .
[14] ISDALE J.D., DYMOND J.H., TREVOR A.B., High Temperature, High Pressure, Vol11, (1979),571-580.
[15] TANAKA Y., HOSOKAWA H., KUBOTA H., MAKITA T., International Journal of

Thermophysics, Vol12,N°2,(1991), 245-264)

[16] DYMOND J.H., ROBERTSON J., ISDALE J.D., International Journal of Thermophysics, Vol12,N°2, (1981),133-154.

[17] ZHOU H., Thèse de doctorat - Université de Pau et des Pays de l'Adour (France). (1988) .

[18] AKHUNDOV R.T., ABDULLAEV F.G., RAMAZANZADE M.G., Ivz. Vyssh. Uchebn, Zaved, Neft Gaz,, Vol26,N°3, (1983),53-54.

[19] KNAPSTAD B., SKJOLSVIK P.A., OYE H.A., Ber. Bunseges Phys. Chem., Vol94, (1990),1156-1165.

[20] LEE A.L., Published by American Petroleum Institute, (1965),7-128.

[21] HUANG E.T.S., SWIFT G.W., KURATA B., A.I.Ch.E. Journal, Vol11,N°2, (1979),571-580

[22] C.K.ZEBERG-MIKKELSEN, A.BAYLAUCQ, G.WATSON, C.BONED , International Journal of Thermophysics, 26 (5) (2005) 1289-1302.

[23] A. BAYLAUCQ, C. BONED, X. CANET, C.K. ZEBERG-MIKKELSEN, S.E. QUIÑONES-CISNEROS, H. ZHOU. Journal of Petroleum Science and Technology 23 (2005) 143-157 .

[24] DILLER D.E., J. Chem. Eng. Dat., Vol29, (1984),215-221.

IOP Conf. Series: Materials Science and Engineering **241** (2017) 012013 doi:10.1088/1757-899X/241/1/012013

A Study on Wet and Dry Tensile Properties of Wood pulp/Lyocell Wetlace Nonwovens

Yinjiang Zhang[1,2,a], Chao Deng[1], Benchen Qu[1], Qu Zhan[1] and Xiangyu Jin[1,a]

[1]Key Laboratory of Textile Science & Technology, Ministry of Education, College of Textiles, Donghua University, No. 2999 North Renmin Road, Songjiang, Shanghai 201620, China
[2]Shaoxing Vocational &Technical College, No.526 Tingshan Road, Yuecheng district, shaoxing 312000, china

[a]Corresponding's author: jinxy@dhu.edu.cn, zyjdhdx2011@hotmail.com

Abstract. A new biodegradable wood pulp/Lyocell moist wipe had been developed, which made from wetlaid/spunlace(wetlace) technology. The dry and wet tensile curve characteristics were described and the relationship between dry and wet strength in both machine direction (MD) and cross-machine direction (CD) were investigated. The results indicate that the fabricated wetlace materials are composed of the entanglements and cohesions of wood pulp/Lyocell fibres. The modulus and tensile strength of the materials were obviously decreased in wet state, and the tensile curves in the dry and wet state both can be divided into two parts. It is noted that there exists a high linear correlation between the dry and wet strength in MD or CD. Meanwhile, the diminished amplitude of wet strength in CD is larger than that of wet strength in MD and the relationship fluctuation between the wet and dry strength in CD is significantly higher than that in MD.

1. Introduction

With the economic development and improvement of life quality, the dispersible moist wipes increasingly were being sought by consumers for their convenience and effectiveness [1, 2]. As of now, wetlace method was an advanced processing technology as a result of 100% degradation or disintegration after use and sufficient dispersion and fabricated materials have became a current trend for moist wipes [3, 4]. Nevertheless, this existing technology still faced several difficulties and need to be examined. It is important to highlight that the wetlace nonwovens are completely made up of cellulose fibres, and MD or CD tensile properties of material will be changed significantly in wet state. As expected, this change is related to the fibre entanglement performances of the materials, and also importantly affects the material characteristics in-use and the product price [5]. But no published literature could be found yet to sufficiently investigate these respects.

In this paper, the wood pulp/Lyocell wetlace nonwovens were fabricated. The apparent structures of wetlace nonwoven were characterized by Scanning Electron Microscopy (SEM). The mechanism of the tensile properties of material dry and wet state was revealed in detail. The relationship between the dry and wet strength in MD or CD based on different basis weights and thicknesses was investigated. Overall, this may guide the dispersible wetlace nonwoven for high value moist wipes.

2. Methods and materials

Preparation of wood pulp/Lyocell wetlace nonwovens: the average lengths of wood pulp and Lyocell were 2.2 mm and 12 mm, respectively. According to required blend ratio (80/20) of wood pulp/Lyocell, two types of fibres both were mixed in water and then drained on an inclined wire to form the homogenous wet-laid nonwoven. Then the fibre web was bonded subsequent multiple passes of jet heads. Eventually, the wetlace nonwovens were dried in an online gas-fired, through-air drum dryer before being wound on to a roll at the end of the production. In this study, 30 samples with different basis weights and thicknesses were prepared at Hezhong Nonwoven Co. Ltd, Zhejiang, China. Characterizations: the prepared wetlace specimens were observed via SEM (TM3000) with 15kV electron. The determination of the dry and wet strength (both MD and CD) of the nonwoven fabrics had been implemented according to the guidelines of ISO 9073-3:1989 strip test method [6]. The width and the test gauge length of samples respectively were 50 and 200 mm and the stretching rate was100 mm/min. In order to determine the wet strength of materials, materials were firstly pre-wetted by water spraying to mimic moist wipes and then measured immediately. At least five samples were tested and the average values for tensile strength at break were recorded.

3. Results and Discussion

As can been seen in Figure 1(a), the wetlace nonwoven includes ribbon-like structure of relatively higher entanglement along the MD. It is noticeable that Lyocell fibre is longer than the wood pulp and is entangled to stride over several stripes. In addition, the Lyocell fibres aggregation forms the three dimensional skeleton structure involving a great number of wide-angle "*U*" shape entanglements and the wood pulp are entangled into it. Figure 1(b) shows pronounced fibre entanglements and cohesions structure of local fibres from Figure 1(a).

Figure 1. SEM images of wetlace nonwoven: (a) apparent structure and (b) fibre entanglements.

The MD or CD tensile curves of wood pulp/Lyocell wetlace nonwoven (65g/cm^2) in dry and wet state show in Figure 2(a) and (b), respectively. Compared with the condition of the dry state, the tensile strength and modulus in MD or CD decreased significantly in wet state. As mentioned previously, Lyocell fibres with longer length are easily entangled and form the entanglement structure to bear stretching. This entanglements are mainly based on the accumulation of basic "*U*" shape entanglement [7] and the bearing force from basic "*U*" shape entanglement depends on the friction coefficient and the wrapped angles between fibres [8, 9]. Although fibres swelling in wet state increase the fibre diameter as well as wrapped angles slightly, the friction coefficient has greater influence on the fibre displacement and the failure of entanglements than the wrapped angles. Therefore, the sharp decrease in friction coefficient in the wet state leads to decreasing the tensile strength greatly as well as modulus.

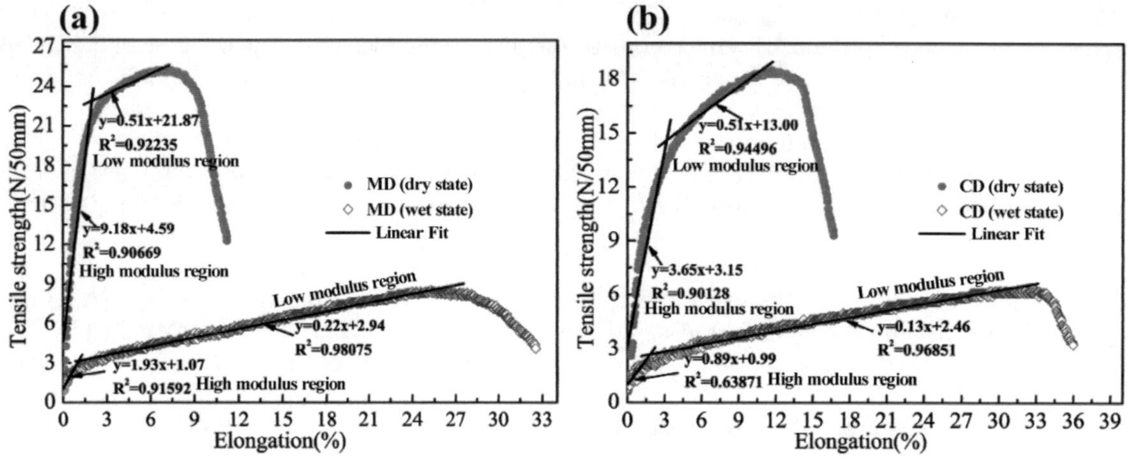

Figure 2. The tensile curves in dry and wet state: (a) MD and (b) CD.

Meanwhile, it is evident that the tensile curves of wetlace nonwovens (see Figure 2) can be divided into high modulus and low modulus regions. Besides, the region of high modulus is reduced substantially and the region of low modulus is increased greatly in wet state. Because the high modulus region attributes to the fibre initial displacement and the low modulus region mainly depends on the fibre straightening and slippage. The decrease of friction coefficient between fibres reduces the initial displacement of fibres and enlarges the fibres stretch time. Notably, in addition to the small correlation degree of CD high modulus region in wet state (see Figure 2(b)), the other parts of tensile curves all have highly linear fitting due to fibres MD arrangement after hydroentanglement [10, 11]. It is noticeable that the MD and CD tensile curves in the wet state have pronounced fluctuation as a result of small friction coefficient between fibres in the wet state. In addition, from the dry state to wet state, the MD high modulus ranges from 9.18 to 1.93 and the rate of decline is 78.98%, and the low modulus changes from 0.51 to 0.22 and the decreasing rate is 56.86%. The CD high modulus decreases from 3.65 to 0.89 and the decrease rate is 75.62%, and the low modulus falls from 0.51 to 0.13 and the reduction rate is 74.51%. The decline amplitude of the MD high modulus is larger than that of the CD high modulus, but the decline amplitude of the MD low modulus is smaller than that of the CD low modulus. This phenomenon is mainly due to the fibre entanglements and ribbon-like structures along MD.

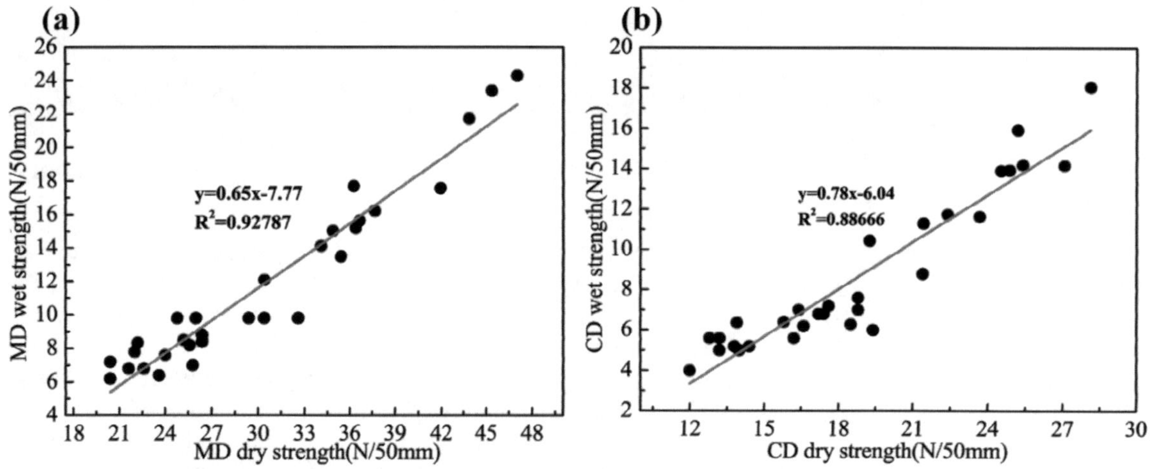

Figure 3. The relationship between the wet and dry strength: (a) MD and (b) CD.

Figure 3(a) and (b) respectively show the relationship between the dry and wet strength of 30 samples in MD and CD. No matter vertical or horizontal direction, it is evident that the increase in dry strength of samples results in increasing the wet strength, which present highly linear correlation. This trend is responsible for no structure change of material in dry or wet state. The fitting linear slope between the dry and wet strength in CD is larger than that in MD, namely that the CD wet strength is reduced relatively larger than that of the MD wet strength. Moreover, the correlation degree between the dry and wet strength in CD is smaller than that in MD, thus the relationship volatility of dry and wet strength in CD is larger than that in MD. This may be due to the results that the MD tensile strength is mainly affected by the fibre entanglement, and the CD tensile strength is primarily caused by the fibre displacement and entanglements. Additionally, the coefficient of friction in the wet state decreases sharply, which greatly weakens the friction force caused by fibre displacement and affects the failure of entanglement.

4. Conclusion

In this research work, the prepared wood pulp/Lyocell wetlace nonwovens have obvious ribbon-like structure. The Lyocell entanglements form the three dimensional skeleton structure, and the wood pulp is entangled into it. No matter wet or dry state, the MD and CD tensile curves both can be divided into high modulus and low modulus regions. The high modulus region significantly decreases but the low modulus region increases substantially in wet state. In addition to the small correlation degree of CD high modulus region in wet state, the other parts have the highly linear correlation. Furthermore, there is highly linear correlation relationship between the dry and wet strength in MD or CD.

Acknowledgments

The authors wish to express their sincere gratitude to Zhejiang Hezhong Non-woven Co., Ltd. for their support in sample production. This work was supported by the Fundamental Research Funds for the Central Universities (Grant No.CUSF-DH-D-2016017).

References

[1] W. Bunyard, K. Branham, M. Lostocco, T. Dyer, J. Hockersmith, K. Possell, U.S.Patent App11/300,967.(2005).
[2] Information on http://www.nonwovens-industry.com.
[3] Zwick KJ, Vogel NJ, Baker JK, U.S.Patent App. 14/068,847. (2013).
[4] K.J. Zwick, N.J. Vogel, W. Lee, J.K. Baker, C. Ackroyd, U.S.Patent 9,320,385.(2015).
[5] S.J. Russell, Handbook of nonwovens, Woodhead Publishing2006.
[6] EN ISO 9073-3:1989 Textiles-Test methods for nonwovens-Part 3: Determination of tensile strength and elongation.
[7] N. Mao, S.J. Russell, A framework for determining the bonding intensity in hydroentangled nonwoven fabrics, Composites Science and Technology 66(1) (2006) 80-91.
[8] H. Wang, L.S. Pang, X.Y. Jin, B.P. Yin, H.B. Wu, The Influences of Hydrophilic Finishing of PET Fibers on the Properties of Hydroentangled Nonwoven Fabrics, Journal of Engineered Fibers and Fabrics 5(4) (2010) 26-32.
[9] Y. Zhang, X. Xu, X. Jin, Z. Zhang, Analysis of elastic recovery property based on the mesh structure of spun-laced nonwovens, Advanced Materials Research, 2013, pp. 19-24.
[10] A. Rawal, D. Moyo, V. Soukupova, R. Anandjiwala, Optimization of Parameters in Hydroentanglement Process, J Ind Text 36(3) (2007) 207-220.
[11] Y. Zhang, Y. ZHAO, M. Latifi, R. Wang, X. Jin. Investigation of the mechanical and dispersible properties of wood pulp/Danufil wetlaid nonwovens with/without hydroentanglement[J]. J TEXT I, 2017. DOI: 10.1080/00405000.2017.1362747.

IOP Conf. Series: Materials Science and Engineering **241** (2017) 012014 doi:10.1088/1757-899X/241/1/012014

Research on wire rope stress distribution of WR-CVT

Wu Zhang, Wei Guo, Chuanwei Zhang, Zhengxiong Lu, Xiaobin Xu

School of Mechanical Engineering, Xi'an University of Science and Technology, Xi'an, Shaanxi , 710054, China

corresponding author's: zhangwu0828@163.com

Abstract. A wire rope continuously variable transmissions (WR-CVT) has been introduced in the paper, in view of its less research, this paper mainly studied the stress distribution of 6×7+IWS bending wire rope. The results shown that the wire stress is layered distribution in each section, the stress at the outer strand center wire and outer strand side wire was the greatest, the stress value of the outer strand side wire and metal block circular notch is second. As the transmission ratio decreases, the wire stress decreases, which is related to the pulley working radius increases. Compared with the section A1, the stress value on the section A2 is smaller, mainly because the section A2 is not in contact with the metal block or the contact pressure is small. This study provides a basis for the study of fatigue and wears failure of WR-CVT components.

1. Introduction

Over the past two decades, continuously variable transmissions (CVT) have aroused a great deal of interest in the automotive sector due to the potential of lower emissions and better performance [1]. Metal pushing V-belt and metal chain CVT have been developed for automotive applications. In a comparison of the fuel consumptions of vehicles with different transmission types on the US market for the US Environmental Protection Agency city cycle [2], it was seen that, generally, vehicles with CVT achieve a higher fuel economy than vehicles with stepped automatic trans-missions do. According to forecasts made by CSM Worldwide [3], it is expected that the market share of CVT equipped vehicles will keep growing in the future. Nowadays CVT are widely used in agricultural vehicles, mini cars and motor-cycles, and more and more attention is placed on vehicle com-fort, fuel economy and low environmental impact. To achieve this objective, a continuously variable transmission could be helpful [4]. Wang, Dagang, et al [5] researched the stress distributions along the contact path (between the rope and friction lining) by using finite element analyses, the result indicate that with an increases of hoisting parameters, the stress concentrations increases. A wire rope CVT (WR-CVT) has been designed by Wu Zhang et al [6], WR-CVT as shown in Figure 1. Traditional wire ropes are commonly used in long-distance and low-speed transmission systems, and the CVT is classed as a high-speed transmission system. Therefore, the application of closed wire rope not only enhances the CVT-product type, but also extends the application of closed wire ropes. The knit method and the manufacturing level of this closed wire rope place higher demands, requiring advanced technologies to be developed. WR-CVT requires suited to the passenger cars and motorcycles.

2 Closed wire rope geometry model

In this paper, the wire rope model is 6×7+IWS. It contains 1 core strand and 6 side strands, one strand contains a center wire and six side wires. 6×7+IWS contains four different geometric forms of wire, they are core strand center wire, core strand side wire, outer strand center wire and outer strand side wire. Core strand center wire is the axis of closed wire rope, core strand side wire and outer strand center wire are single helix, Outer strand side wire is second helix. 6×7+IWS wire rope section is shown in **Figure. 2.**

According to the right-hand rules to establish a Cartesian coordinate system O-xyz as a closed wire rope mathematical model of the global coordinate system[7,8], the centerline equation for each wire in a closed wire rope is given [9,10], Wire rope specific parameters References [11]. The 3D solid model of 6×7+IWS bending wire rope was established by using parametric modeling and surface modeling of Pro/E, as shown in **Figure**.3.

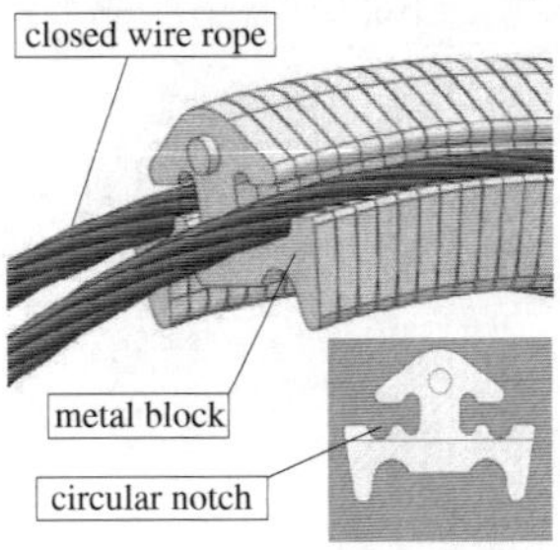

Fig. 1. WR-CVT

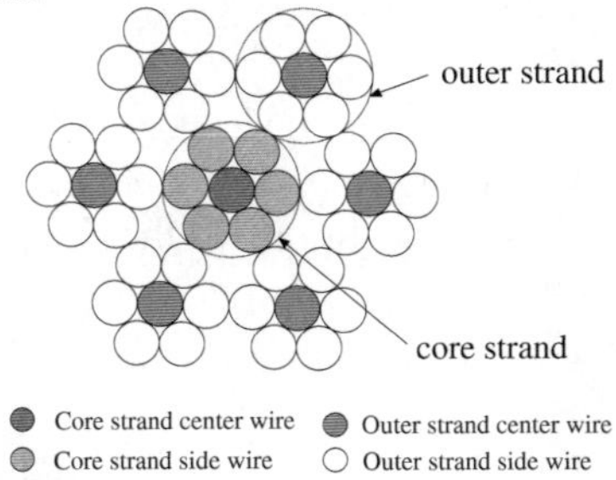

Fig. 2. 6×7+IWS wire rope section

Fig.3 6×7+IWS 3D solid model

3 Wire rope mesh generation and Section definition

3.1 Wire rope mesh generation

The selection of the wire rope length is very important for the finite element analysis. Wire rope is too short can not fully contact between the wire rope and too long will affect the calculation efficiency. In order to obtain higher analytical precision and computational efficiency, we select one-twist length steel rope loop as the analysis model in this paper. The ABAQUS software was used for the finite element modeling. The wire material model is defined by the following parameters: Modulus of elasticity E= 1.96×10^{11}Pa, Yield limit: $\sigma_s=1.54\times10^{9}$Pa, Shear modulus $G=8.53\times10^{10}$Pa, Poisson ratio $\mu=0.3$, Friction coefficient f=0.115. Definition of closed wire rope and metal block unit property type C3D8R hexahedron. According to the calculation requirements, select the appropriate grid size, complete the meshing in Hypermesh, the number of divisions is 67982, the number of nodes is 94994, as shown in Figure 4. In the finite element model of the wire rope, one end applies the surface constraint, restrains all degrees of freedom in three directions, x, y, z, and applies the surface load at the other end.

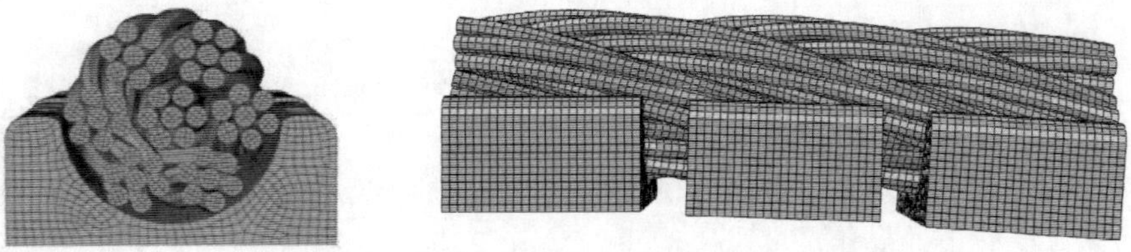
Fig.4 Wire rope mesh generation

3.2 Section definition

When the transmission ratio i=0.42, the angle between the two cross sections is 1.6 ° and 2.6 °, respectively, and is defined as the section A_1 and the section A_2. The following main analysis of the two sections of the steel wire stress distribution. Figure 5 is the section position; Figure 6 is the wire number and path schematic.

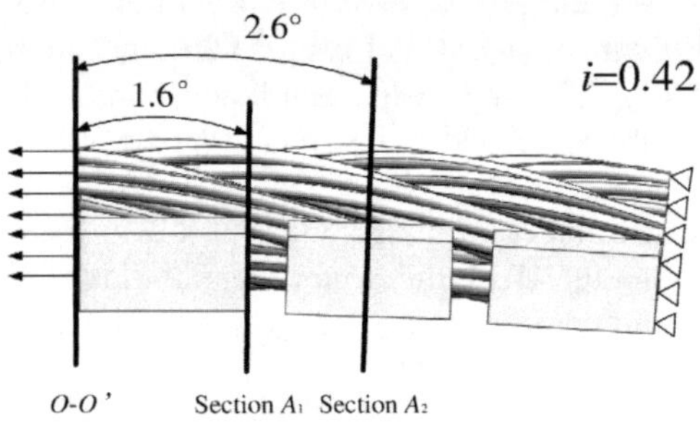

Fig.5 Section position

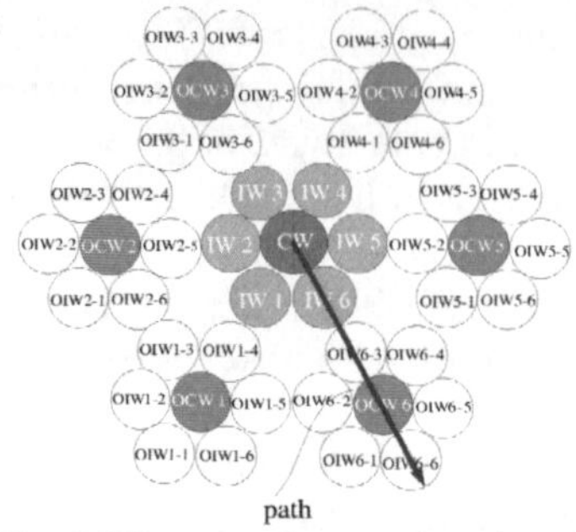

Fig.6 Wire number and path schematic

4 Wire rope stress analysis

Figure 7 shows the transmission ratio i=0.42, the driving pulley bending section of the overall stress distribution of the rope ring. It can be seen from the figure 7: the wire stress along the direction of rotation of the wire rope staggered distribution.

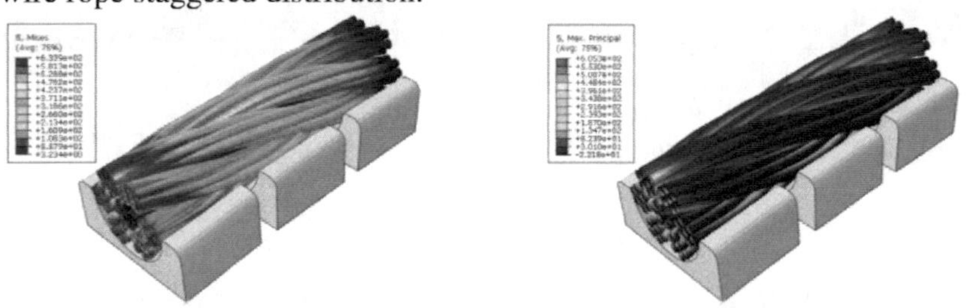

(*a*) Equivalent stress (*b*) The maximum stress

Fig.7 Wire rope stress distribution (i=0.42)

Figure 8 is the wire maximum stress distribution of section A_1 and A_2 in. It can be seen from the figure that the stress distribution of the steel wire is symmetrical in the left and right, and asymmetrical in the up and down direction. The wire stress on the side of the metal block circular notch is less than that on the other side. The stress in each section of the wire is in the form of a layer distribution, the middle stress is small, and the stress on the contact side with other steel wires is larger. In section A_1, the maximum stress value is located in the wire contact with the edge of the metal block circular notch, and the stress value is much higher than other wire, indicating that the two wire stress concentration is the most serious, and thus more prone to early failure. In Section A_2, although the value of the stress of the wire near the side of the rope circular notch is high, the stress difference of each wire is small and the distribution is relatively uniform.

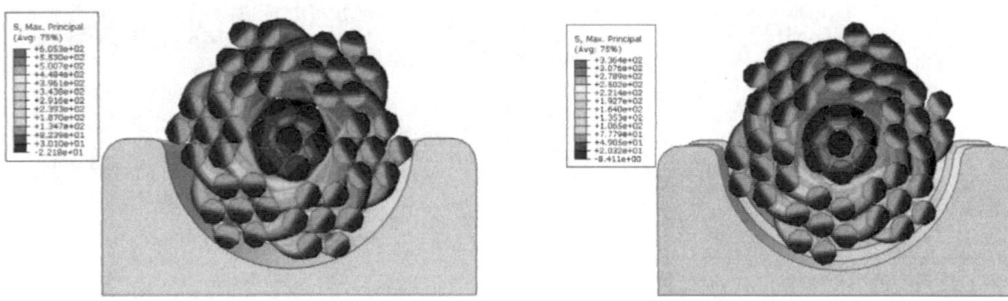

(*a*) The maximum stress of section A_1 (*b*) The maximum stress of section A_2

Fig.8 wire stress (i=0.42)

When the transmission ratio for other values, the stress profile on each section is similar to Figure 8, here is no longer shown, only the value extracted.

Figure 9 shows the wire stress distribution along the path in sections A_1 and A_1 for transmission ratios i=2.35, 1.0 and 0.42. Figure 9 (*a*) shows that with the decrease of the transmission ratio, the wire stress is decreasing. Among them, the core strand side wire in the different transmission ratio stress is basically the same. The lateral stress of the outer strands was the highest, and the maximum value of the stress appeared at the contact point between the outer strand center wire and the outer strand side wire because of the influence of the metal block circular notch. The wire stress at the corresponding position decreases as the transmission ratio decreases.

Figure 9 (*b*) shows that as the transmission ratio decreases, the overall stress of steel wire are still decreasing. The stress difference of the upper side steel wire is smaller and the stress distribution is more uniform. The stress of the side wire in contact with the metal block circular notch is larger, and the maximum value is in contact with the center filament of the side strands, which is similar to the section A_1.

Since the section A_2 is not in contact with the metal block or the contact pressure is small, the stress value on the section A_2 is generally smaller in comparison with the section A_1.

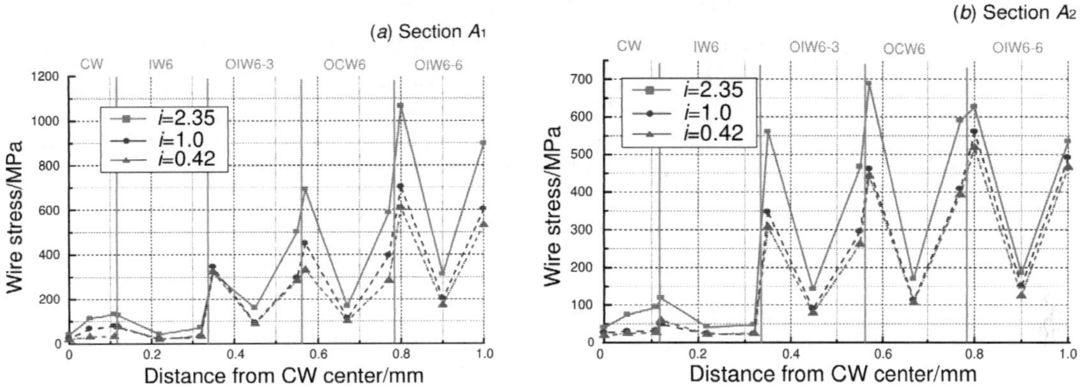

Fig.9 Stress distribution of wire along path

5. Conclusion

In this paper, the stress distribution of WR-CVT bending wire rope is mainly studied. The geometrical model of 6x7+IWS bending wire rope is established by Pro/E software. The finite element model of bending wire rope is completed by ABAQUS software. The stress distribution of the wires on the two sections is analyzed. The results shown that the stress in each section of the wire is in the form of layered distribution, the middle stress is small, and the stress in contact with other wires is larger. The stress at the outer strand center wire and outer strand side wire was the greatest, the stress value of the outer strand side wire and metal block circular notch is second. As the transmission ratio decreases, the wire stress decreases, which is related to the pulley working radius increases. Compared with the section A_1, the stress value on the section A_2 is smaller, mainly because the section A_2 is not in contact with the metal block or the contact pressure is small.

Acknowledgement

The research presented here was supported by the National Natural Science Foundation of China (Grant No.51505373). The authors are grateful for the support.

References

[1] Nilabh Srivastava, Imtiaz Haque. Clearance and friction-induced dynamics of chain CVT drives[J]. *Multibody System Dynamics*, 2008, 19(3): 255-280

[2] FVD Sluis, AVD Velde, TV Dongen, GJV Spijk, AV Heeswijk. Efficiency optimization of the push belt CVT, *SAE paper* 2007-01-1457, 2007.

[3] Doi T. New compact, lightweight, low friction CVT with wide ratio coverage[C]. 2010 international continuously variable and hybrid transmission congress, Maastricht, The Netherlands, 17–19 November 2010. Eindhoven: Eindhoven University of Technology: 146–151.

[4] G. Carbone, L. Mangialardi, G. Mantriota. Influence of clearance between plates in metal pushing V-Belt dynamics[J]. *Journal of Mechanical Design*, 2002, 124(9): 543-557

[5] Wang, Dagang; Li, Xiaowu; Wang, Xiangru; Shi, Ganyu; Mao, Xianbiao; Wang, Dao'Ai. Effects of

hoisting parameters on dynamic contact characteristics between the rope and friction lining in a deep coal mine[J]. *Tribology International*, 2016, 96(4): 31-42.
[6] Zhang Wu, Guo Wei, Lu Zhengxiong, Cao Jianbo. Design and Analysis of a New Metal Belt Continuously Variable Transmission[J]. *Journal of Mechanical Transmission*, 2015, 39(9): 59-61, 88
[7] Lee W K. An insight into wire rope geometry [J]. *International Journal of Solids & Structures*, 1991, 28(4):471-490.
[8] Erdönmez C. n-tuple complex helical geometry modeling using parametric equations [J]. *Engineering with Computers*, 2013, 30(4):715-726.
[9] Stanova, E. Fedorko, G.; Fabian, M.; Kmet, S. Computer modelling of wire strands and ropes Part I: Theory and computer implementation [J]. *Advances in Engineering Software*, 2011, 42(6): 305-315.
[10] Stanova, E. Fedorko, G.; Fabian, M.; Kmet, S. Computer modelling of wire strands and ropes Part II: Finite element based applications [J]. *Advances in Engineering Software*, 2011, 42(6): 322-331.
[11] Guo Wei, Lu Zhengxiong, Zhang Wu. Geometric modeling theory of bent wire rope based on Pro/E [J]. *China Mechanical Engineering*, 2015, 26(17): 2363-2368

IOP Conf. Series: Materials Science and Engineering **241** (2017) 012015 doi:10.1088/1757-899X/241/1/012015

Stress analysis of composite wind turbine blade by finite element method

Meng-Kao Yeh and Chen-Hsu Wang

Department of Power Mechanical Engineering, National Tsing Hua University, Hsinchu 30013 Taiwan, ROC

mkyeh@pme.nthu.edu.tw

Abstract. In this study, the finite element analysis software ANSYS was used to analyze the composite wind turbine blade. The wind turbine blade model used is adopted from the 5 MW model of US National Renewable Energy Laboratory (NREL). The wind turbine blade is a sandwich structure, comprising outermost carbon fiber cloth/epoxy composites, the inner glass fiber/vinylester layers, and PVC foam core, together with stiffeners. The wind pressure is converted into the load on the blade structure. The stress distribution and deformation of wind turbine blade were obtained by considering different pitch angles and at different angular positions. The Tsai-Hill criterion was used to determine the failure of wind turbine blade. The results show that at the 0° pitch angle, the wind turbine blade is subjected to the largest combined load and therefore the stress is the largest; with the increasing pitch angle, the load gradually decreases and the stress is also smaller. The stress and displacement are the greatest when the wind blade is located at 120° angular position from its highest vertex.

1. Introduction

In recent years, the wind power industry is booming, the wind turbine market has also been expanding. Most of the wind turbine blades are made of composite materials, which with light weight can effectively enhance the structural strength and general performance for wind power generation. Daniel and Abot [1] used three-point and four-point bending tests to obtain the bending behavior of sandwich composite beams and found that since the Young's modulus of the core material was much smaller than that of the surface material, the influence of the core material was negligible. Kim and Swanson [2] performed the experiment for the carbon fiber/epoxy composite sandwich beam structure and found that when the core density was small, the failure resulted from shear damage of the core material; when the core was thicker with high density, the failure came from the fiber fracture of surface layers. Steevens and Flect [3] found that from three-point bending test, the foam core could be used for relatively lower load applied on the sandwich structure. Duan and Zhao [4] used the finite element method to analyze the stress distribution of a 600 kW wind turbine blade and found that the maximum stress occurred at the root of wind blade. Yeh et al. [5] evaluated the stress distribution and deflection of NREL 5MW wind blade with different structural stiffeners.

In this study, the stress and deformation analyses of the NREL 5MW wind blade with different pitch angles, 0°, 5°, 15° at different angular positions under wind load were carried out by finite element analysis. The wind turbine blade, comprising outermost carbon fiber cloth/epoxy composites, the inner glass fiber/vinylester layers, and PVC foam core, is a sandwich structure. The Tsai-Hill failure criterion was used to determine the failure of wind blade.

IOP Conf. Series: Materials Science and Engineering **241** (2017) 012015 doi:10.1088/1757-899X/241/1/012015

2. Finite element analysis

The finite element analysis software ANSYS was used to analyze the stress distribution and deformation of the wind turbine blades under the wind load.

2.1 Wind turbine blade

The wind turbine blade model is adopted from the 5 MW model of US National Renewable Energy Laboratory (NREL). The model has a length of 61.5 m, using 19 airfoil sections. The circular part connecting the hub is 6.83 m long, 14% of total length; the remaining part uses DU99W405, DU99W350, DU97W300, DU91W2250, DU93W210, and NACA64-618 airfoils [6]. The longest chord length is 4.652 m, located at position about 23% of total length [7]. Two vertical stiffeners are placed inside the wind blade to improve the blade strength, as shown in Figure 1. The outer carbon fiber cloth/epoxy, the inner glass fiber/vinylester composites, and PVC foam core were adopted as the blade material to form a sandwich structure. The material parameters are referred to Yeh et al. [5]. The carbon fiber cloth/epoxy composites has total of 20 layers including both upper and lower parts. The glass fiber/vinylester composites has stacking sequence, $[0_n/\pm45_n/90_n]_s$, to make a minimum of 16 layers and a maximum of 80 layers including both upper and lower parts. The single layer thickness for both materials is 0.28 mm, together with central PVC foam to form a sandwich structure in the wind blade. Figure 2 shows the thickness variation of the three materials in the sandwiched wind blade.

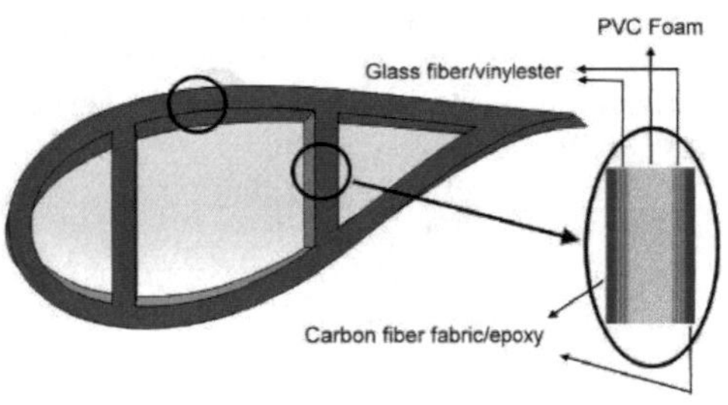

Figure 1 Stiffened wind turbine blade

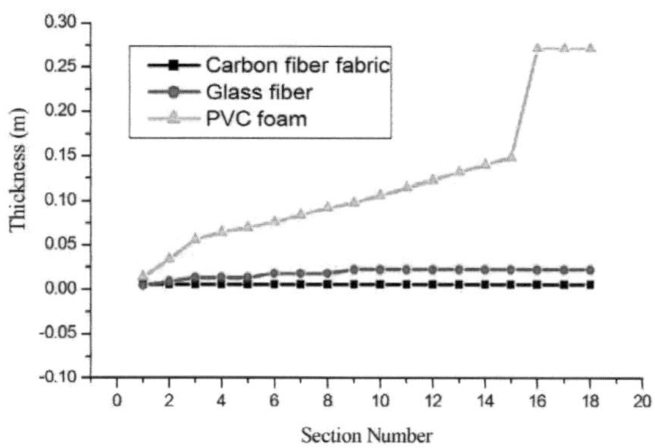

Figure 2 Thickness variation in wind blade

2.2 Analysis model of wind turbine blade

The ANSYS software was used to analyze the deformation and stress distribution of the wind turbine blade. Mesh convergence was carried out for the element number from 7000 to 60000 and the optimal model was obtained when the blade model reaches more than 25000 elements, as shown in Figure 3, and the maximum von Mises stress approaches a convergent value. In this study, the clockwise rotation speed of the wind blade was set at 12.1 rpm, which corresponding to the maximum designed wind speed of 25 m/s. The three cases analyzed were for three pitch angles 0°, 5°, 15° of wind blade. The wind blade was subjected to the gravity, the wind load, and centrifugal force. Figure 4 shows various load and boundary conditions for the wind blade at the vertical position and the gravity force is acting through the wind blade from its tip to the root end. The root end of the wind blade is fixed on the wind turbine hub, so the petiole degrees of freedom are all fixed.

3. Results and discussion

The stress analysis results of NREL 5MW wind turbine blade by finite element software ANSYS were obtained. The stress distribution and deformation of three angular positions and four pitch angles of wind blade on are discussed.

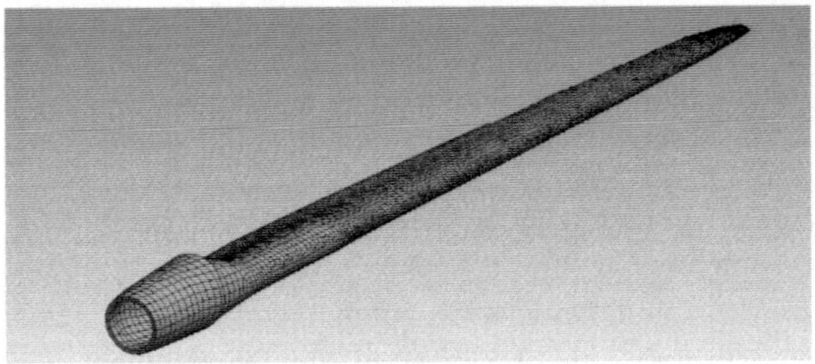

Figure 3 Finite element model of wind turbine blade

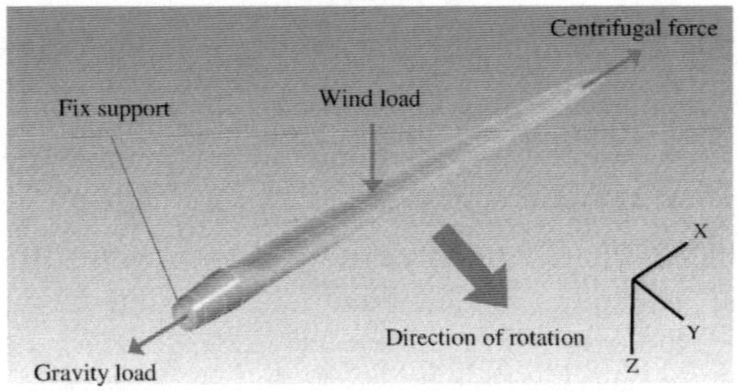

Figure 4 Load on wind turbine blade

3.1 Different angular position

The wind turbine blade under wind load at three angular positions is investigated. With the pitch angle 0°, the wind blade is subjected to a maximum wind pressure load to bend the blade structure. The maximum stress occurs at the junction of circular airfoil and DU99W405 airfoil near the stiffener edge, where the blade shape changes a lot. When the wind blade is located at vertical position, the maximum von Mises stress value is 86.11 MPa. When the blade is at 120° counterclockwise angular position from its highest vertex, the combination of the wind load, the gravity and the centrifugal force resulted in a maximum von Mises stress of 72.95 MPa at the junction of the circular airfoil and DU 99W405 airfoil near the stiffener edge. When the wind blade is at 120° clockwise angular position from its highest vertex, the load on the blade is added by the wind load, the gravity and the centrifugal force and resulted in a maximum value of von Mises stress 108.79 MPa at the junction of the circular airfoil and DU99W405 airfoil near the stiffener edge.

Figure 5 shows the stress distribution and the deformation of the wind blade at 120° clockwise angular position. The maximum tip displacements at three positions are 2.55 m, 2.27 m and 2.54 m, respectively. The deflection of wind blade is similar to that of a cantilever beam as shown in Figure 5(b), in the flapwise direction of wind blade. The maximum von Mises stresses and maximum tip displacements of NREL 5MW wind turbine blade at different angular position are shown in Table 1.

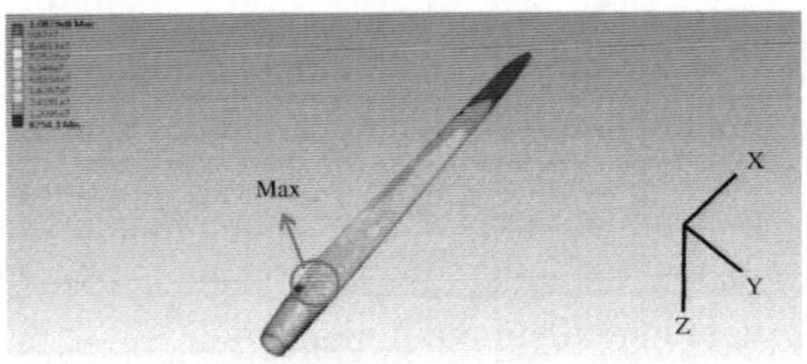

(a) von Mises stress distribution

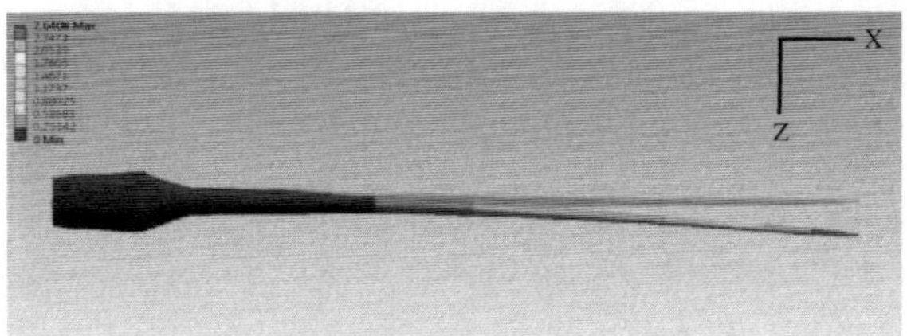

(b) Deflection of wind blade

Figure 5 Results of NREL 5MW wind blade at 120° clockwise angular position

Table 1 Results of wind blade at different angular positions and for different pitch angles

Item	Blade angular position			Pitch angle		
	Vertical	-120°	120°	0°	5°	15°
$\sigma_{vM,max}$ (MPa)	86.11	72.95	108.79	108.79	69.10	60.92
Tip disp.(m)	2.55	2.27	2.64	2.64	2.38	2.12

3.2 Different pitch angle

This section explores the results of three different pitch angles at 0°, 5°, and 15° for wind blade. The results of the wind blade with 0° pitch angle are described in the previous section for different angular position; the results of the wind blades with 5° and 15° pitch angles, with decreasing wind load, the smaller maximum stresses are all located at the junction of the circular airfoil and DU99W405 airfoil near the stiffener edge, same as those in the wind blade with 0° pitch angle. For the wind blade with larger pitch angle 15°, the values of von Mises stress and tip displacement are smaller than the case with pitch angles 0° and 5°. The maximum von Mises stress values for the wind blade with pitch angle

5° and 15° were 69.1 MPa and 60.92 MPa, respectively. The simulated maximum von Mises stresses and maximum tip displacements of NREL 5MW wind turbine blade at different pitch angle are shown in Table 1.

3.3 Failure criterion of wind blade

The Tsai-Hill failure criterion was used to determine the failure of wind blade. The maximum von Mises cases, at 120° clockwise angular position in Table 1, are summarized in Table 2 for different pitch angles. It can be seen from Table 2, for the wind blade with pitch angle 0°, the glass fiber/vinylester composite layers fails for fiber orientation 45° and 90°. Reinforcements are needed for the wind blade with pitch angle 0° and increasing the pitch angle is an alternative way to prevent the wind blade from failure.

Table 2 Simulated results by Tsai Hill failure criteria

Material	Fiber orientation	Pitch angle 0°	Pitch angle 5°	Pitch angle 15°
Carbon fiber fabric/epoxy	0°/90°	0.447	0.568	0.235
Glass fiber/vinylester	0°	0.470	0.723	0.281
	45°	1.435	0.106	0.254
	90°	1.611	0.639	0.530

4. Conclusions

In this study, the stress distribution and deformation of the NREL 5MW wind turbine blade under various loads was analyzed using finite element analysis software ANSYS. The results for wind turbine blade with different angular position and different pitch angle are discussed. For the wind blade with 0° pitch angle, the combined load on the wind blade is a maximum. With increasing pitch angle, the load on the wind blade is decreasing at a specified angular position. When the blade is located at 120° clockwise angular position, maximum von Mises stress occurs at the junction of the circular airfoil and DU99W405 airfoil near the stiffener edge, resulted from the combined load by the wind load, the gravity and the centrifugal force. The use of stiffener improves the rigidity of the blade to resist the bending moment of the blade, and also bears the combined load of wind blade. Increasing the pitch angle can reduce the stresses in the wind blade and can prevent the wind blade from failure.

Acknowledgements

The authors would like to thank the support by Ministry of Science and Technology, Taiwan, ROC, under the Grant MOST 105-2221-E-007-031-MY3. The support is greatly acknowledged.

References

[1] Daniel I M and Abot J 2000 *Compos. Sci. Technol.* **60** 2455
[2] Kim J and Swanson S R 2001 *Compos. Struct.* **52** 365
[3] Steevens C A and Flect N A 2004 *Script Mater.* **50** 1335
[4] Duan W and Zhao F 2010 Loading analysis and strength calculation of wind turbine blade based on blade element momentum theory and finite element method. *Power and Energy Engineering Conference* (APPEEC), Chengdu, p 1
[5] Yeh M K, Cheng Y C and Wang C H 2015 Finite element stress analysis of wind blade structure under wind pressure. *Taiwan Wind Energy Conference*, Taipei, Taiwan, Paper No. SI_07 (in Chinese)
[6] Jonkman J, Butterfield S, Musial W and Scott G 2009 *Definition of a 5-MW reference wind turbine for offshore system development*, Technical Report, NREL/TP-500-38060
[7] Yeh M K and Wang C H 2017 Stress analysis of wind turbine blade with different stacking angle and different skin thickness, *Joint Conferences of 2017 International Conference on Materials Science and Engineering Application (ICMSEA 2017) and 2017 International Conference on Mechanics, Civil Engineering and Building Materials (MCEBM 2017)*, April 21-23, Nanjing, China, p 558.

IOP Conf. Series: Materials Science and Engineering **241** (2017) 012016 doi:10.1088/1757-899X/241/1/012016

Acoustic Modeling of Lightweight Structures: A Literature Review

Shasha Yang[1], Cheng Shen[2]

[1]Nanjing Institute of Industry Technology, Nanjing 210046, P. R. China
[2] College of Energy and Power Engineering, Nanjing University of Aeronautics and Astronautics, Nanjing 210016, P. R. China

Corresponding author's email: cshen@nuaa.edu.cn

Abstract. This paper gives an overview of acoustic modeling for three kinds of typical lightweight structures including double-leaf plate system, stiffened single (or double) plate and porous material. Classical models are citied to provide frame work of theoretical modeling for acoustic property of lightweight structures; important research advances derived by our research group and other authors are introduced to describe the current state of art for acoustic research. Finally, remaining problems and future research directions are concluded and prospected briefly

1. Introduction

The content of this section will be organized in the way of explaining every important phrase in the topic of this paper, and it intends to help readers to know quickly about the subject and scope of this paper involved.

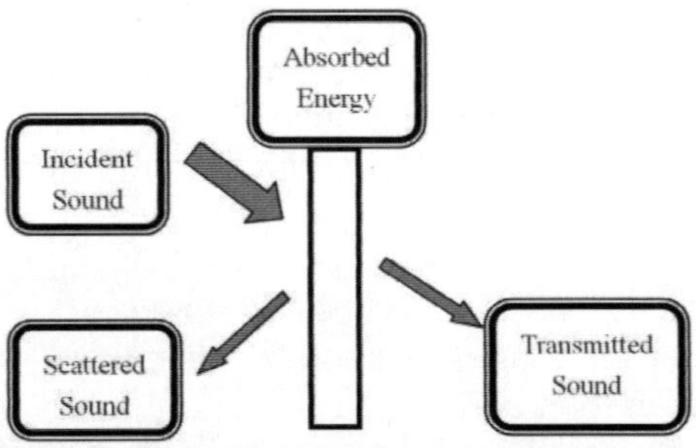

Figure 1. Acoustic behavior of sound wave striking a structural surface

Acoustic modeling: As a branch of physics, acoustic is an extensional subject including quite different contents across with other fields. In the present paper, we will mainly discuss the acoustic modeling of beam-plate structure and sound absorption of porous material. The total incident sound energy is divided into three parts (as shown in Fig.1) including scattering sound, absorbed energy and transmitted sound when a sound wave strikes a structural surface. Among them, how to predict scattered sound and transmitted sound is a problem involving vibroacoustic field. The contents of this paper are arranged as follows: The main body of this paper is constituted of three parts including double-leaf plate system, stiffened plate system and porous material, respectively.

2. Double-Leaf Plate Systems

As one of the most basic theories in acoustics, mass law tells that the transmission loss of double-leaf plate system can add 6 dB over one of single-leaf plate in ideal case, so double-leaf plates have been widely applied in industries due to their better sound insulation performance.

In the earliest period, a few models have been proposed to explore the physic mechanism of sound transmission problems. Cremer [1] underlined the influence of critical frequency on sound transmission through single panel. Under the assumption of two identical walls, London introduced standing wave resonance and mass-air-mass resonance phenomenon of double-leaf structure using wave based model [2]. These two kinds of resonances greatly reduce the sound insulation property of double-leaf plate system and make the problem more complex. Beside classic wave-based approach, transfer matrix method [3] were also employed to solve the problem of sound transmission through layered structures. However, all the above works are limited to structure in infinite fluid domain. If finite room backing panel system is considered [4], the model will consist of five coupling oscillators including 'room-panel-cavity-panel-room', and it is shown that modes of room play a significant effect especially at low frequency [5].

The double panel structure is really rather complicated because there exist many variables though it seems to be a simple problem at first glance. In fact, many sophisticated models have been developed with the consideration of different structural variables. For instance, Ng and Zheng [6] have extended sharp's model [7] to predict sound transmission through double-leaf panel with corrugated core;; Among series of parameters, the viscosity of air play a significant role when the gap is small especially. The results of Dijckmans [8] et al showed that the thin air layer could improve the sound transmission loss of the structure remarkably; Lee and Ih [9] stated that non-diffuse of test room is an important reason resulting in the discrepancies between theories and experiments. The concept of limited angle introduced by Beranek [10] can favorably evaluate the diffuse field sound transmission [11]. A more detailed discussion for the wave-based modeling of diffuse field sound transmission can refer to the review paper [12]; It maybe a simplest way to improve the sound insulation of double panel by filling porous material (e.g. mineral wool) which has two kinds effect [13] including damping and fictive increase of volume in the cavity. Unfortunately, it is not an easy work to model the acoustic property of porous material, detailed discussion can be found in the following sections.

3. Stiffened Plate System

Stiffeners are used extensively in order to improve the static mechanical strength of plate-like structure, which make the original structure to be not uniform and homogeneous and bring the complexity of analyzing. In fact, the stiffeners not only influence the vibration behavior of structures directly, but also affect the radiated sound field especially in the case of double panel with structural links.

3.1. Overall dispose of periodic problem

Stiffened single panel has been widely used in aircraft and marine structures [14], while stiffened double plate is a typical structure used in building field [13] especially in Nordic countries. Mead [15] comprehensively reviewed the research developments for wave propagation in continuous periodic structure concentrating on the contributions of Southampton. Only space harmonic method and energy method can be used to analyze two-dimensional structure, Moreover, space harmonic method is suitable for predicting sound radiation since transverse displacement has been expressed as a series of sinusoidal travelling waves. Brunskog and Hammer have also given a comprehensive review [16] relevant to present work.

3.2. Overall dispose of stiffeners

A most straightforward way [17] is to equivalent the whole structure to be an orthotropic uniform plate if the bending wave length is greater than the stiffeners spacing. Lee and Kim [18] predicted the transmission loss of periodically stiffened plates just considering the equivalent tension and torsion stiffness of supports. If the geometry of stiffeners has been taken into account, the stiffeners are usually equivalent to be independent beams. In most cases, only normal force between beam and base panel are included in theoretical model for simplicity, but moment coupling should be embodied to achieve more accurate results [19]. Moreover, more accurate beam theories such as Timoshenko theory has been used to model base panel or stiffeners to include the effect of shear motion. Note that the rigid coupling between base panel and stiffeners has been simplified in all the above models, because the practical connections may be much complex such as spot-weld or fixing through screws. To solve this problem, Arruda et al[20] introduced elastic points to approximate the coupling condition. Most stiffeners are placed parallelly or orthogonally on base panel, but Xu et al[21] found that the effect of orientation of stiffeners play an important role in composite structure system. Besides, most papers have ignored the eccentric effect of stiffeners, which means that the base panel and stiffeners are jointed along middle surface, but the stiffeners are set on the based panel along side surface actually. A more accurate theoretical model has been developed by several authors[22] with this eccentric effect taken into account.

4. Porous Material

When sound propagates in porous material, the inside space and matrix provide two paths, which leads to a complex problem. There exist many models[23] from different viewpoints, and three kinds of main models and existing problems are summarized in the following.

4.1. The rigid framed models

The basic assumption of the following rigid framed models is that the frame of material is motionless, and then the fluid inside porous material can be equivalent to be free fluid. This assumption is valid when the stiffness of matrix of porous material (e.g. metal foam) is much larger than the fluid inside [24] or porous material has been set on rigid wall [25].

The first conceptual model for porous media may be the Rayleigh model [26], which describes exact sound propagation in parallel circular capillaries, and many theories are evolved from the Rayleigh model. Later, viscous and thermal effect were assumed to be independent by Zwikker and Kosten [27], they found that such treatment was at least efficient for fluid motion in a cylindrical channel. Moreover, Stinson [28] revisited the validation of Zwikker and Kosten's assumption and concluded that this approximation could be applicable for a broad range of tube radius from 10-3 cm to several centimeters. Following the way of this analytical modeling, Lu et al [29] proposed a model to predict acoustic wave through various cell shapes by combining three kind of drag forces including parallel, perpendicular to propagation direction forces and drag force caused by joints, microstructural optimization [30] have been subsequently conducted based on this model.

4.2. The rigid framed models

If the elastic frame of porous material is taken into account, many wave types play significant roles due to the coupling between matrix and fluid inside the pores, which thus leads to a more complex problem. In this case, a more complete model Biot's theory [31] was developed, which is the most popular model so far. In the context of Biot's theory, both fluid and matrix are assumed to be continuum ignoring the microscopic level when wavelength is longer than volume element.

Detailed derivation process can be found in Atalla's book [23]. Besides, the Biot's theory can predict three kinds of waves propagating in porous material including two compressional waves and one transverse wave. Plona's experiments [32] have supported the existence of two compressional waves. Based on Biot's theory, other authors have developed new models (e.g. Atalla's work [33]) by modifying the representation of stress-strain relation or other relation.

5. Conclusions

A number of theoretical methods for vibroacoustic property of three kinds of lightweight structures have been reviewed, which tries to present the whole frame work of acoustic modeling with recent advances also introduced. In general, there must emerge new problems with the development of other fields such as material production.

Acknowledgments

This work is supported by National Natural Science Foundation of China (Grant No. 11502110),

Natural Science Fund project in Jiangsu Province(BK20150737) and funded by Hong Kong Scholars Program.

References

[1] L. Cremer, Akust. Z, vol. 7, pp. 81-104, 1942.
[2] A. London, *The Journal of the Acoustical Society of America,* vol. 22, p. 270, 1950.
[3] W. Lauriks, P. Mees, and J. F. Allard, *Journal of Sound and Vibration*, vol. 155, pp. 125-132, 1992.
[4] K. Mulholland and R. Lyon, *The Journal of the Acoustical Society of America*, vol. 54, pp. 867-878, 1973.
[5] L. Gagliardini, J. Roland, and J. L. Guyader, *Journal of Sound and Vibration*, vol. 145, pp. 457-478, 1991.
[6] C. F. Ng and H. Zheng, *Applied Acoustics*, vol. 53, pp. 15-34, Jan-Mar 1998.
[7] B. H. Sharp, *Noise Control Engineering,* vol. 11, pp. 53-63, 1978.
[8] A. Dijckmans, G. Vermeir, and W. Lauriks, *The Journal of the Acoustical Society of America*, vol. 128, pp. 3513-3524, 2010.
[9] J. Lee and J. Ih, *Journal of Sound and Vibration,* vol. 277, pp. 881-893, 2004.
[10] L. L. Beranek, Ed., Noise reduction. New York: McGraw-Hill, 1971, p.^pp. Pages.
[11] M. Villot, C. Guigou, and L. Gagliardini, *Journal of Sound and Vibration,* vol. 245, pp. 433-455, 2001.
[12] A. Pellicier and N. Trompette, *Applied Acoustics*, vol. 68, pp. 1192-1212, 2007.
[13] J. Brunskog, PhD, Engineering Acoustics, Lund University, Lund, 2002.
[14] B. R. Mace, *Journal of Sound and Vibration*, vol. 73, pp. 473-486, 1980.
[15] D. J. Mead, *Journal of Sound and Vibration*, vol. 190, pp. 495-524,1996.
[16] J. Brunskog and P. Hammer, *Building Acoustics*, vol. 7, pp. 89-112, 2000.
[17] L. Cremer, Heckl, M. and Ungar, E. E. , Structure-borne sound, Second ed. Berlin: Springer-Verlag, 1988.
[18] J. H. Lee and J. Kim, *Journal of Sound and Vibration*, vol. 251, pp. 349-366, 2002.
[19] D. Takahashi, *Journal of Sound and Vibration*, vol. 90, pp. 541-557, 1983.
[20] J. Arruda, F. Gautier, and L. Donadon, *Journal of Sound and Vibration*, vol. 307, pp. 564-577, 2007.
[21] X. Xu, H. Lee, and C. Lu, *Journal of Sound and Vibration*, vol. 282, pp. 1264-1272, 2005.
[22] A. Mejdi and N. Atalla, *Journal of Sound and Vibration,* 2010.
[23] J. F. Allard and N. Atalla, John Wiley & Sons Inc, 2009.
[24] T. J. Lu, D. P. He, C. Q. Chen, C. Zhao, D. Fang, and X. Wang, *Advances in Mechanic,* vol. 36, pp. 517-535, 2006.
[25] T. J. Cox and P. D'antonio, Acoustic absorbers and diffusers: theory, design and application: Taylor & Francis Group, 2009.
[26] B. J. W. S. Rayleigh, The theory of sound vol. 2: Macmillan, 1896.
[27] C. Zwikker and C. W. Kosten, Sound absorbing materials: Elsevier Pub. Co., 1949.
[28] M. R. Stinson, *The Journal of the Acoustical Society of America*, vol. 89, pp. 550-558, 1991.
[29] I. D. J. Dup¨"re, A. P. Dowling, and T. J. Lu, ASME International Mechanical Engineering Congress, 2004.
[30] Iain D.J.Dupère, Lu Tianjian, and A. P.Dowling, *Journal of Xi'an Jiaotong University,* vol. 41, pp. 1251-1256, 2007.
[31] M. A. Biot, *The Journal of the Acoustical Society of America*, vol. 28, pp. 168-178, 1956.
[32] T. Plona, *IEEE*, pp. 1044-1048.1982
[33] N. Atalla, R. Panneton, and P. Debergue, *The Journal of the Acoustical Society of America,* vol. 104, p. 1444, 1998.

IOP Conf. Series: Materials Science and Engineering **241** (2017) 012017 doi:10.1088/1757-899X/241/1/012017

Calculation of the end form the rotating electrode in the liquid environment

I V Chumanov[1], A N Anikeev[1]

[1] Department of engineering and technology of materials, South Ural State University (national research university), 16 Turgenev Street, Zlatoust, Russia

Abstract. Electroslag Remelting Process (further – ESR) is an advanced technology of ingots' production, which are used in applications such as aviation, power generation, medicine etc. [1, 2]. However, the purity of the resulting metal and its structural homogeneity depend on the selected remelting regime. This article provides mathematical calculation, which allows estimating the optimal shape of the electrode's end and this will contribute to the creation of optimal conditions for obtaining the maximum performance values of electroslag remelting. Such conditions can be provided at radial flow of molten metal on the even melted-off surface, that is reached by rotation of the melted electrode with an optimum speed.

1. Introduction

The passage of an alternating current from the electrode to the base plate creates Joule heating in the highly resistive calcium fluoride based slag which is sufficient to melt the electrode. Electrode metal drops while passing a layer of less dense slag, from a thub of liquid metal, followed by the formation of an ingot in a water-cooled crystallizer. Due to the heat loss solidifies the metal gets crystallized, forming an ingot with dendrites structure and maintaining a shallow liquid metal pool (depth of 1/3 to 1 diameter) throughout the process [3, 4] 'figure 1'.

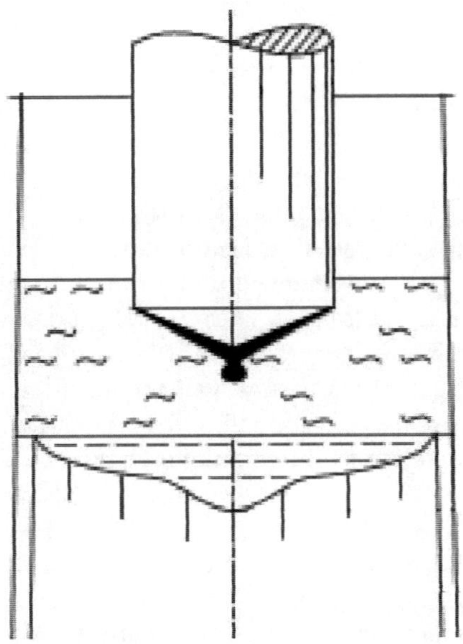

Figure 1. Diagram of the ESR.

The main disadvantages of the electroslag process are a relatively low output, high specific energy consumption, scarce materials used for flux production, and the high cost of a remelted metal. It restricts the field of its application. Therefore, the problems of a further improvement in ESR are relevant and important [5]. To ensure continuous energy saving, the heat transfer must be optimized to a heated metal by improving the thermal operation of a furnace unit [6]. Novel commercial processes and technological regimes imply an enhancement of the set of thermal, mechanical, and chemical actions on metallic and slag melts. New solutions can be obtained during ESR with the rotation of a consumable electrode about its axis, which is a rather simple technical procedure [7], Electrode rotation mechanism 'figure 2'.

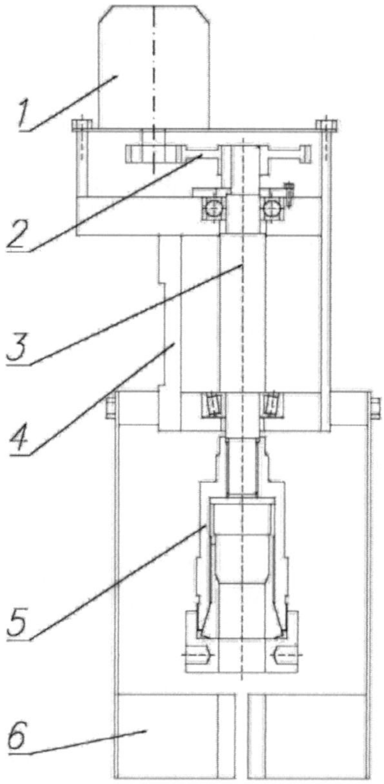

Figure 2. Electrode rotation mechanism: 1 – engine; 2 – transmission; 3 – shaft; 4 – body; 5 – collet; 6 – contact brushes.

In this process, instead of just moving downward, additionally the electrode is revolving around its axis. As a result, an ascending slag flow is induced under the electrode zone. The slag heats up while rising and that leads to an enhanced heat transfer with the electrode, which can improve the efficiency with nearly the same power.

Optimum conditions for obtaining the maximum values of efficiency and the refining ability of process will be reached when can be ensured of a film on all melted-off surfaces, the uniform thickness what is possible in the presence of the flat melted-off electrode end.

2. Mathematical formulation

Formation of the melted-off part of the consumable electrode in case of its rotation is of interest not only from hydrodynamics positions, as it was shown above, but also process from thermal physics. The distribution of flows in the slag bath, shown in 'figure 3'. For this purpose the mathematical model of formation of a fluid film at a end of the rotating cylinder has developed and investigated, and also a liquid thermal physics in this film.

Let's accept the following conditions: thick liquid, incompressible, uniphase; stable process liquid, from the cylinder enters all throw the section (I), driving in a film lamina; the cylinder rotates in liquid

(II) at a speed ω. The strimming of thick incompressible liquid under the rotating disk is considered in work [8]. Numerical results are received, without gravity for the corresponding values of speed liquid's on a cylindrical coordinate (r,θ, t). They are provided in type of schedules.

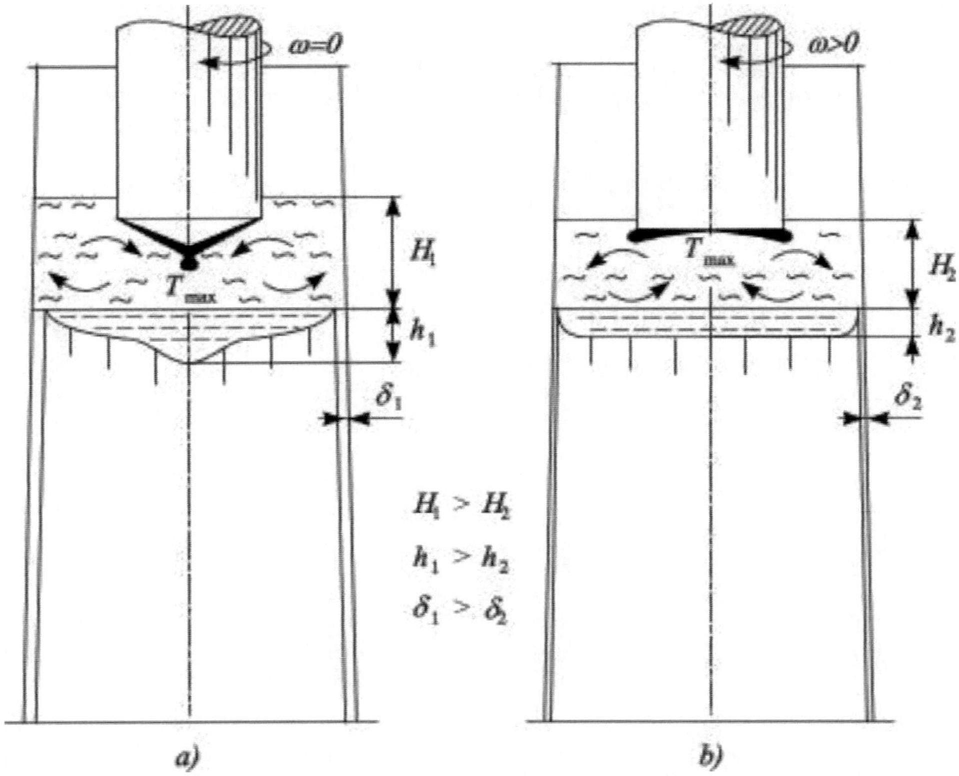

Figure 3. The distribution of flows in the slag bath in the accepted scheme of melting (a) and with rotation of the consumable electrode (b) in electroslag process: ω – the rotation speed of the consumable electrode, turn/min; T_{max} – the epicenter of the zone with the maximum temperature in the slag bath, °C; H_1, h_1 – the height of the slag and metal bath in the accepted scheme of melting, respectively, mm; H_2, h_2 – the height of the slag and metal bath in the implementation of the scheme remelting with rotation of the consumable electrode around its own axis, respectively, mm; δ_1, δ_2 – the thickness of slag skull in the implementation of the scheme remelting without rotation and with rotation of the consumable electrode, respectively, mm.

Assuming that film thickness $\delta(r)$ under end of the rotating cylinder is small in comparison with its radius R, we apply process of alignment, which is provided in work [9] to curves for speed components (Vr, Vz) with method of least squares with the subsequent correcting of results, using an equation of continuity [10].

$$\frac{\partial Vr}{\partial r} + \frac{Vr}{r} + \frac{\partial Vz}{\partial z} = 0. \tag{1}$$

After that we will receive the following expressions for speed components near the formed cylinder:

$$Vr = 0.21Kr\sqrt{Z};$$
$$Vz = -0.28Kz\sqrt{Z}; \tag{2}$$

where $K = \sqrt[4]{\frac{\omega^5}{\nu}}$; ω – angular velocity of cylinder rotation; ν – entering liquid kinematic viscosity (I) from the cylinder.

Minus sign at the making speed Vz shows that liquid in a film and liquid, adjacent to a film (II) move to a cylinder end, to fill a rate of flow in the radial direction.
Taking into account symmetry of rotation and a smallness $\delta\,(\delta << R)$ the heat conduction equation will look like [10]:

$$0.21Kr\sqrt{Z}\frac{\partial T}{\partial r} - 0.28Kz\sqrt{Z}\frac{\partial T}{\partial z} = a\frac{\partial^2 T}{\partial z^2}, \quad (3)$$

where a – liquid thermal diffusivity (I); T – temperature in a fluid film.

Boundary conditions for a liquid film (I) at the end face of the cylinder we will set as following:

$$Z = 0,\ T = T_c; \quad (4)$$

$$Z = \delta(r),\ T = T_l; \quad (5)$$

$$r = R,\ \frac{T - T_c}{T_1 - T_c} = \frac{Z}{\delta}, \quad (6)$$

where T_c – cylinder temperature; T_l –environmental liquid temperature (II).

We will find dependence $\delta(r)$ using kinematic condition

$$Vr\frac{d\delta}{dr} = Vz, \quad (7)$$

which expresses a equality condition of normal components of speed of liquid (I) and liquid (II) on limit of their section, that is at $z = \delta(r)$.

Substituting (2) in (7), we will receive:

$$\delta = \delta_{max}\left(\frac{r}{R}\right)^{4/3}. \quad (8)$$

Here δ_{max} – the maximum value of thickness of a film on cylinder's end face edge, received in account of inertial forces.
Influence of the edge on processes in a film is not considered. In (3) and (6) let's pass to the dimensionless variables:

$$\eta = \frac{r}{R},\ \xi = \frac{z}{\delta},\ u = \frac{T - T_c}{T_1 - T_c} - \xi. \quad (9)$$

Then we will have the following heat conduction equation in the dimensionless parameters:

$$3\eta^{\frac{13}{3}}\xi^{\frac{1}{2}}\frac{\partial u}{\partial \eta} - 8\xi^{\frac{2}{3}}\eta^{\frac{10}{3}}\frac{\partial u}{\partial \xi} - A\frac{\partial^2 u}{\partial \xi^2} = 8\xi^{\frac{2}{3}}\eta^{\frac{10}{3}}, \quad (10)$$

at the homogeneous boundary conditions of a look:

$$\xi = 0,\ u = 0; \quad (11)$$

$$\xi = 1,\ u = 0; \quad (12)$$

$$\eta = 1,\ u = 0, \tag{13}$$

where $A = \dfrac{100}{7\delta_{\max}^{2.5} K}$ - the dimensionless parameter.

For the solution of the (10) under (11)...(13) we will use Galerkin's method [11, 12], meaning that we will look for the decision in a look:

$$u = \xi(1-\eta)(1-\xi)\sum_{i=0}^{\infty} a_1 (1-\eta)^i . \tag{14}$$

For $i = 2$ we will receive the following approximate expression of distribution of temperature in the liquid film, entering from the cylinder:

$$u^* = \frac{\xi(1-\eta)(1-\xi)}{A} \cdot \left[0.88 + 1.28(1-\eta) - 2.34(1-\eta)^2\right] + \xi . \tag{15}$$

Temperature gradient at the cylinder end at $\xi = 0$ takes in this case a form:

$$g(n) = 1 + \frac{(1-\delta)}{A} \cdot \left[0.88 + 1.28(1-\eta) - 2.34(1-\eta^2)\right], \tag{16}$$

where $g(\eta) = \dfrac{\partial u}{\partial \xi_{(\xi=0)}}$, $A = \dfrac{1000a}{7\delta_{\max}^{2.5}\omega} \cdot \sqrt[4]{\dfrac{\nu}{\omega}}$.

3. Results and discussion

Given below is the qualitative picture of influence of cylinder's rotation speed ω on the form of its end face at constant ν and a. For example, for values ω = 6,28 rad/s ('figure 4', curve 1), and ω = 10,42 rad/s ('figure 4', curve 2).

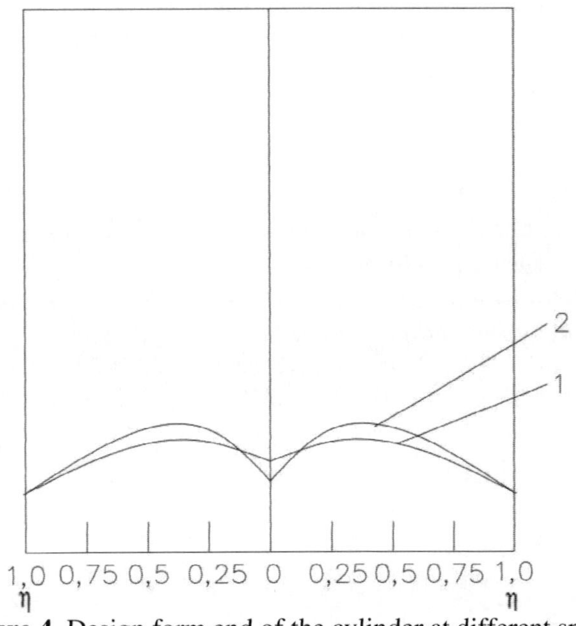

Figure 4. Design form end of the cylinder at different speed

If we assume, that the gradient of temperature is proportional to liquid thickness under the cylinder, we can define that the cylinder end form depending on an angular velocity of rotation, liquid's viscosity and coefficient of thermal diffusivity.

In that way, knowing the speed of rotation of the cylinder, it is possible to predict a form of its end, and also to calculate a temperature profile in the entering liquid.

4. Conclusion

The intention to receive a minimum thickness, and evenly distributed on the melted-off electrode surface a fluid metal film is dictated by the fact, that the main refinement of the melted metal, as it was noted above, happens exactly in this layer, and the less thickness of a layer is, the cleaning process is more effective.

The beneficial factor is that technology of manufacturing electrodes for secondary remelting is already quite advanced.

Intake of electrode metal from top of its conic part in a fluid metal tub leads to increase in depth of its central part; in return that promotes violation of configurationally regularity of an ingot. In this regard creation of such conditions at electroslag remelting, at which thickness of a layer of molten metal on an electrode would be minimum, and metal tub, the front of a crystallization and end face of an electrode in the course of remelting would remain flat, would allow to increase effectiveness of refinement of metal. Such conditions can be provided at radial flow of molten metal on the even melted-off surface, that is reached by rotation of the melted electrode with an optimum speed. Based on the above calculation we found out that it is possible to provide optimal conditions for carrying out the electroslag remelting.

Acknowledgments

The work was performed under the state order of the Ministry of Education and Science 14.577.21.0185 (unique number RFMEFI57715X0185).

References

[1] Johnson A J D and Hellawell A 1972 *Metall. Trans.* **3** 1016
[2] Li Z B 2010 *Electroslag Metallurgy Theory and Practice* (Beijing: Metallurgical Industry Press)
[3] Jardy A, Ablitzer D and Wadier J F 1991 *Metallurgical and Materials Transactions B* **22B** 111
[4] Paton B E and Medovar L B 2008 *Steel in Translation* **38** 1028
[5] Dub A V, Levkov L Ya, Shurigin D A, Kriger Yu N, Markov S I, Orlov S V, Dub V S, Volobuyev Y S, Kuznetsova K N and Kisselman M A 2013 *Questions of atomic science and technology. Series: Security of NPP* **34** 11
[6] Tilman R and Crainberg I 1987 *Electroslag remelting process* **9** 99
[7] Chumanov I V and Chumanov V I 2001 *Metallurgist* **45** 125
[8] Landau L D and Livhec E M 1954 *Mechanics of continua* (Moscow: Gosizdat)
[9] Shvidsky V S, Ladygichev M G and Shavrin V S 2005 *Mathematical methods of a thermal physics* (Moscow: Heating engineer)
[10] Loycanski L G 1970 *Mechanics of liquid and gases* (Moscow: Science)
[11] Mihlin S Sh 1970 *Variation methods in mathematical physics* (Moscow: Science)
[12] Ryabtsev A D, Pashinsky V V, Carter W T, Benz M G 2005 *Proc. of the 2005 Int. Symposium on Liquid Metal Processing and Casting "LMPC 2005"* (Santa Fe: Omnipress) p 116

Analysis on the workspace of palletizing robot based on AutoCAD

Jin-quan LI[1], Rui ZHANG[1, a], Qi GUAN[1], Fang CUI[1] and Kuan CHEN[1]

[1]Automation School, Beijing University of Posts and Telecommunications, Beijing 100876, China

Email: [a]zhangrui0908@bupt.edu.cn

Abstract. In this paper, a four-degree-of-freedom articulated palletizing robot is used as the object of research. Based on the analysis of the overall configuration of the robot, the kinematic mathematical model is established by D-H method to figure out the workspace of the robot. In order to meet the needs of design and analysis, using AutoCAD secondary development technology and AutoLisp language to develop AutoCAD-based 2D and 3D workspace simulation interface program of palletizing robot. At last, using AutoCAD plugin, the influence of structural parameters on the shape and position of the working space is analyzed when the structure parameters of the robot are changed separately. This study laid the foundation for the design, control and planning of palletizing robots.

1. Introduction

In recent years, considerable achievements have been made in the research of the workspace of palletizing robot[1]. G. Shanmugasundar[2] et al make the robot mechanical configuration design and modeling and simulation of the workspace of robot in CATIA 3D modelling software, which proved that the robot is totally capable of perform the given task. Zhang Ming[3] et al analyze the workspace of joint type stacking robot to get the reachable workspace diagram. Xu Chengyi[4] et al give a simple method of collision checking through simulation of joint driving by rotate3d function based on AutoCAD. Yin Feng[5] et al Advanced a new method based on the local point quadrant pattern which shown that it is accurate enough for extracting the workspace boundary of the robot. In this paper, using AutoLisp language the workspace of the four-degree-of-freedom joint stacking palletizing robot is drawn and simulation is constructed. The influence of structural parameters on the shape and position of workspace is analyzed intuitively. It also provides a powerful help for the design and research of palletizing robot.

2. Workspace of Palletizing Robot

According to the D-H parameter, the four-degree-of-freedom palletizing robot link coordinate system is established. See Figure 1. Then the position and attitude of the end of manipulator can be obtained:

$$\boldsymbol{T} = \begin{bmatrix} n_x & o_x & a_x & p_x \\ n_y & o_y & a_y & p_y \\ n_z & o_z & a_z & p_z \\ 0 & 0 & 0 & 1 \end{bmatrix} \quad (1)$$

According to the D-H parameter, the four-degree-of-freedom palletizing robot link coordinate system

Where: $\begin{bmatrix} n_x & o_x & a_x \\ n_y & o_y & a_y \\ n_z & o_z & a_z \end{bmatrix}$ is the rotation matrix (attitude) of the robot end coordinate system with respect to base coordinate system; $\begin{bmatrix} p_x & p_y & p_z \end{bmatrix}^{\mathrm{T}}$ is the center of the end effector of the robot. D-H parameters are shown in Table 1.

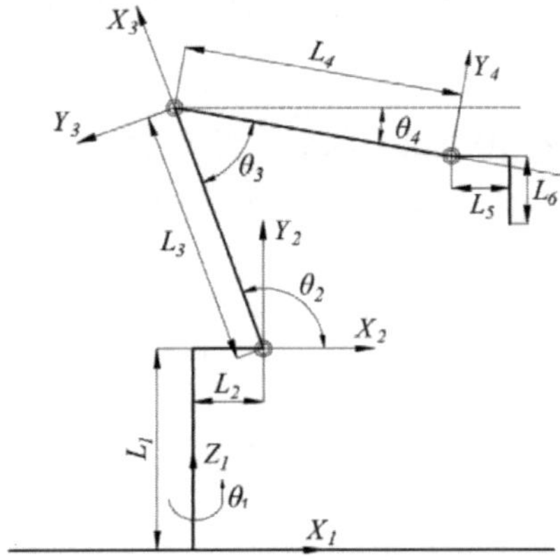

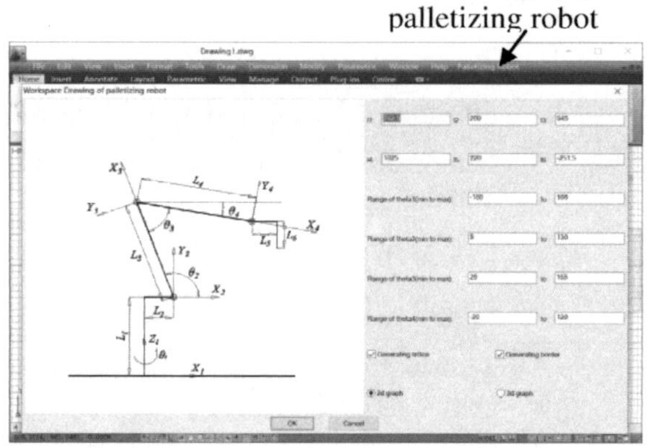

Figure 1. Mechanism coordinate system. **Figure 2.** Palletizing robot workspace software interface.

Table 1. Palletizing robot D-H parameter table.

i	a_{i-1} (mm)	α_{i-1} (°)	d_i (mm)	θ_i. (°)
1	0	0	0	θ_1
2	l_2	$-90°$	l_1	0
3	l_3	0	0	θ_3
4	l_4	0	0	θ_4

3. The AutoCAD plug-in in the workspace of palletizing robot

In the AutoCAD development environment Visual Lisp, using AutoLisp language the file"palletizing robot.lsp" is written to realize the simulation of the workspace of palletizing robot. And generating the format dialog interface file "*.dcl". The dialog interface file achieves the loading and display of dialog box, After the user clicks on the "OK" button, the data is gotten from the edit box and check box. At the same time closing the dialog box to complete the following procedures for the drawing command of the workspace. Adding a custom menu "palletizing robot" in AutoCAD menu bar. The custom command macro loads the " palletizing robot.lsp" source file and the "*.dcl" dialog box. The final software interface is shown in Figure 2.

4. Drawing the workspace of palletizing robot

When the bottom rotary angle θ_1 keeps constant, the workspace of palletizing robot in the plane XOZ is analyzed. Figure 1 shows that in the XOZ plane the ranges of θ_2、θ_3、θ_4 have the relation as follow:

$$\theta_2 + \theta_3 + \theta_4 = 180° \tag{2}$$

Where any two angles increase from the minimum value of the angle to the maximum value, which is constituting a dual loop. In this loop, the third angles are judged within the scope of the request. If consistent, the result of the coordinate point of the equation (1) is obtained, as shown in Figure 3.

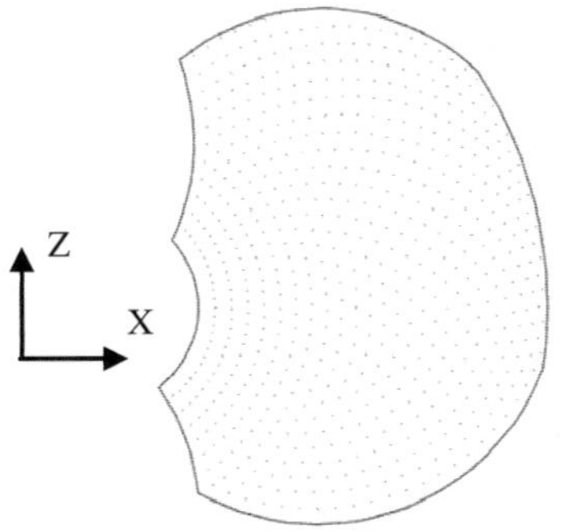

Figure 3. Lattice and boundary diagram of 2D workspace.

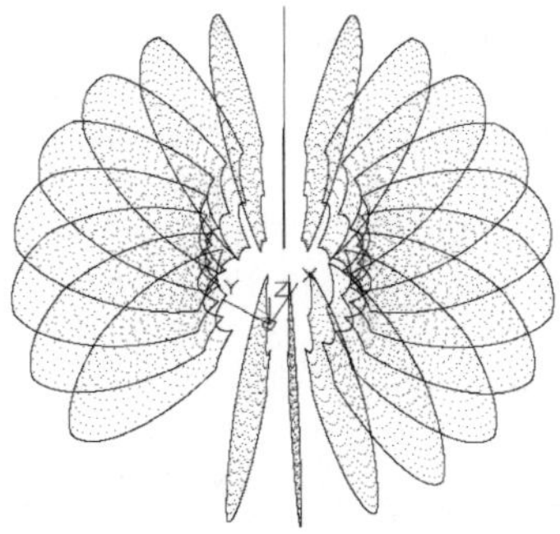

Figure 4. Lattice and boundary diagram of 3D workspace.

Analysis on the formation conditions of the workspace boundary can be seen: Firstly, it can be seen that each boundary is an arc, because one angle is fixed and the other angle is gradually increased when drawing each boundary. Secondly, when one of θ_2 、 θ_3 、 θ_4 is equal to the upper limit and lower limit of the angle range, one of workspace boundary is formed, so the maximum number of boundary of workspace should be 6; when any two angles equal to the upper limit and the lower limit of the respective angle range, one vertex of the workspace is formed. Thus, the method of three points to determine the arc can be used to draw the workspace boundary. The center point of the arc can be determined by taking the intermediate value of the angle range.

Therefore, running the AutoLisp program to draw the lattice and boundary diagram of 2D workspace, as shown in Figure 3. 3D workspace can be rotated by the 2D workspace along the Z axis, as shown in Figure 4.

5. The Influence of Structural Parameters on the Workspace of the Palletizing Robot

The influence of structural parameters on the workspace is analyzed by simulation and drawing. The original value of the parameter changes as shown in the spacing in Figure 5-7. The original workspace is represented by thick lines; the changing figure boundary is represented by thin lines.

5.1. The Influence of the Changing Rod

As can be seen from Figure 5, l_1 and l_2 only affect the position of the workspace, and the workspace moves toward the forward direction of the Z axis and the X axis.

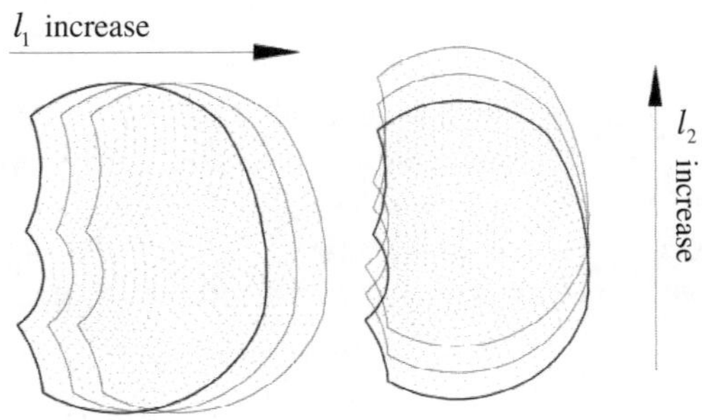

Figure 5. When the length of l_1 and l_2 change, the variation of the workspace.

As shown in Figure 6, when the other parameters are constant and l_3 increase or decrease, 2D workspace is deformed along the direction of the arrow. But the workspace is always composed of six arcs. l_4 is also shown in Figure 6.

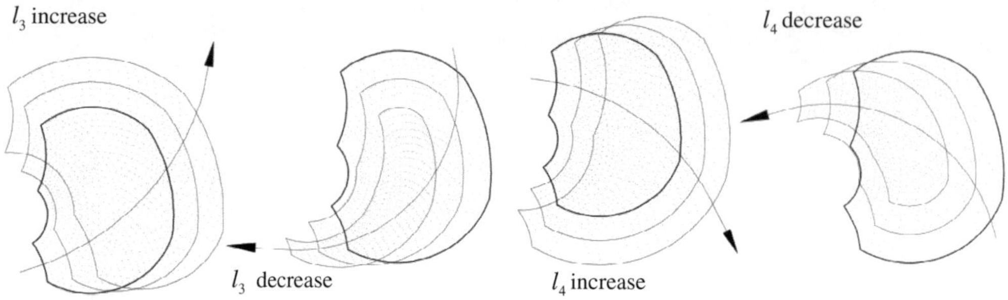

Figure 6. When the length of l_3 and l_4 changes, the variation of the workspace.

5.2. The Influence of the Changing Angle

Because the angle θ_1 is the bottom rotary angle, only the changes of θ_2 and θ_3 are analyzed. the 3D workspace changes the rotation range as the angle θ_1 range changes.

When the range of θ_2 is enlarged, the workspace is deformed along the direction of the arrow shown in Figure 7, and the workspace is gradually increased, otherwise, the reverse. In the both cases, it is possible that a boundary of the workspace is degraded into a point so that the resulting workspace boundary arc is less than 6. It is the same for θ_3.

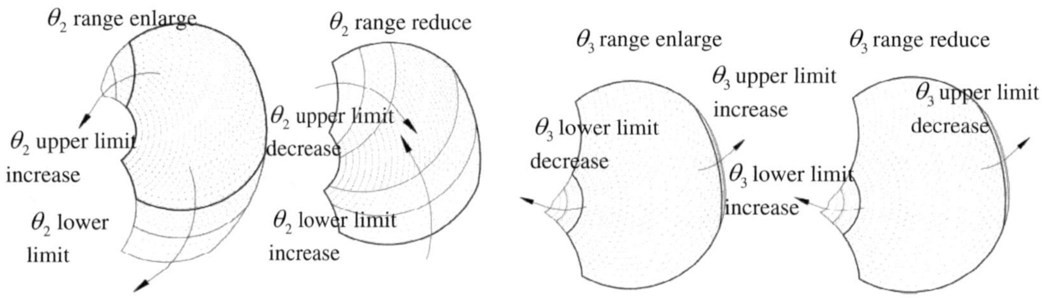

Figure 7. When the length of θ_2 and θ_3 changes, the variation of the workspace.

6. Conclusion

This paper studies the workspace of four-degree-of-freedom articulated palletizing robot by a secondary development on the AutoCAD platform. The robot is modeled according to D-H parameter method. This paper develops a simulation interface platform of workspace of palletizing robot in which the drawing of 2D and 3D workspace of robot is realized. The workspace is drawn by changing the input parameters. The influence of structural parameter change on workspace can be analyzed. This is a clear, simple and effective method for determining the workspace of palletizing robots, and it lays a foundation for the study of robot dynamics, control and planning.

References

[1] Li J, Duan B, Li Z. *Journal of Beijing University of Posts and Telecommunications*. **2011,34(6)** 78-81

[2] G. Shanmugasundar, R. Sivaramakrishnan, M.Rajmohan, et,al. *IEEE International Conference on Computational Intelligence and Computing Research, IEEE ICCIC 2014*. **2014** 1-6

[3] Zhang Ming, He Qingzhong, Wang Zhipen, et.al. *Modular Machine Tool & Automatic Manufacturing Technique*. **2013,(7)** 68-70,74

[4] Xu C, Ding H, Ji Z, et.al .*Modem Manufacturing Engineering*. **2013,(3)** 28-31,67

[5] Yin F, Wang Y, Yu H. *Control Theory and Applications*. **2010,(6)** 891-896

IOP Conf. Series: Materials Science and Engineering **241** (2017) 012019 doi:10.1088/1757-899X/241/1/012019

Temperature Control of Autothermal Reformer System with Coefficient Diagram Method

N Srisiriwat[1] and C Wutthithanyawat[2]

[1]Department of Petrochemicals and Environmental Management, Faculty of Engineering, Pathumwan Institute of Technology, Bangkok 10330, Thailand
[2]Department of Instrumentation and Control Engineering, Faculty of Engineering, Pathumwan Institute of Technology, Bangkok 10330, Thailand

navadee@hotmail.com

Abstract. The objective of this paper is to design the autothermal reformer (ATR) temperature control by using a coefficient diagram method (CDM). The adiabatic temperature is a main controlled variable of the ATR which is a combination of endothermic and exothermic reactions. The simulation results of control parameters were calculated to maintain the ATR reaction temperature by manipulating air feed flow rate. In this work, two strategies of ATR temperature controller system with and without the feed temperature control of a preheater unit are compared to investigate the appropriate controller system when the change of surrounding temperature is considered as a key disturbance. The results showed that by using the CDM, the stability and robustness for controlling the ATR temperature system were considered to offer the proper control parameters and the designed temperature control of ATR system gave a good performance to maintain the controlled variables and reject the disturbance. Moreover, the ATR control system design with the feed temperature controller can compensate the surrounding temperature better than that without the feed temperature control.

1. Introduction

An autothermal reformer (ATR) is one of hydrogen processing options for on-board hydrogen storage applying for the fuel cell to generate electricity. The moderate size of ATR, combining the exothermic reaction of partial oxidation (POX) with the endothermic reaction of steam reforming (SR) in a single process, is an advantage for the transportation application because of its high thermal efficiency and dynamics during transient operation as well as its lower system complexity [1]. The temperature control system design of ATR is important for controlling the reaction temperature to sustain the performance of hydrogen production. In our previous work, the modeling and simulation for ATR system was carried out to investigate the open loop response for designing the control system strategy [2] and an internal model control (IMC) method was performed to control the ATR temperature by manipulating the air flow rate [3-4]. Furthermore, there are many studies about the control design techniques for ATR process to produce hydrogen from gasoline [5-6], methanol [7-10], and other hydrocarbon fuels [11-13] whereas our researches have concentrated on the hydrogen production from ethanol [2-4]. By using several techniques of controller design, most of the studies were proposed to regulate the ATR reaction temperature and hydrogen yield by manipulating the air flow rate, water or fuel.

In this work, the controller design for ATR system is performed by using a coefficient diagram method (CDM) introduced by Manabe [14] when the change of surrounding temperature, which is a major effect for countries having cold weather, is also focused as the main disturbance. In order to maintain the near adiabatic reaction temperature of the hydrogen production from ethanol, two strategies of controller design with and without the feed temperature control of the preheater unit are proposed and compared.

2. Mathematical Model and Control Strategy of Autothermal Reforming System

In the ATR system, hydrogen-rich gas can be produced from the reaction between ethanol fuel, water and air that they are mixed and fed together into a preheater before entering the ATR reactor. Two control loops of ATR process include a feed temperature control loop and an ATR temperature control loop as illustrated in figure 1. The details of nominal parameters and operating points of this ATR system were proposed in the previous work [2]. The feed loop controller employs the heating power fraction of preheater (X, [%]) as the manipulated variable (MV) and the feed temperature of fluid entering ATR (T_{feed}, [K]) as the controlled variable (CV). In the ATR loop controller, the air flow rate ($\dot{N}_{Air}$, [mol/s]) is the MV for controlling the adiabatic temperature of ATR reactor (T_{ATR}, [K]) as the CV. Consequently, two transfer functions of the match of the CVs and the MVs for the ATR control system can be written as equation (1) and equation (2).

$$T_{feed}(s) = \frac{1}{(2.0s+1)}[320.5X(s) - 33.1\dot{N}_{Air}(s)] \tag{1}$$

$$T_{ATR}(s) = \frac{1}{(41.4s+1)}[137.3\dot{N}_{Air}(s) + 0.8T_{feed}(s)] \tag{2}$$

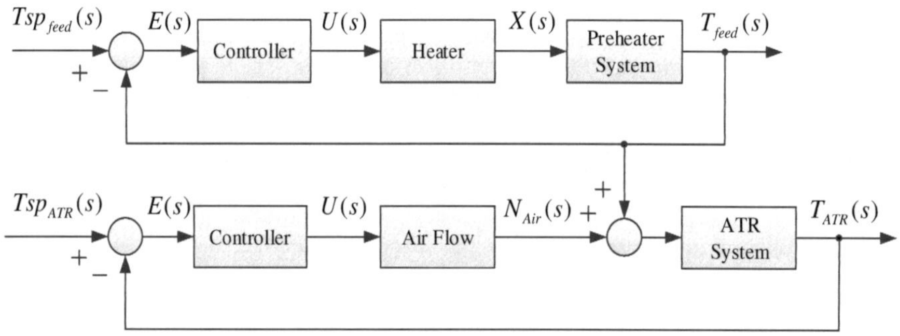

Figure 1. Block diagram of two control loops for ATR process control [3].

3. Control Design Using CDM

As a concept of coefficient diagram method (CDM) was described in a previous work [15], the control parameters of two control loops were designed by using CDM. In the ATR system control loop design, the desired settling time (t_s) of 50 sec is specified; therefore the equivalent time constant (τ) can be determined as 20 sec according to the standard Manabe form [14] given by $\tau = t_s / (2.5\sim3)$. In order to compare the effect of stability index on the controlled temperature response and control signal, the stability index were varied as 2.5, 5.0, 7.5, and 10.0 that the controller parameters of CDM-PI controller can be calculated as demonstrated in table 1.

In the feed temperature control loop design, table 2 presents the controller parameters of CDM-PI for different values of stability index of 2.5, 5.0, 7.5, and 10.0 when the desired settling time (t_s) is fixed at 2 sec; consequently, the equivalent time constant (τ) can be calculated as 0.8 sec.

4. Simulation Results

In this work, to regulate the ATR temperature that has an effect on an amount of hydrogen production, two strategies of the ATR temperature control system with and without the feed temperature control were designed and compared. According to the calculated control parameters using CDM at different values of stability index, the simulation results of the responses of ATR temperature and feed temperature were presented when an ambient temperature changed -25 K from a nominal condition is a key disturbance to consider the suitable control system for applying in countries that have cold weather nearly the water freezing point.

Table 1. Controller parameters of CDM-PI for various values of stability index of ATR control loop.

Stability index, γ_1	**K_p**	**K_i**	τ_i
2.5	0.0304	0.0019	16.1353
5.0	0.0681	0.0038	18.0677
7.5	0.1058	0.0057	18.7118
10.0	0.1435	0.0075	19.0338

Table 2. Controller parameters of CDM-PI for various values of stability index of feed control loop.

Stability index, γ_1	**K_p**	**K_i**	τ_i
2.5	0.0164	0.0244	0.6723
5.0	0.0360	0.0489	0.7362
7.5	0.0555	0.0733	0.7574
10.0	0.0751	0.0977	0.7681

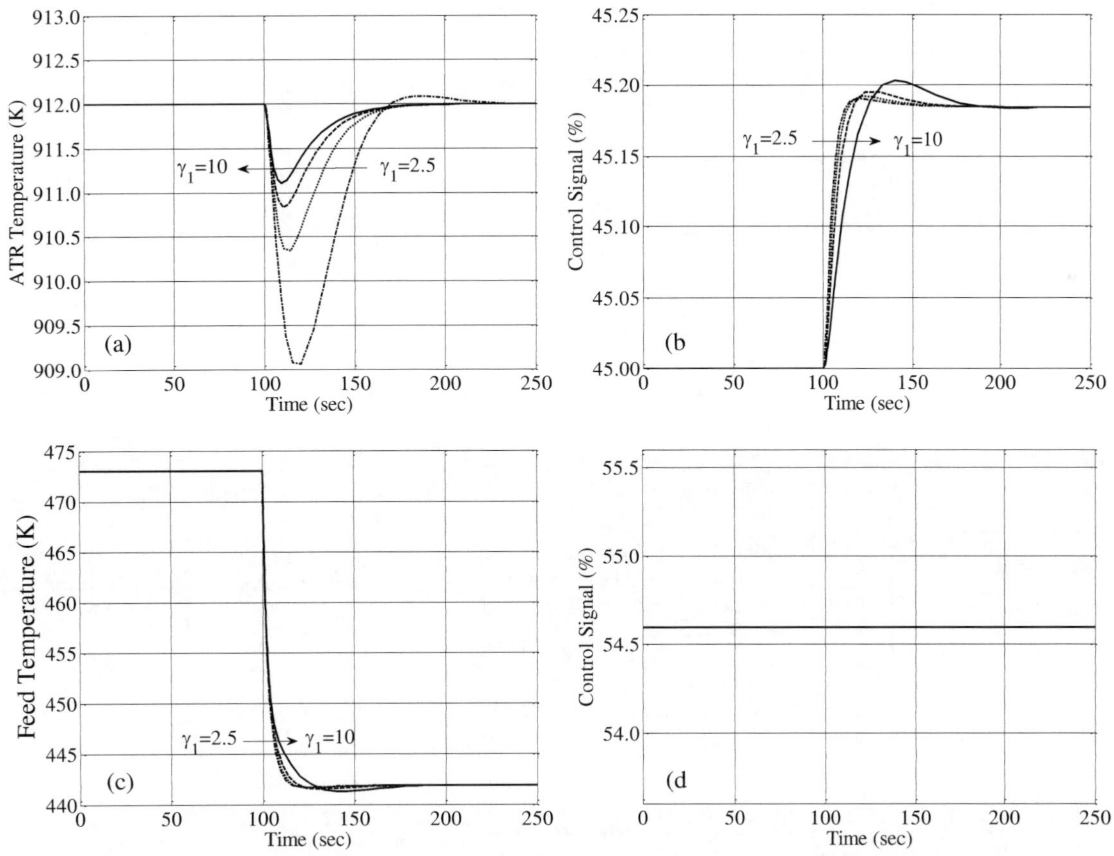

Figure 2. Response of ATR system without feed temperature control (a) Response of ATR temperature, (b) Control signal of air valve, (c) Response of feed temperature and (d) Control signal of heating power fraction of preheater.

IOP Conf. Series: Materials Science and Engineering **241** (2017) 012019 doi:10.1088/1757-899X/241/1/012019

In figure 2, the ATR temperature as the CV (figure 2 (a)) was controlled by changing the control signal as the MV (figure 2 (b)) for the ATR system without feed temperature control loop whereas the change of air flow rate affected the change of feed temperature (figure 2 (c)) with fixed control signal of a feed preheater (figure 2 (d)). When the stability index were varied between 2.5 and 10.0, the results showed that an increase of stability index decreases both of the deviation of ATR temperature from a set point of 912 K and the settling time but increases the control signal of air valve. In case of the ATR temperature control system with the feed temperature control loop, the control system designed by CDM showed a good performance that the ATR temperature is very well controlled in the range between 911.94 and 912.02 K in the whole range of varied stability index between 2.5 and 10 as shown in figure 3 (a) when the air flow rate was manipulated by control signal in the range between 44.9995 and 45.0015 % as shown in figure 3 (b). Additionally, the feed temperature was rapidly maintained (figure 3 (c)) by changing the control signal between 54.60 and 54.69 % (figure 3 (d)).
As shown in figure 3, by varying the stability index between 2.5 and 10.0, the difference between the ATR temperature and its set point can be decreased by increasing the stability index and the increase of stability index also decreases the change of feed temperature from its set point. Moreover, the ATR and feed temperatures can speedily reach their set points as the stability index increase. It can be found that the designed ATR control system with the feed temperature control is robust to the fast change in ambient temperature which is considered as the key disturbance for this work. In addition, the deviation of ATR and feed temperatures of two control loops are smaller than that of ATR temperature control loop without the feed temperature control loop. Consequently, the proposed CDM method for the ATR control system achieved a good performance to reject the disturbance.

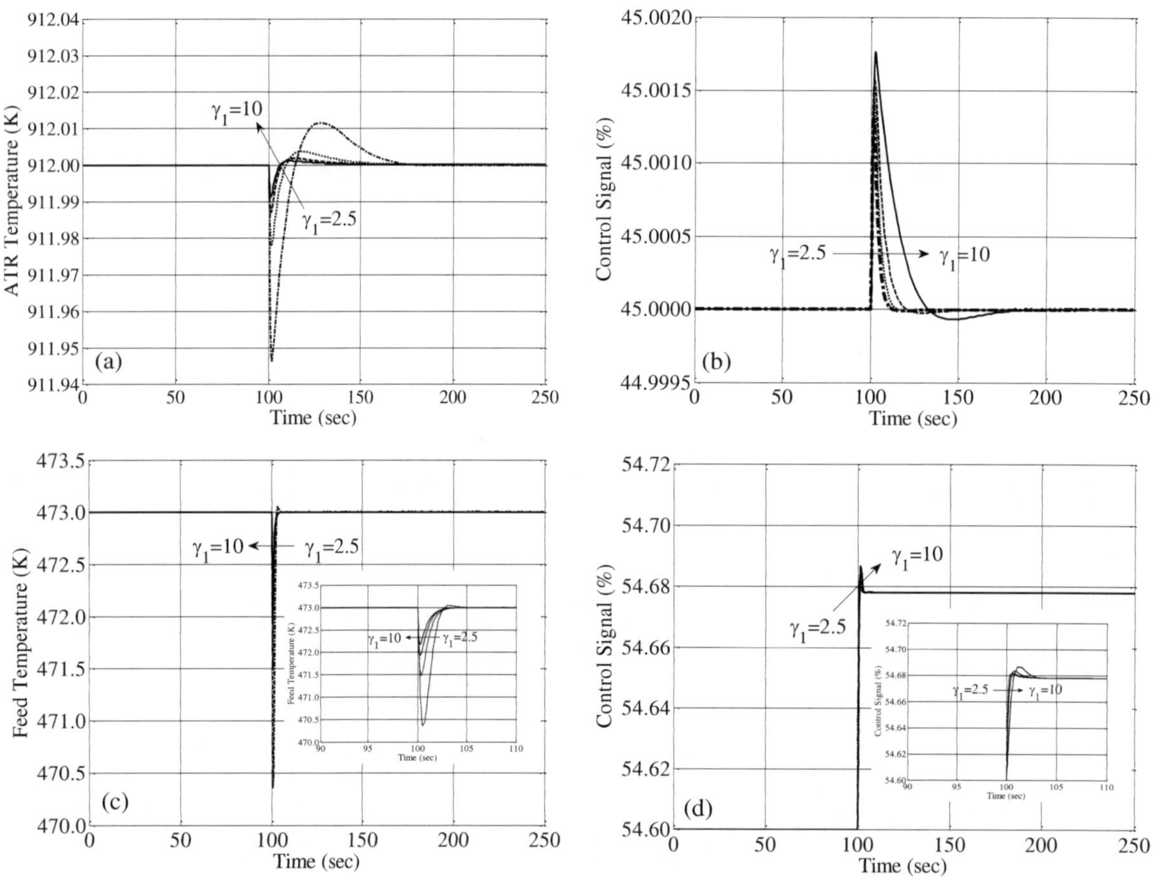

Figure 3. Response of ATR system with feed temperature control (a) Response of ATR temperature, (b) Control signal of air valve, (c) Response of feed temperature and (d) Control signal of heating power fraction of preheater.

5. Summary

In this work, a temperature control of ATR system for hydrogen production from ethanol was investigated by using CDM to regulate the adiabatic ATR temperature when two control loops of the ATR temperature control and the feed temperature control were proposed. In a feed control loop, the feed temperature is the CV and the heating power fraction is the MV. The ATR loop controller employs air feed rate as the MV and ATR temperature as the CV. For transportation application, because of the change of weather especially the countries having cold weather, a main change of ambient temperature was considered to be a major disturbance. It can be concluded that the ATR temperature control system designed by CDM achieves required performance. Moreover, the ATR control system design with the feed temperature controller can compensate the surrounding temperature better than that without the feed temperature control.

References

[1] Ipsakis D, Voutetakis S, Seferlis P, Papadopoulou S and Stoukides M 2009 *Proc. 17th Mediterranean Conference on Control & Automation* pp 1421-1426
[2] Wutthithanyawat C and Srisiriwat N 2014 *Applied Mechanics and Materials* **541-542** 108
[3] Wutthithanyawat C and Srisiriwat N 2014 *Applied Mechanics and Materials* **627** 236
[4] Srisiriwat N and Wutthithanyawat C 2016 *Key Engineering Materials* **689** 121
[5] Hu Y, Chmielewski D J and Papadias D 2008 *J. Power Sources* **182** 298
[6] Hu Y and Chmielewski D J 2009 *Proc. 2009 American Control Conference* pp 659-664
[7] Zheng J,Yang Q,Lu J and Sun Y 2012 *Advanced Materials and Research* **418-420** 377
[8] Wang X, Lu J, Zhuang H, Zheng J, Yang Q, Chen J and Sun Y 2013 *Proc. 10th IEEE International Conference on Control and Automation* pp 218–222
[9] Lu J, Zhuang H, Yang Q,Wang X,Zheng J, Chen J and Sun Y 2014 *ISA Transactions* pp 1470–1475
[10] Tang H Y 2009 *PhD Dissertation University of California, Davis*
[11] Chen J and Sun J 2007 *Proc. 46th IEEE Conference on Decision and Control* pp 4608–4613
[12] Chen J and Sun J 2010 *IEEE Transactions on Control Systems Technology* **18** 779
[13] Chen J and Sun J 2009 *Joint 48th IEEE Conference on Decision and Control and 28th Chinese Control Conference* pp 5188-5193
[14] Manabe S 1998 *Proc. of the 14th IFAC Symposium on Automatic Control in Aerospace* pp 199-210
[15] Srisiriwat N and Wutthithanyawat C 2016 *Advances in Engineering Research* **90** 38

A study for prevent theft of the bike design and analysis

Huang, Yu-Che [1], Huang, Tai-Shen [2],

168, Jifeng E. Rd., Wufeng District, Taichung, 41349 Taiwan(R.O.C)[1]
168, Jifeng E. Rd., Wufeng District, Taichung, 41349 Taiwan(R.O.C)[2]

che59965996@gmail.com[1] , tshuang@mail.cyut.edu.tw[2]

Abstract. This research will deal with the analysis and design of an anti-theft bicycle. At the first, the structure of the bicycle lock device will be focused. Therefore, all related academic journals, dissertations, and patent were collected, ordered, classified, and analysis in order to understand the phase of the developing tendency towards the bicycle lock apparatus presently. Nowadays, a wave of riding bicycles all over the world is quite prevailing. Also, the bicycles incline to be high-price sporting and recreational equipment. Due to this, the rate of stolen bicycles has increased as well. Even though there are many different types of bicycle locks sold on the market, they are unable to work out effectively to prevent from being stolen because they are easy to be broken. In this case, the target design in this research will be mainly put on the simple structure and facile use enabling users feeling secure to park their bicycles. The design also includes a lock that works together with the bicycle frame to provide a new option of anti-theft design for bicycles. We defined the functional requirements by the process of an innovative mechanical design, before designing the fundamental bicycle structure of the present anti-theft bicycle and finally integrate all the requirements of design. The SolidWorks software was used to draw a 3D model of the anti-theft bicycle; the ANSYS software was also used to do CAE analysis, to understand the distribution of stress, so that the design and the requirements of use are met.

1. Introduction

As bicycles provide people to move around in the outdoor or a park at leisure, they are increasingly used as a transportation instrument in the era with emerging environmental awareness. Yet, as the prices of bicycle keep increasing, bicycle theft has become a problem that cannot be taken lightly. Funds and human power are thus pumped into research, development and production in advanced countries, which helps the markets for bicycle develop well. However, as the materials used as the bike frame are becoming more light-weight thanks to technological breakthrough in recent years, the number of bicycle thefts keeps increasing. Anti-theft design thus is at present an important issue taken seriously by R&D personnel. How to make bicycles hard to steal and easy to use is a pressing question. When the body per of a bicycle is also an anti-theft device, the theft rate can be decreased as a result of the body structure and strength being compromised at the same time when the lock is sabotaged, which causes a safety problem with riding.

2. Patent analysis and classification

According to the documents, academic periodicals and patent gazettes from which this study gathered data, the devices with the structures for bicycle lock are largely divided by the locking method in five categories:

(1) Locking rod device

This locking rod device is a fixed block that is fixed to either fork or either tubing of the seat stay of a bicycle, comprising a lock which can be controlled by a key to be turnable or be fixed. Patents [1] and [2] that were found by the search both disclose a way of theft prevention by locking with a rod, which is hereby discussed. As Fig. 1 shows, [1] is a method that inserts a rod through the spokes, making the wheel unable to turn, and in turn locking the bicycle.

The title is set 17 point Times Bold, flush left, unjustified. The first letter of the title should be capitalized with the rest in lower case. It should not be indented. Leave 28 mm of space above the title and 10 mm after the title.

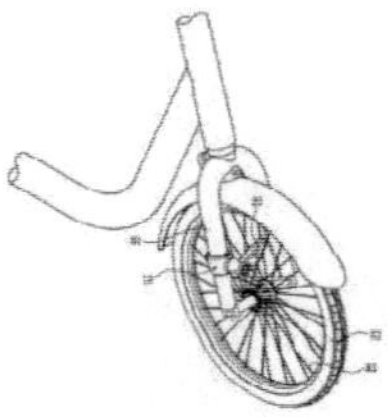

Fig. 1 Application of locking by a rod [1]

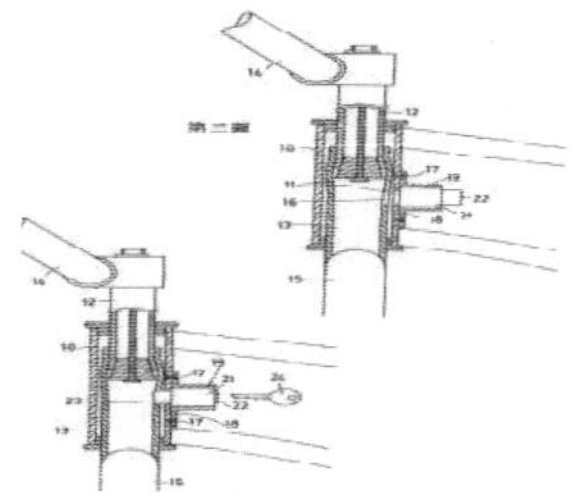

Fig. 2 Fork with a lock [3]

(2) Fork-locking

The fork-locking device provides a lock mechanism in a fork, in which the lock mechanism is enabled when the bicycle is parked, making the fork unable to turn. Patents [3], [4] and [5] all use a fork-locking device to lock a bicycle; here Patent [3] is particularly discussed. As Fig. 2 shows, [3] discloses a fork that is provided with an exterior hole, and a link, which is disposed, fixed inside said fork, having an internal hole that corresponds to said exterior hole; a positioning plate is fixed to said fork on said exterior hole, a bushing is provided on the positioning plate at the position corresponding to said exterior hole, and a lock is fixed to said bushing; by means of the locking of said lock, it is possible to make the handle bar unable to turn, achieving the effect of locking the bicycle.

(3) Rigid-lock

A rigid-lock device forms a close lock with rigid rods or tubing that go through the wheel axel, which is a device so strong that it is difficult to break, and thus achieves the effect of theft prevention. Patents [6], [7] and [8] were found; here we elaborate on [6] and [7]. Fig. 3 shows [6] as a rigid-lock device for bicycles, which is a locking device installed in front of the forks, which includes a steel wire; when the wire is pulled up, it pulls two L-shape rods on both sides so that they go through the wheel and form a closed locking device. Fig. 4 shows that [7] provides a lock on the rack of a bicycle, said rack comprising a fixed rod, a pivot rod disposed fixed to the end of said fixed rod, and two support rods that are pivotably joined to said pivot rod at both ends and are supported by the rod that runs through the wheel's axel; said lock comprises a U-shape rod and a lock head, characterized by that said U-shape rod is disposed with its ends pivoting on said pivot rod of the rack and, when the lock head is removed, can swivel to a position where its front end is by the sides of the wheel allowing the lock head to lock the wheel or swivel to a position parallel to the fixed rod of the rack and be fixed with the lock head.

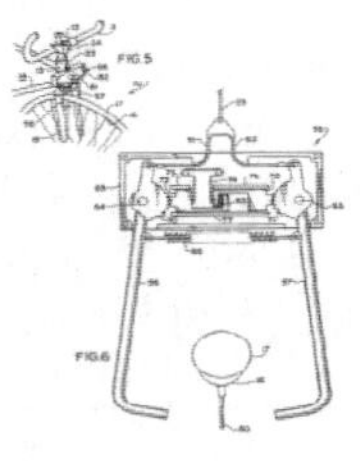

Fig. 3 Structure of a rigid lock for bicycles [6]

Fig. 4 Structure of bicycle lock [7]

(4) Flexible lock device

A flexible lock device uses flexible materials, such as rope and steel cable, for locking device. When locking, the flexible lock is handy in going around any object that is attached to ground, such as railing. Patents [9], [10], [11] and [12] all disclose a means of locking using flexible materials; we discuss particularly [9], as shown in Fig. 5. It is a lock device built with the bicycle frame as an integral body, it being a lock disposed at one end of the bicycle rack and having a first end and a second end; a cable, which is fixed to said first end of the lock on one end and the other end goes around the frame and back to be removably attached to said send end of the lock, and which can be in a folded status and fit against the frame, whereby the cable is free from being pendulous, and thus ensures of the safety during riding.

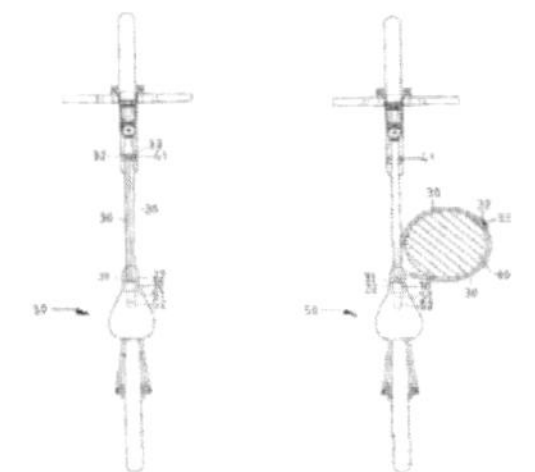

Fig. 5 Locking by flexible lock [9]

Fig. 6 Locking device by bike parking rack [13]

(5) Bike parking rack and lock device

Bike parking racks are a device disposed by roadsides; this bike parking rack is a special lock device, with its ring freely rotatable to any position and any orientation to lock a bicycle on the forks, the top tube, the rear brake or the wheel; it is an omni-directional lock device, as [13] shown in Fig. 6.

3. Design conditions for theft bikes

We analyzed and sorted the available documents and patents on bicycle lock structures and devices to understand the advantages and disadvantages of the anti-theft devices available in the market. Through regulations and market survey, we set out the design conditions for new bicycle anti-theft devices before designing them, so as to render this bicycle locking structure or device with maximum functions, letting users be assured to use this device [14] [15] [16].

(1) The bicycle can structurally withstand a load up to 100kg (980N) to ensure of user safety.
(2) For handiness, safety and versatility, this study makes the simplest design of locks.
(3) The design herein enables easy locking when in the outdoor.
(4) A minimum of two methods of locking is required: locking with the tire and locking onto utility poles.
(5) The innovative design herein aims at decreased theft rates and thus consideration is given to combining the lock with the bike frame, whereas they are joined in a fixed manner, such that once the lock is damaged, the bike frame is also damaged.

4. Research and design results

The review herein of existed patents revealed that most bicycle locks are portable and some are installed on the bicycle, but the locks, being of steel wires, are easy to cut. Out design thus was made for the specific requirements of an anti-theft bicycle, which are materials that are difficulty to cut and using roadside bike parking racks as locking means. Based on these two design requirements, this study came up with two design concepts, one being that the lock is partly the bicycle's frame, thus when the lock is damaged from sabotage, the bike will collapse due to under-strength to support the load. The other concept is to design locks that can be mounted on bicycles of all kinds.

4.1. Dedicated anti-theft type of bike

For this anti-theft bicycle, we use part of its top tube as a part of the lock; the lock, which, as Fig. 7 and Fig. 8 show, is consisted of three tubes as a locking device with five degrees of freedom, is installed on the seat tube with rivets in a fixed manner. Fig. 9 is the perspective view of this lock device. In the event that it is damaged, the bicycle will suffer from the loss of support by the top tube and tends to collapse. As such, this design can discourage a thief to steal the bike. In addition, this lock device is designed to allow the rider to lock the wheel when he wishes to park the bicycle or when there is not a parking rack nearby, so as to protect the bicycle. Fig. 10 is the perspective view of our exclusive anti-theft bicycle.

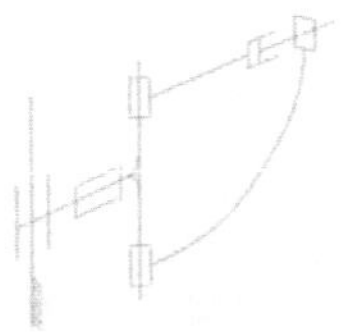

Fig. 7 3D diagram

Fig. 8 Exploded view

Fig. 9 Assembly diagram for lock device

4.2 Lock bike

This lock device is a stand-alone device, as Fig. 11 shows. It can be installed on the seat tube, the top tube, the forks or the chain stay of a bicycle. When installed on a seat tube, a top tube or a down tube, the lock device can lock the wheel or lock with a utility pole or a bike parking rack against theft, as Figs. 12 and 13 show. If installed on the fork or the down tube, though it cannot lock the wheel, it can lock with a utility pole or a bike parking rack.

Fig.10 Perspective view

Fig.11 Device installed on seat tube

Fig.12 Lock wheel

Fig.13 Lock stop lever

4.3 Innovative anti-theft bike design

With the above-mentioned two design concepts and the required design criteria defined, we can begin to make detailed design of the anti-theft bicycle. To create a 3D model in diagrams of parts and assembly, the SolidWorks 2016 software was used. Fig. 14 shows the diagram of locking function of the lock device.

IOP Conf. Series: Materials Science and Engineering **241** (2017) 012020 doi:10.1088/1757-899X/241/1/012020

Fig. 14 Diagram for locking method

4.4 Analysis of Frame Structure Strength

Following the completion of model creation was a stress analysis to ensure the bicycle is free of safety concern under loads. To do this, an ANSYS Workbench module was run on the ANSYS 11.0 software. As the first concept herein features taking part of the top tube of a bicycle as a rod piece of the lock device, while the second concept features the adoption of existed bicycles available in market, the stress analysis will be applied only to the exclusive anti-theft bicycle of the former. We applied user's body weight on the bike frame, ignoring the stresses, distribution and variation thereof caused by the weight of the frame itself, whereby to understand whether each rod is adequately strong, as reference for designing.

Fig.15 Bike frame under load

Fig.16 Bike frame under load

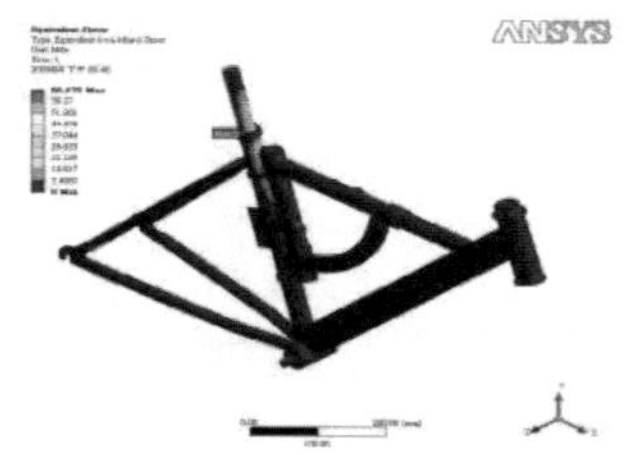

Fig.17 Diagram for stresses

We began by creating a static analytic model, which was used on the bike frame and the lock device, of the material of 7005 aluminum, in two applications of forces exerted on the frame, as Fig. 15 and Fig. 16 show. In the first application, a 980N force is exerted on the seat tube, and in the second, a 980N force is divided in 490N forces, each of which is exerted on the seat tube and the bottom bracket, respectively, for analysis. The analytic results were put to stress diagrams herein by the Von Mises Stress, as Figs. 17, 18 and 19 show, which are the diagrams for stress distribution on bike frame and for the Von Mises Stress. Such results indicated that the area where greater stresses focus was at the top of the seat tube in both the first and the second applications, with the maximum stresses at 66.679MPa and 33.571MPa, respectively. As according to the material database, the 7005 aluminum has a yield strength at 280MPa, because the stresses on the current bike frame under analysis in the three applications were all within the tolerable range, the anti-theft bicycle with our exclusive frame meets the strength requirement. In addition, through the industrial design to use this result to do the concept of design, as Figs. 20 show .

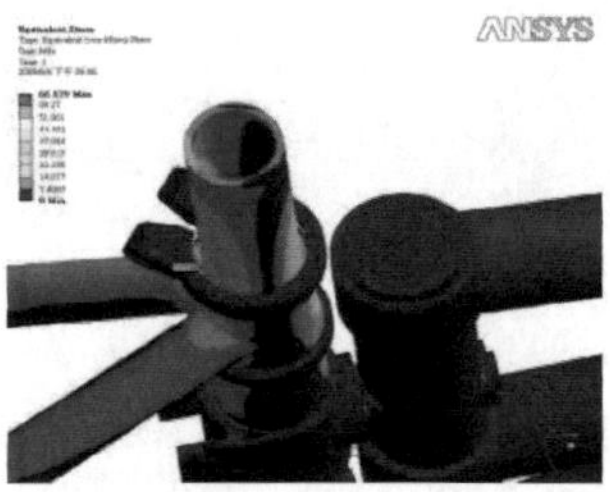

Fig. 18 Diagram for Max.

Fig. 19 Diagram for stress

Fig. 20 Future conceptual design

5. Conclusion

This study developed, by searching for extensive types of bicycle lock structure and device and sorting and analyzing various locking means, a new type of anti-theft device for bicycles. In the process, consideration was given to locking methods in different situations and the purpose of increased handiness and reduced mechanical complexity; hence, compared with other products, the design herein is characterized by the following:

IOP Conf. Series: Materials Science and Engineering **241** (2017) 012020 doi:10.1088/1757-899X/241/1/012020

(1) The use of kit-type lock device, wherein the current lock device is characterized by being able to installed on the top tube, seat tube or the fork of any kind of bicycle and able to lock the bicycle wheel itself as well as with the utility pole or bike parking racks.
(2) The adoption of an exclusive bike rack-incorporated lock device, wherein part of the lock device is also the top tube, such that any damage to the lock will also damage the bicycle, thus making it less liable to theft. Also, the locking methods include locking the wheel of the bicycle itself as well as locking with a utility pole or bike parking racks on road side.

References

[1] Ahu Zhe hoe in: The structure of the bicycle lock (Taiwanese patent) number 219028
[2] Liao Ben Liu in: Bicycle lock structure (Taiwanese patent) number M346551
[3] Li Jiyuan in: The structure of the bicycle lock (Taiwanese patent) number 431341
[4] Lhen Guotai,in: Bike lock structure improvement (Taiwanese patent) number 397000
[5] Li Jiyuan in: The lock in the fork shaft of the bicycle (Taiwanese patent) number 487070
[6] Mark L All n ,in: BICYCLE LOCK (US Patent Number) 4708004 1987
[7] Mou Xiuyu in: lock light device (Taiwanese patent) number M345024
[8] Penneth Andersson Gronegatan in: DEVICE FOR LOCKING A BIBYCLE (US Patent) Number 4794770 1989
[9] Ruang Shimin, in: with the bicycle frame integrated lock Taiwanese patent number I224063
[10] Su Fangxiong, in: Bike lock device improved, Taiwanese patent number 506377
[11] Sai Kam-hui, in: Can be used with the horseshoe lock the use of the bike cable its bolt seat improved, Taiwanese patent number 553125
[12] Shih-Min Li 2016 esearch on Patent Analysis of Taiwan Bicycle Industry National Institute of Information Management (Qinchi University of Science and Technology Master thesis Taiwan)
[13] Suai En and Syu Meng-Hui and Hsu 2014 Design Analysis New Folding Bikes (Kunshan University of Science and Technology Institute of Mechanical Engineering Master thesis Taiwan)
[14] Shen Zhaojun Applying 2102 TRIZ method to increase comfort and anti-theft performance of Power-Assist Electric Bicycle (Jianguo University of Science and Technology Department of Automation Engineering and Electrical and Mechanical Systems Master thesis Taiwan)
[15] Yi-Lung Chiu and Jan-Nan Chen 2014 Applying TRIZ Theory to Integrated Design of Bicycle Parts and Anti-Theft Device (Chaoyang University of Science and Technology Industrial Design Institute Master thesis Taiwan)
[16] Yu-Chun Huang and Jenq Huey Shyu 2010 Design and Analysis of Anti-theft Bicycles (Taipei University of Science and Technology Institute of Electrical and Mechanical Integration Taiwan)

IOP Conf. Series: Materials Science and Engineering **241** (2017) 012021 doi:10.1088/1757-899X/241/1/012021

Bi-fuel System - Gasoline/LPG in A Used 4-Stroke Motorcycle - Fuel Injection Type

Tongchit Suthisripok, Nachaphat Phusakol and Nuttapol Sawetkittirut

Automotive Engineering Department, College of Engineering, Rangsit University, Phaholyothin Road, Lakhok, Pathumthani 10200 Thailand

E-mail: tongchit.s@rsu.ac.th

Abstract. Bi-fuel-Gasoline/LPG system has been effectively and efficiently used in gasoline vehicles with less pollutants emission. The motorcycle tested was a used Honda AirBlade i110 - fuel injection type. A 3-litre LPG storage tank, an electronic fuel control unit, a 1-mm LPG injector and a regulator were securely installed. The converted motorcycle can be started with either gasoline or LPG. The safety relief valve was set below 48 kPa and over 110 kPa. The motorcycle was tuned at the relative rich air-fuel ratio (λ) of 0.85-0.90 to attain the best power output. From dynamometer tests over the speed range of 65-100 km/h, the average power output when fuelling LPG was 5.16 hp; dropped 3.9% from the use of gasoline91. The average LPG consumption rate from the city road test at the average speed of 60 km/h was 40.1 km/l, about 17.7% more. This corresponded to lower LPG's energy density of about 16.2%. In emission, the CO and HC concentrations were 44.4% and 26.5% lower. Once a standard gas equipment set with ECU and LPG injector were securely installed and the engine was properly tuned up to suit LPG's characteristics, the converted bi-fuel motorcycle offers efficiently, safely and economically performance with environmental friendly emission.

1. Introduction

Since all vehicles heavily depend upon petroleum fuels, the fluctuations of fuel price are of great concern. In Asian countries like Thailand, Vietnam and Myanmar, motorcycles are popular and widely used because of affordable prices and not well support countries' infrastructure. Interestingly, Thailand, Vietnam and Myanmar are gas production countries [1,2]. Therefore clean gaseous fuels, NG and LPG, would be potential alternate fuels for gasoline vehicles. This would benefit both environment improvement and national energy development. In Thailand, fast growing car and motorcycle population cause traffic congestion and pollution problems in the capital and major cities. About two-fold lower pricing than gasoline and more gas filling stations available than NG, LPG becomes very attractive alternate fuel for vehicles nationwide. Though a standard gas equipment set for vehicle and safely installation are costly, much lower prices of both LPG and NG are justifiable for vehicle owners; passenger, light and heavy duty vehicles, to modify their vehicles for gaseous fuel use.

In Thailand, over 20 million registered motorcycles have been used nationwide, about 54.2% of total registered vehicles [3]. Motorcycle was the highest share among the registered vehicles in the country. They are needed not only for personal uses but also for public services. The bi-fuel system of gasoline/LPG fitted in a used motorcycle with fuel injection and water cooling system was therefore conducted and tested for engine performance, fuel consumption and emission. This converted bi-fuel motorcycle will be a prototype providing safe, efficient and smooth riding. Furthermore it can be developed for transporting agricultural products in community and rural areas. However law and regulation applied to modified gaseous fuel vehicles has not yet included gaseous motorcycles.

Combustion in spark ignition engines
In spark ignition (SI) engines, the fuel-air mixture in the engine intake system enters the engine cylinder, reacts and combustion occurs inside the cylinder. Combustion of the fuel-air mixture controls engine power, efficiency, and emissions. The theoretical or stoichiometric fuel-air ratio is the ratio of fuel completely reacted with enough oxygen to produce CO_2 and H_2O. Engines can be operated under the condition of either fuel-rich mixture or fuel-lean mixture. In case of excess air or fuel-lean combustion, the non-reacted air appears in the products stream. Under the fuel-rich mixtures or insufficient oxygen environment, the incomplete combustion takes place and yields a mixture of CO_2 and H_2O and carbon monoxide CO and hydrogen H_2 and nitrogen N_2. The composition of the combustion products is significantly difference from fuel-lean and fuel-rich mixture conditions and the stoichiometric fuel-air ratio depends upon fuel composition [4-5]. It is therefore to define an informative parameter for mixture compositions. That is, the fuel-air equivalence ratio, ϕ and its inverse, the relative air-fuel ratio, λ are as follows: [6]

$$\phi = \frac{\left(\frac{A}{F}\right)_{stoic}}{\left(\frac{A}{F}\right)_{actual}} = \frac{1}{\lambda} \quad (1)$$

For fuel-lean mixture: $\phi < 1; \lambda > 1$
For stoichiometric mixture: $\phi = \lambda = 1$
For fuel-rich mixture: $\phi > 1; \lambda < 1$

To ensure smooth and reliable operation, the SI engines have normally been operated close to stoichiometric or slight fuel-rich condition. Best torque and power will be produced at about 10-15% fuel-rich mixture and best economy will be produced at a similar amount of fuel-lean mixture. The amount of CO in the exhaust is a measure of fuel-rich condition whereas the amount of O_2 in the exhaust is a measure of fuel-lean condition. Oxides of nitrogen NOx depend on temperature and pressure [5] and it reaches a peak with slightly fuel-lean condition.

The stoichiometric combustion reaction of C_8H_{16} is [4]

$$C_8H_{16} + 12(O_2 + 3.776N_2) \rightarrow 8CO_2 + 8H_2O + 45.312N_2 \quad (2)$$

and $\left(\frac{A}{F}\right)_{stoic} = 14.8.$

The stoichiometric combustion reaction of LPG, a mixture of 70% propane and 30% butane, is

$$(0.7C_3H_8 + 0.3C_4H_{10}) + 5.45(O_2 + 3.776N_2) \rightarrow 3.3CO_2 + 4.3H_2O + 20.58N_2 \quad (3)$$

and $\left(\frac{A}{F}\right)_{stoic} = 15.7.$

In reality the combustion processes are complex and often incomplete, even in the considerable excess air environment. It is therefore more practical to analyse the products experimentally. As shown in Figure 1, the emission curves indicate the complexities of emission control. Exhaust gas composition depends on the relative proportions of fuel and air fed to the engine, fuel composition, and completeness of combustion.

IOP Conf. Series: Materials Science and Engineering **241** (2017) 012021 doi:10.1088/1757-899X/241/1/012021

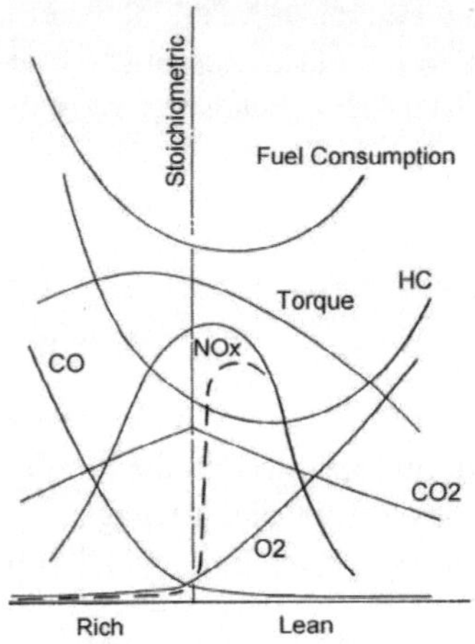

Figure 1. Relationship between torque, fuel consumption and gaseous compositions inside a SI engine with relative to air-fuel ratio [7]

Liquefied Petroleum Gas (LPG) is a non-renewable gaseous fossil fuel of propane and butane blends depending on the season - more propane in winter specification and more butane in summer. LPG is a by-product of oil refining, natural gas processing and petrochemical processes. LPG fuelled vehicles provides comparable power, acceleration and cruise speed to those of an equivalent gasoline fuelled vehicle. To be combustible, LPG must be in between 1.5% and 9.6% of the total LPG/air mixture. That is, the lower flammable limit (LFL) and the upper flammable limit (UFL) of LPG [8] is 1.5 and 9.6%. LPG is water insoluble; LPG does not pollute underground water sources. Its advantages are non-toxic, non-corrosive, having high octane rating (RON) of 102-108 depending on local specifications and, more importantly, environmental friendly fuel which produces less CO and toxics, especially benzene and butadiene emission, as well as regulated emissions. Carbon dioxide (CO_2) emission levels are reduced by up to 40% over those of gasoline vehicles [2]. With less carbon build-up, spark plugs often last longer and less frequently oil changes are needed. However there are some LPG drawbacks when fuelling vehicles which are starting problem in cold condition at below $0^{O}C$ because of the low vapour pressure of propane at low temperatures, and having lower energy density than gasoline resulting in 14% consumption more than a comparable gasoline vehicle. In addition the LPG filling stations are much less than the gasoline gas station and needs more frequent refuelling, it will be more practical to keep some amount of gasoline as a back-up fuel. Gasoline and LPG characteristics are given in Table 1 [9,10].

Table 1. Fuel characteristics [9,10]

Properties	LPG	Gasoline
Chemical formula	$0.7C_3H_8/0.3C_4H_{10}$	$C_{8.26}H_{15.5}$
Molecular weight	48.2	114.8
Specific gravity	0.5-0.58	0.72-0.78
Reid Vapor Pressure @$38^{O}C$ (kPa)	-	48-108
Boiling point (^{O}C)	-42	27-225
RON	105	92-98
MON	97	80-90
Heat of Vaporization (kJ/kg)	435	350
Lower Heating Value (MJ/kg)	50	44
Flash point (^{O}C)	-104	-43 to -39
Auto-ignition Temperature (^{O}C)	360	260-460
Flame temperature (^{O}C)	1920	2030
Combustible Limit (%gas in air)	1.5-9.6	-

2. Testing methods

The motorcycle tested, Honda AirBlade i110, was used over 70,000 kilometres; the engine conditions i.e. fuel system, transmission system, and lubrication system must be thoroughly inspected before conducting the tests [11-12].

Equipment List

1. The used 4-stroke motorcycle, Honda AirBlade i110 - fuel injection, water cooling system and CVT system
2. Test fuels: Gasoline91 and LPG (C_3:C_4 = 70:30)
3. Dynamometer [13]
4. VERSUS ECU (Electronic Control Unit) Gas Tuning Set
5. Emission Gas Analyzer – Gena 2000

.

Performance Test on Dynamometer

1. Tight the motorcycle tested firmly on the dynamometer and then calibrates the engine speed (rpm) see Figure 2.
2. Start the engine and accelerate up to full throttle.
3. Cut off the transmission system. Torque and power output were automatically computed and graphically displayed.
4. Apply brake and turn off the engine.

Fuel Consumption

Mileage and fuel consumption were collected from the following test conditions [11-12]:

1. City road test at the average speed of 60 km/h. The test data were collected from daily riding in the city at the average speed of 60 km/h.
2. Long distance riding test at the constant speed of 60 km/h. This was tested on the certain road during 16:00-18:00 hr. due to light traffic

Emission Measurement

1. While running the engine, insert the oxygen sensor right in the middle of its exhaust pipe referred to Figure 3.
2. Record data shown on the screen and then remove the sensor and turn off the engine.

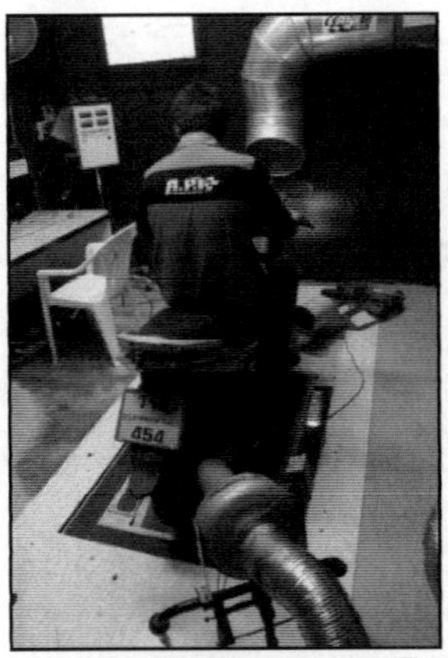

Figure 2. Performance Test on Dynamometer

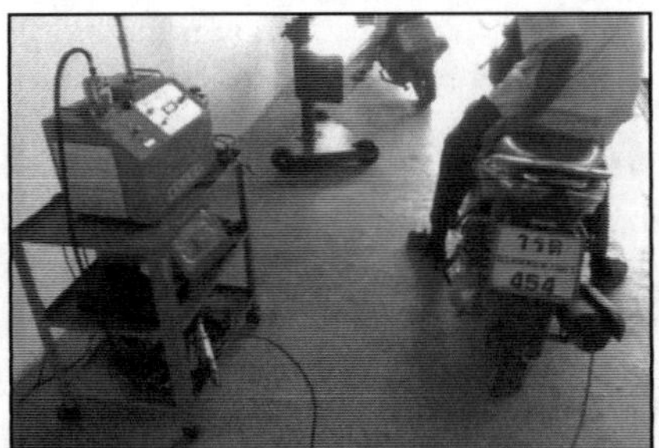

Figure 3. Emission Measurement

Engine Modifications

To utilize the original engine arrangement system, the bi-fuel converted vehicle was fitted with fuel-injected engine modified to control the gas flow and revised ignition timing or alternatively, fitted with a standard gaseous control system [14-17]. Similarly, the motorcycle tested which was a fuel-injection type with water cooling system was fitted with the standard LPG equipment set for vehicle which was modified to control LPG firing properly.

1. To obtain the performance data base of the motorcycle tested using gasoline91, the motorcycle was properly tuned on the dynamometer at the relative fuel-rich ratio (λ) of 0.85-0.90 which theoretically was in the best giving power output range.

2. For the use of bi-fuel system of LPG/gasoline in the motorcycle whose compression ratio was maintained, the specific 3-litre LPG storage tank was installed and wired to the VERSUS ECU kit of 4-cylinder-4-stroke engine. Note that it is wired to only 1 cylinder and left out the others. All was in the U-box beneath the seat see Figure 4. The 1 mm diameter LPG injector was inserted close to the original fuel injector and a regulator was tightly placed near the cooling over the frame bar. The engine was tuned up on the dynamometer at the λ range of 0.85-0.90. The stoichiometric air-fuel ratio required for gasoline combustion was 14.8:1 and 15.7:1 for LPG. Therefore LPG combustion is required to be operated in the fuel-leaner condition. To be combustible, the total LPG/air mixture must be present in between 1.5 and 9.6 %.

Based on VERSUS Hi-end Technology Gas for Car manual, the calculated nozzle size for LPG used in a 1-cylinder-4-stroke 110 cc engine was approximately 1 mm and the LPG injection control diagram was shown in Figure 5. The fuel nozzle injector geometry produces optimum fuel air mixing and increase the volumetric efficiency of the engine which will promote a comparable engine performance [15]. Due to having low energy density and low engine volumetric efficiency, LPG must be compressed and stored in pressure vessel. To safely use LPG for the motorcycle tested, the relief pressure was set below 48 kPa and over 110 kPa while the operating pressure and temperature was 78-80 kPa and 35°C. If the system pressure was below or above the set pressure, ECU will automatically cut off LPG feed and switch to fuel with gasoline instead. In Table 1, heat of vaporization of LPG was 435 kJ/kg and 350 kJ/kg for gasoline, the temperature of water cooling system must be kept at 35°C or slightly above for smoothly LPG vaporization. After tuning for the use of LPG, the ECU controlled the proper air/fuel ratio injection to be within the λ range of 0.85-0.90 so as to achieve the best performance with smooth riding and low emission of CO and HC.

The motorcycle tested can be either started with gasoline or LPG. Typically, it is started with gasoline and then manually changed to run with LPG from the fuel selector switch (Figure 6). In case of having trouble with fuel pump, the engine can be started with LPG. If running out of LPG while riding, a rider can just switch to use gasoline from the fuel selector switch without any interruption. Beside that refuelling LPG gas stations are not widely available; it is more practical to keep some gasoline in the fuel tank. This was a fuel back-up arrangement. Typically, a gas-fuelled vehicle needs refuelling twice or three-times as often as similar petrol [14].

Figure 4. 3-litre LPG storage tank and ECU kit

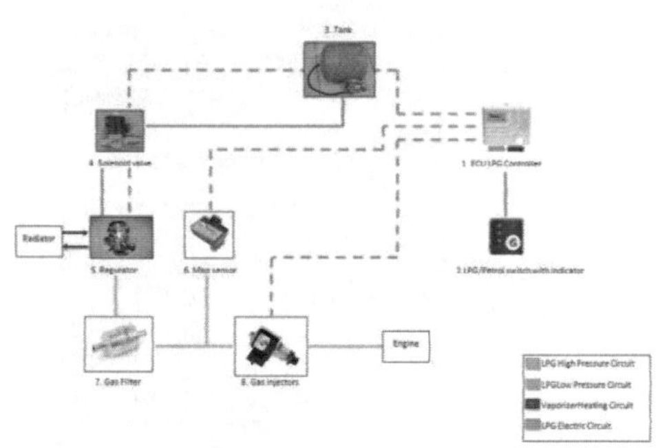

Figure 5. LPG injection control diagram

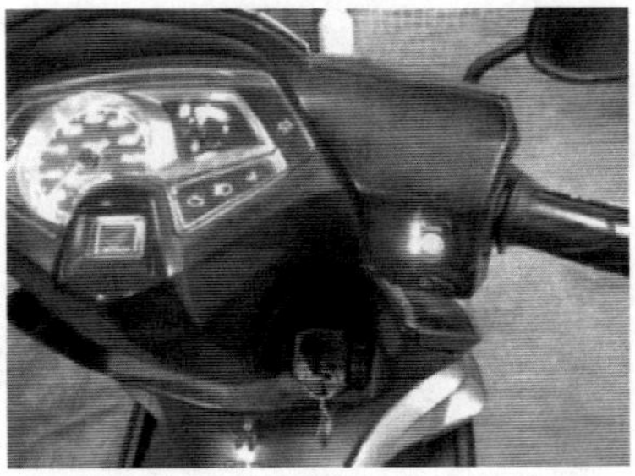

Figure 6. Fuel selector switch

3. Results and Discussion

Performance Test on Dynamometer

For the use of bi-fuel system-gasoline/LPG in the motorcycle tested, the engine was tuned up on the dynamometer at the relative fuel-rich ratio range of 0.85-0.90. When fuelling with gasoline, the engine gave the average torque 5.81 N-m over the speed range of 65-100 km/h and 5.59 N-m while using LPG (Figure 7). The average power output from burning LPG over the speed range of 65-100 km/h was 5.16 hp which was about 3.9% lower than that from using gasoline91 of 5.37 hp. According to Semin et al [14-17], the performance of the converted bi-fuel engine will generally drop 15-20% in maximum power from the petrol. In general, bi-fuel modified vehicles suffer power loss and encounter drive ability problems as from the design and/or the installation of the retrofit conversion packages. A bi-fuel system was arranged while retaining the petrol fuel system; the engine was prevented to fully optimize the high-octane gas. Such arrangement provides a back-up fuel where gaseous refuelling facilities are not well developed.

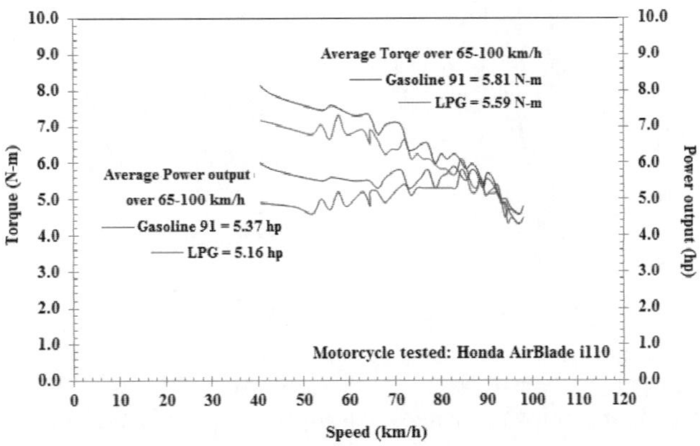

Figure 7. Comparison of Gasoline 91 and LPG performance test on SuperFlow CycleDyn Dynamometer

Gaseous fuels have low energy density resulting in low engine performance, low engine volumetric efficiency, and need of pressurized storage vessel [14,16]. However large pressure fuel tank could be utilized but weight, space, performance and cost must be compromised. For safe operation of gas-fuelled vehicles, an appropriate pressure fuel storage carried on board must be installed. The standard 3-litre LPG-pressure-storage tank was therefore installed (Figure 4). The operating pressure and temperature was 78-80 kPa and 35°C, respectively, while the safety relief valve pressure was set at below 48 kPa and above 110 kPa. It is confirmed that the bi-fuel system-gasoline/LPG can be securely and efficiently used if the engine was properly modified and tuned up. Moreover, the engine can be started with either gasoline or LPG. It is normally started with gasoline and then changed to use LPG from the fuel selector switch. In case of having problem with fuel system or run-out of gasoline, the engine can also be started with LPG. This was because the LPG vaporizer or pressure reducer was kept at 48 kPa which was high enough to vaporize LPG. After that LPG injection is controlled by cooling water temperature, 35°C or slightly above. When switching from firing gasoline to LPG, shorter driving range was experienced see Figure 8.

Fuel Consumption Test

The comparative results between the consumption rate of gasoline91 and LPG collected from both conditions for over 2,000 kilometres was graphically displayed (Figure 8). After tuning at the λ range of 0.85-0.90, it showed that the consumption rate of LPG was about 17.7% higher than that of gasoline91. Note that LPG has heating value of 50 MJ/kg and density of 0.56 kg/l where gasoline has lower heating value of 44 MJ/kg with higher density of 0.76 kg/l. Therefore, energy density per unit volume of LPG was 28 MJ/l and 33.4 MJ/l for gasoline. Theoretically, it required more LPG about 16.2% for combustion. Like natural gas used as an engine fuel [14-17], the disadvantages of LPG were low energy density, low engine volumetric volume as a gaseous fuel, and need of pressurized fuel tank. Economically, LPG fuelled vehicles have a significant advantage over petroleum vehicles due to lower cost per energy unit. Moreover, LPG price was about 50% lower than gasoline/gasohol prices. Additional maintenance cost incurs less than USD 80; for fuel line, fuel filter required to be scheduled change at 6-80,000 kilometres and LPG nozzles to be replaced every 3 to 5 years. The fuel flexible line is required frequent inspection, typically every year and replaced every 3 to 5 years. However, conversion kits and required accessories and pressure storage tank are costly, but safety must not be compromised.

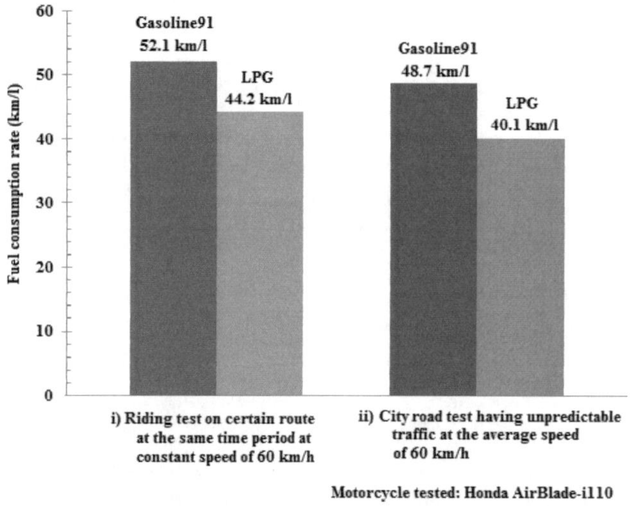

Figure 8. Comparison of the fuel consumption rate of gasoline 91 and LPG

Emission Measurement

The motorcycle tested was purposely designed for the use of gasoline91 and has been used over 70,000 kilometres. After being thoroughly inspected, the LPG equipment set with ECU was added while maintaining the compression ratio. The engine was then properly tuned up on dynamometer aiming at the best power output which was in the range of 0.85-0.90. Lower CO and HC concentrations in the exhaust were observed as shown in Figure 9 and Figure 10. Those were much below the regulations which required <4.5%vol for CO concentration and 10,000 ppm for HC concentration. This corresponded to most often cited advantage of gaseous fuels on pollutants; the percentages vary depending on the source [14,16]. It was likely that natural gas and LPG burning vehicles substantially produces lesser amounts of pollutants than vehicles burning petroleum fuel. More benefits to engine components and effective reduction on maintenance requirements were also claimed [14-17]. Since lubricating oil was not contaminated or diluted with gaseous fuels, NG and LPG; less deposits were formed in combustion chamber and on spark plugs, consequently, prolong piston ring and spark plugs life. Therefore, LPG is an alternative clean fuel for vehicles and provides better combustion than gasoline.

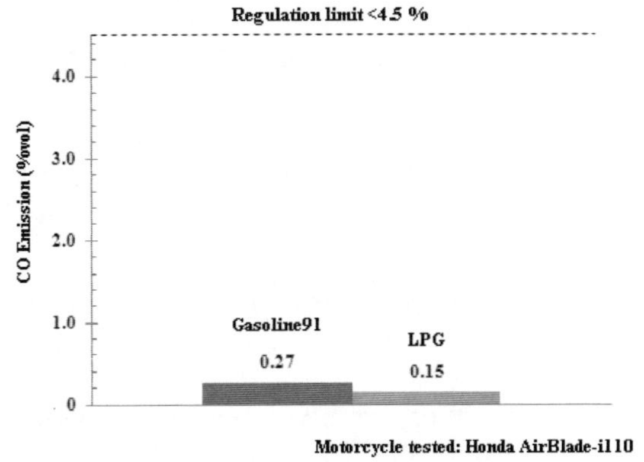

Figure 9. Comparison of CO concentration in the exhaust of gasoline 91 and of LPG.

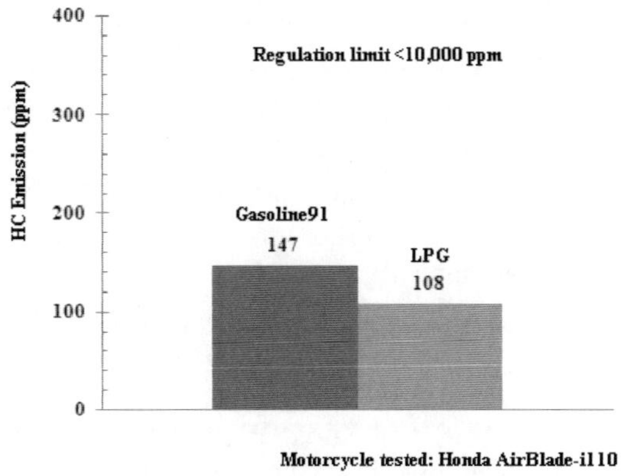

Figure 10. Comparison of HC concentration in the exhaust of gasoline 91 and of LPG.

4. Conclusion

It clearly confirmed that bi-fuel system – gasoline/LPG can be safely and effectively used in a 4-stroke motorcycle equipped with fuel injection and water cooling system. The motorcycle tested was the used 4-stroke Honda AirBlade i110 which was used over 70,000 kilometres. After being thoroughly inspected, the LPG equipment set and ECU were securely installed in the U-box under the seat. For properly LPG fuel injection, the calculated 1-mm diameter injector was inserted close to the gasoline fuel injector and the regulator was tightly placed over frame bar near the cooling system. After that, the engine was tuned up at the relative rich air-fuel ratio range of 0.85-0.90 which theoretically gave the best power output for SI engines. The operating pressure and temperature were 78-80 kPa and 35°C. For smooth LPG vaporization, cooling water temperature was controlled at 35°C while the relief valve pressure was set at below 48 kPa and above 110 kPa. If the system pressure is below or above the set pressure, ECU will automatically cut off LPG and switch to use gasoline instead.

Comparing to the use of gasoline91, the dynamometer performance test result of firing LPG showed that the average power output over the engine speed range of 65-100 km/h was 5.16 hp which was only 3.9% lower. From the road tests of both long riding and city riding at the average speed of 60 km/h, the average LPG consumption rate was 17.7% more which corresponded to LPG's energy density of about 16.2% lower than gasoline's. It therefore required more LPG to burn. The average concentrations of CO and HC in the exhaust were 0.15%vol and 108 ppm. Those were lower than in the exhaust of burning gasoline91, 0.27%vol CO concentration and 147 ppm HC concentration, but much below the regulations. In conclusion, LPG is clearly a clean fuel and safe to use as fuel for motorcycles once the engine is properly modified and tuned up to suit the fuel's characteristics, such as combustible limit. To assure smooth operation, the cooling water temperature must be regularly checked. For safety purpose, all gas equipment must be inspected as scheduled.

Acknowledgement

The study was conducted successfully with the great supports from College of Engineering, Rangsit University and A.P.I. Tech Shop Company. The A.P.I. Tech Shop Company provided valuable technical advice along with standard testing equipment, dynamometer.

References

[1] Energy Statistics, Energy Policy and Planning, Office, Ministry of Energy http://www.eppo.go.th/index.php/en/en-energystatistics/value-energy-statistic?orders[publishUp]=publishUp&issearch=1# (Accessed on January 2017)

[2] Bui V G Tran V N Ho T Q and Pham T D P 2004 Application of LPG on Motorcycles and small sized bus, Proc. of Clean Fuels and Vehicles in Vietnam, Hanoi, 13-14 May, 2004

[3] Transport Statistics Sub-Division, Planning Division, Department of Land Transport, Ministry of Transport. http://apps.dlt.go.th/statistics_web/statistics.html (Accessed on January 2017)

[4] Cengel Y A and Boles M A 2010 *Thermodynamics: An Engineering Approach* 7th ed New York McGraw-Hill ISBN10: 9780073529325 p 1024

[5] Pulkrabek W W 2004 *Engineering Fundamentals of the Internal Combustion Engine*. 2nd ed. New Jersey Pearson Prentice Hall ISBN10: 0131405705 p 478

[6] Heywood J B 1988 *Internal Combustion Engine Fundamentals* International ed New York McGraw-Hill p 71

[7] AJ6 Engineering *Introductory Technicalities: A guide to the general principles for increasing the performance of modern engines*, http://www.jagweb.com/aj6eng/technics.php (Accessed on December 2016)

[8] Sirisawat S and Supat P 2012 Learning Equipment for the Flammability Limits of Liquefied Petroleum Gas *Am J Applied Sci* **9**(8) 1316-20 doi: 10.3844/ajassp.2012.1316.1320

[9] The Properties & Composition of LPG, http://www.elgas.com.au/blog/453-the-science-a-properties-of-lpg (Accessed on October 2016)

[10] Alternate Energy Systems Inc. *LPG Data - Definitions and Technical Data for LPG, Natural Gas and LPG/Air Mixtures*, http://www.altenergy.com/technology/lpgproperties.htm (Accessed on October 2016)

[11] Suthisripok T Promjun T and Rangron A 2011 A Comparative Study of a Used 4-Stroke Motorcycle Engine Performance Using E85 and Gasoline 91, Proc. of The 7th International Conference on Automotive Engineering (ICAE-7), March 28-April 1, 2011, Bangkok, Thailand

[12] Suthisripok T 2012 The use of E100 to fuel a used 4-stroke motorcycle *Am J Applied Sci.* **9**(5) 647-653 doi: 10.3844/ajassp.2012.647.653

[13] Martyr A and Plint M A 2007 *Engine Testing: Theory and Practice* 3rd ed Oxford Butterworth-Heineman ISBN-10: 9780750684392 p 442

[14] Semin and Rosli A B 2008 A Technical Review of Compressed Natural Gas as an Alternative Fuel for Internal Combustion Engines *Am J Eng and Applied Sci* **1**(4) 302-311 doi: 10.3844/ajeassp.2008.302.311

[15] Semin Ismail A R and Rosli A B 2009 Diesel Engine Convert to Port Injection CNG Engine Using Gaseous Injector Nozzle Multi Holes Geometries Improvement: A Review *Am J Eng and Applied Sci* **2**(2) 268-278 doi: 10.3844/ajeassp.2009.268.278

[16] Semin Bakar R A and Ismail A R 2009 Green Engines Development Using Compressed Natural Gas as an Alternative Fuel: A Review *Am J Environ Sci* **5**(3) 371-381 doi: 10.3844/ajessp.2009.371.381

[17] Semin Ismail A R and Rosli A B 2009 Effect of Diesel Engine Converted to Sequential Port Injection Compressed Natural Gas Engine on the Cylinder Pressure vs Crank Angle in Variation Engine Speeds *Am J Eng and Applied Sci* **2**(1) 154-9 doi: 10.3844/ajeassp.2009.154.159

IOP Conf. Series: Materials Science and Engineering **241** (2017) 012022 doi:10.1088/1757-899X/241/1/012022

A Gradient Taguchi Method for Engineering Optimization

Shun-Fa Hwang[1, a], **Jen-Chih Wu**[1] **and Rong-Song He** [2]

[1]Department of mechanical engineering, National Yunlin University of Science and Technology, 123 University Road, Sec. 3, Touliu, Taiwan 640, R. O. C.

[2]Department of mechanical engineering, WuFeng University, 117 Chian-Kuo Road, Sec. 2, Ming-Hsiung, Chia-Yi, Taiwan 621, R.O.C

[a]hwangsf@yuntech.edu.tw

Abstract. To balance the robustness and the convergence speed of optimization, a novel hybrid algorithm consisting of Taguchi method and the steepest descent method is proposed in this work. Taguchi method using orthogonal arrays could quickly find the optimum combination of the levels of various factors, even when the number of level and/or factor is quite large. This algorithm is applied to the inverse determination of elastic constants of three composite plates by combining numerical method and vibration testing. For these problems, the proposed algorithm could find better elastic constants in less computation cost. Therefore, the proposed algorithm has nice robustness and fast convergence speed as compared to some hybrid genetic algorithms.

1. Introduction

In engineering applications, optimization is often required to enhance the performance of a material or structure and to reduce its weight or cost at the same time. Recently, stochastic methods such as genetic algorithm (GA), simulated annealing (SA), particle swarm optimization, ant colony optimization, and immune algorithm have been widely and successfully applied in various engineering problems. However, one drawback of stochastic methods is the large times of function evaluation. Taguchi method is a statistical methodology to analyze tested data for determining the effects of various factors and optimizing their levels. Taguchi method using orthogonal arrays could reduce the number of tests even when the number of factors is quite large. However, since Taguchi method is concerned with the discrete level of factors, it is troublesome to handle the continuous values of factors.

Among different algorithms for optimization, fast convergence speed and robustness in finding the global optimum are always concerned. However, it is not easy to be satisfied simultaneously by just a single algorithm. Therefore, there is an increasing interest to propose hybrid approaches [1-4] to avoid premature convergence towards a local minimum and to reach the global optimum.

Similar to the above attempts, it is desired to propose a method to balance the robustness and the convergence speed. Hence, a novel hybrid algorithm is proposed in this work. This novel hybrid algorithm consists of Taguchi method and the steepest descent method. Three composite plates discussed by other researchers are chosen to test the applicability of the proposed algorithm for the inverse determination of elastic constants of materials by combining numerical method and vibration testing.

2. Gradient Taguchi Method

By using orthogonal arrays, Taguchi method could quickly find the optimum combination of the levels of various factors, even when the number of level and/or factor is quite large. In real applications, the number of level is common to be two or three. Taguchi orthogonal array is usually represented as $L_a(Q^b)$, where a denotes the number of experiments, Q is the number of levels, and b is the number of factors. In the present study, the number of level is always chosen as two. For example, the Taguchi orthogonal array for three factors is chosen to be $L_4(2^3)$. The main effect of each level j of each factor i, f_{ij}, is calculated according to Eq. (1), and the difference of the main effect of each factor i, fd_i, is calculated according to Eq. (2).

$$f_{ij} = \frac{1}{N}\sum_{k=1}^{a} F_k \delta_k \quad (1)$$

$$fd_i = \left| f_{i1} - f_{i2} \right| \quad (2)$$

In these equations, F_k is the experiment value. The Kronecker delta $\delta_k = 1$ when the F_k value is from the factor i and the level j, otherwise its value is zero. Also, N is the number of experiments that are accounted during the summation. In general, when the minimum is the better, smaller f_{ij} represents better level of j for factor i, and larger fd_i denotes that factor i has profound effect as compared to other factors.

From the values of f_{ij}, the optimum combination of the level of each factor could be obtained. However, in some circumstances, this combination is not the best, and some modifications need to be done to get the optimum combination. By using the two levels of factor i, its gradient $\nabla f_{i,0}$ is calculated as

$$\nabla f_{i,0} = \frac{f_{i2} - f_{i1}}{x_i^{0,U} - x_i^{0,L}} \quad (3)$$

where $x_i^{0,L}$ and $x_i^{0,U}$ denote the first and second levels. Hence, the gradient of all factors could be represented as

$$\nabla f_0 = \left\{ \nabla f_{1,0}, \nabla f_{2,0}, \ldots\ldots, \nabla f_{n,0} \right\} \quad (4)$$

The index 0 in Eqs. (3) and (4) is used to represent the number of iteration. Then, the point of next generation is obtained as

$$\tilde{x}^{i+1} = \tilde{x}^i \pm \lambda \nabla f_i \quad (5)$$

where λ is the step length, "+" is for the maximum problem, and "-" is for the minimum problem. To obtain the appropriate step length, the steepest descent is used. An initial step length, H, could be guessed or set to be 1 in this work. Then, three different step lengths are created as 0, 1, and 2 times of H, and the corresponding values of objective function are Y_0, Y_1, Y_2, respectively. Also, define a value D as

$$D = 4Y_1 - 2Y_0 - 2Y_2 \quad (6)$$

If $D \geq 0$, let λ be equal to $H/3$. Otherwise, for the maximum problem,

$$\lambda = \frac{H(4Y_1 - 3Y_0 - Y_2)}{D} \quad (7)$$

For the minimum problem,

$$\lambda = \frac{H(4Y_1 - Y_0 - 3Y_2)}{D} \quad (8)$$

From the Taguchi method, the best combination of the level of each factor, $\tilde{b}$, could be obtained according to the main effects and the modification. However, there is one point named base point for each generation, $\tilde{x}$, which could be further used to accelerate the searching. At this point, the value of each factor is equal to the average of the two levels. In some circumstances, the base point may provide even better choice than the best value from main effect analysis. By considering both the best point from

Taguchi method and the base point for iteration number i, the possible values for the next iteration are $\tilde{b}^i \pm \lambda \nabla f_i$ and $\tilde{x}^i \pm \lambda \nabla f_i$. Here, "+" is for the maximum problem, and "-" is for the minimum problem.

3. Inverse Determination Examples

To apply the present optimization method to engineering problems, the inverse determination of elastic constants of materials by combining numerical method and vibration testing is considered. Three composite plates discussed by other researchers are chosen. Plate A is a unidirectional glass/epoxy laminate with 12 layers, its dimensions are 268x150x1.43 mm and its density is 1932 kg/m^3 [5]. Seven natural frequencies measured are 72.3, 97.6, 176.7, 197.4, 238.3, 277.5, 348.0. Plate B is a symmetric cross-ply laminate with 14 layers, its dimensions are 191x282x2.87 mm and its density is 1624 kg/m^3 [6]. Eight natural frequencies measured are 72.2, 108.6, 182.7, 293.9, 308.7, 343.5, 369.4, 442.6. Plate C is a commercially pressed, woven-glass epoxy board having 18 fabric layers. Its dimensions are 120x90x1.19 mm and its density is 1755 kg/m^3 [7]. Eight natural frequencies measured are 138.7, 261.1, 374.3, 463.0, 536.5, 713.3, 771.0, 834.1.

For all these three plates, the optimization problem could be stated as follows:

$$\text{Minimize } F(X) = \sum_{i=1}^{n_0} \left(\frac{f_i - \overline{f_i}}{\overline{f_i}} \right)^2 \qquad (9)$$

where F is the objective function that is a function of the design variable set $X = [x_1, x_2, x_3, ..., x_m]$. Here m is the number of the design variables x_i. The natural frequency obtained by vibration testing is denoted as $\overline{f_i}$ and that obtained by finite element analysis is represented as f_i. The number of natural frequencies used in the objective function is n_0. The elastic constants are assigned as design variables and restrained to be positive. Since all these three plates are considered to be thin and each layer is transversely isotopic, only four elastic constants are necessary and they are the design variables. In the finite element analysis to calculate the natural frequencies, 100 shell elements were used for all three plates.

Table 1. Elastic constants obtained for composite plate A.

	Number	E_1(GPa)	E_2(GPa)	υ_{12}	G_{12}(GPa)	Obj. Fun.
Ref [5]		44.20	17.70	0.195	7.20	0.000956
ARSAGA	100	45.30	17.91	0.202	7.42	0.000085
GTM	102	45.22	17.83	0.215	7.44	0.000076

Table 2. Elastic constants obtained for composite plate B.

	Number	E_x(GPa)	E_y(GPa)	υ_{xy}	G_{xy}(GPa)	Obj. Fun.
Theory		23.30	13.80	0.217	2.70	0.000513
ARSAGA	97	23.24	14.22	0.229	2.71	0.000101
GTM	171	23.44	14.20	0.241	2.71	0.000030

For plate A and under the condition of plane stress, the obtained elastic constants are listed in Table 1. Since the present algorithm may be dependent on the initial guess, different runs may result in different results. Hence, five runs are executed to check its repeatability, and only the best one is shown. The reference results in the first row were obtained by Rikards et al. [5]. The best results denoted as

ARSAGA were obtained by an adaptive genetic algorithm combined with simulated annealing [1]. The five runs by the present method obtain very close results and its repeatability is excellent. As shown in the table, from the viewpoint of the objective function, the present method has the lowest value. In the table, the column denoted as "Number" represent the generation number for ARSAGA and the iteration number for the present method. Since the population size of the present method is much less than that of ARSAGA, the convergence speed of the present method is much superior that that of ARSAGA.

Table 3. Elastic constants obtained for composite plate C.

	Number	E_1(GPa)	E_2(GPa)	υ_{12}	G_{12}(GPa)	Obj. Fun.
Ref [7]		16.47	16.49	0.207	2.44	0.001197
ARSAGA	71	16.74	16.72	0.039	2.48	0.000759
GTM	59	16.92	16.57	0.174	2.46	0.000798

Since plate B is a cross-ply laminate, the effective elastic constants for the whole plate are sought in this work. Five runs by the present method are executed and the repeatability of the obtained effective elastic constants is excellent. As shown in Table 2, the row denoted by "Theory" represents the results calculated by the classic lamination theory. As the value of the objective function is utilized to compare the present method with the theory and ARSAGA, it is evident that the best result of present method is superior to the others. Table 3 shows the elastic constant obtained for the woven-glass epoxy layer from plate C. The reference values in the first row were from the results of Hutapea and Grenestedt [7]. Similarly, these results indicate that the present method have excellent repeatability as well as fast convergence speed and could find the best results.

4. Summary

In this work, a novel hybrid algorithm consisting of Taguchi method and the steepest descent method is proposed. The inverse determination of elastic constants of three composite plates by combining numerical method and vibration testing is used to test this novel hybrid algorithm. The results imply that the present method have excellent repeatability and fast convergence speed to find the best results as compared to other methods.

References

[1] S.F. Hwang, R.S. He: Adv. Eng. Inform. Vol. 20 (2006), p. 7–21.

[2] J.T. Tsai, T.K. Liu, J.H. Chou: IEEE Trans. Evol. Comput. Vol. 8 (2004), p. 365-377.

[3] A.R. Yildiz: Comput. Ind. Vol. 60 (2009), p. 613-620.

[4] P.W. Tsai, J.S. Pan, S.M. Chen, B.Y. Liao: Expert Syst. Appl. Vol. 39 (2012), p. 6309–6319.

[5] R. Rikards, A. Chate, W. Steinchen, A. Kessler, A.K. Bledzki: Compos. Part B-Eng. Vol. 30 (1999), p. 279-89.

[6] A.L. Araujo, C.M. Mota Soares, M.J. Moreira de Freitas, P. Pedersen, J. Herskovits: Compos. Struct. Vol. 50 (2000), p. 363-72.

[7] P. Hutapea, J.L. Grenestedt: J. Electron Mater. Vol. 32 (2003), p. 221-227.

Investigation the Amplitude Uniformity on the Surface of the Wide-Blade Ultrasonic Plastic Welding Horn

Thanh Hai Nguyen[1,a], Quang Thanh Le[2,b], Cong Luat Tran[1,c] and Huu Loc Nguyen[1,d]

[1]Faculty of Mechanical Engineering, Ho Chi Minh City University of Technology, No. 286 Ly Thuong Kiet Street, Ward 14, District 10, Ho Chi Minh City, Viet Nam
[2]Faculty of Mechanical Engineering, Ho Chi Minh City University of Transport, No. 2, D3 Street, Ward 25, Binh Thanh District, Ho Chi Minh City, Viet Nam

[a]haint@hcmut.edu.vn, [b]lequangthanh@hcmutrans.edu.vn, [c]luat.vietsonic@gmail.com and [d]nhloc@hcmut.edu.vn

Abstract. Ultrasonic welding has been applied for joining thermoplastic components due to their advantages such as clean, fast and reliable. The basic principle is to use the mechanical energy of ultrasonic frequency vibration to produce the molten pool at the interface of the joined components under high pressure to create solid-state welding joints. Depending on the specific application, the ultrasonic horn is designed to generate suitable amplitudes on the surface of the welding zone. Uniformity of the amplitudes can be a challenge as the welding area increases. Therefore, design a welding horn in order to obtain the uniform amplitudes at the large area is significant difficult. This work presents a method for obtaining the uniform amplitudes at the working surface of the stepped wide-blade horn. Finite element method is used to analyze the amplitude distribution at the horn surface of 250 x 34 mm^2 with working frequency of 15 kHz and aluminum alloy 7075. The uniformity of amplitude is obtained by changing the shape of the horn.

1. Introduction

Ultrasonic stack involves converter, booster and horn. The converter is used to convert the electrical frequency into mechanical frequency by Langevin type. Booster increases the small amplitude from the ending surface of converter. Ultrasonic horn is designed to cover the welding area and increases the amplitude from the booster to the welding interface. The quality of the welding joints depends on power, frequency, welding time, holding time, mechanical force and the amplitude at the working surface of the horn. The surface of the horn is large, the uniformity of the amplitudes is difficult to control. Therefore, the horn shape must be designed carefully in order to obtain the uniformity of the amplitudes. Some researchers have been investigated the effect of the amplitude for some decades. Lucas and Smith [1] tried to measure the vibration of block and wide-bodied horns using laser Doppler velocimetry. Slotting configuration effects significantly the vibration modes and amplitude performance. Stanasel and Ardelean [2] designed an exponential horns using Matlab application with assumption of 120 μm in amplitude on the working surface. The amplitude of different kinds of horns such as stepped, exponential and conical horns was investigated by using CARD software and calculation in theory [3]. Kim et at [4] used design of experiments (DOE) to obtain amplitude uniformity of the cylinder, bar and spool horns. Good design of grooves and slots increased the uniformity in bar and cylinder horns. Behera et al [5] studied the amplitude at each point on the top of the horn using ANSYS environment. The amplitude depended on the horn materials where aluminum horn gave more amplitude vibration than carbon steel horn. Amplitude distribution of the aluminum block horn [6] was investigated using FEA and homemade measurement apparatus. The top surface of the horn was experimentally divided into smaller areas and each amplitude of the center of each area was measured to build the distribution. The experimental amplitude was distributed roughly due to limited number of divided areas. Both frequency and amplitude of stepped horn [7] was calculated in theory and measured by COMSOL multiphysics with the change of corner radius. The amplitude distribution was measured to optimize the shape of the ultrasonic amplifier using laser vibrometer [8]. This work discussed the amplitude distribution on the top of the aluminum wide-blade horns using finite element method. The change of the shape of the wide-blade horn improved significantly the amplitude on the working surface.

2. Horn Design Methods

The ultrasonic wide-blade horn, Fig. 1, is used for bonding large area on the two joined objects. The width W, length L and thickness R2 of the horn are the important parameters. The design of stepped wide-blade horn is derived from uniform bar horn. Slots separate each bar element and join each other by bridging section. Half-wavelength of the ultrasonic horn is designed in longitudinal resonance. The total length of the horn is calculated using Equation (1).

$$L = \frac{c}{2f} + 2a - 2\frac{c}{\omega}\tan^{-1}\frac{\mathrm{W}}{\mathrm{W}-(n+1)h}\tan\frac{\omega a}{c} \tag{1}$$

Where c is velocity of sound and f is working frequency.

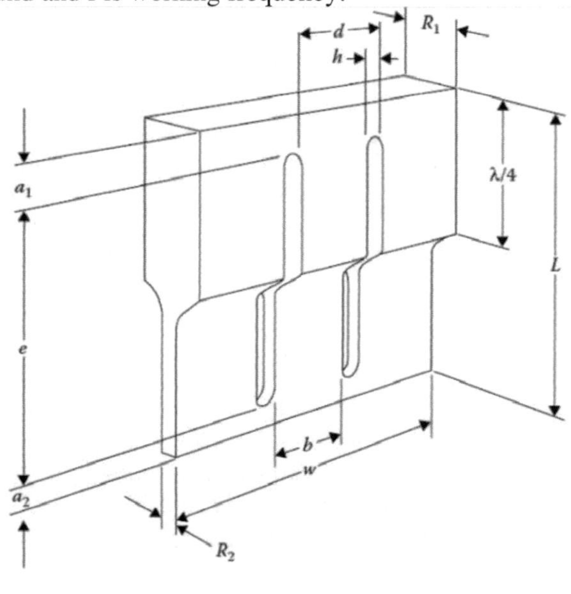

Fig. 1: Stepped wide-blade horn

Computer Aided Resonator Design (CARD) is roughly used to design the stepped wide-blade horn for 7075 aluminum alloy. The area of working surface is 250 mm and 34 mm in width and thickness, respectively. Velocity of aluminum alloy, number of slots, and frequency are 5100 m/s, 4 and 15 kHz, respectively. The option of auto-tune length is used to tune the length of the horn, Fig. 2. The input surface of the horn is 250 x 45 mm while the output surface of the horn is 250 x 34 mm. The radius at the step is 30 mm. The magnified gain of the wide-blade horn is calculated at 1.47. The auto-tune length is tuned at 164 mm and the maximum stress occurred at 101 mm.

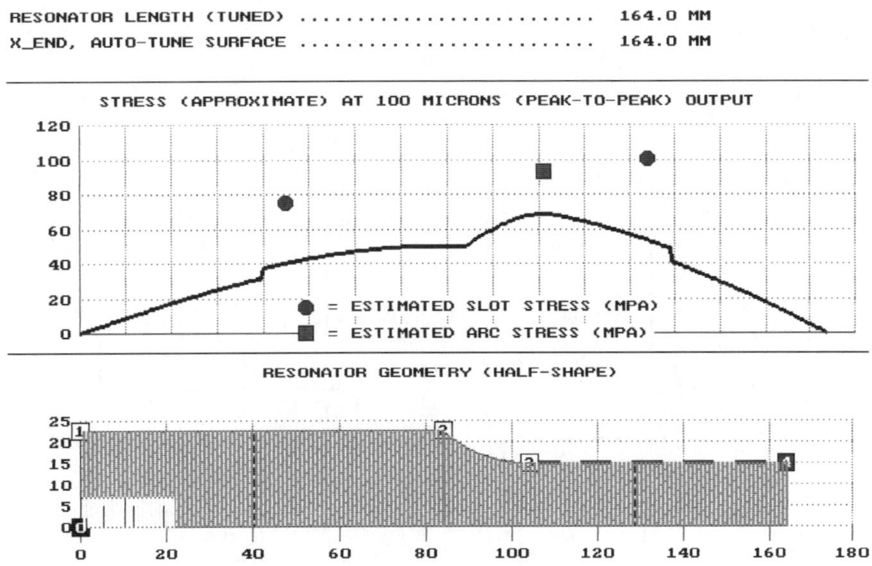

Fig. 2: Auto-tune length of the resonator

3. Simulation Results and Discussion

Using CARD software, the simple shape of the stepped wide-blade horn is obtained. In order to obtain the performance of the horn, amplitude distribution, and stress distribution, Abaqus environment is applied. The ultrasonic stack involving of converter, booster and horn is mechanically modelled in Fig. 3. The Langevin transducer with four PZT rings (6) and four copper electrodes (7) are clamped by bolt (1) between back mass (2) and front mass (3). Booster (4) connecting with front mass and horn (5) by bolts (8) and (9).

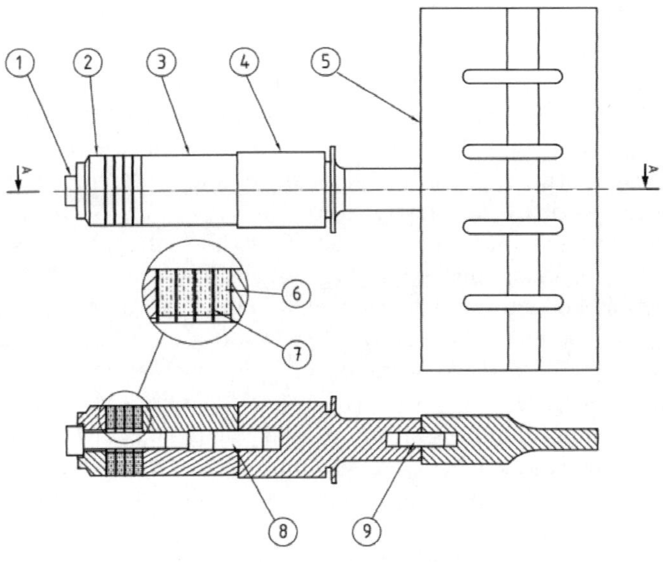

Fig. 3: Modelling of ultrasonic stack

IOP Conf. Series: Materials Science and Engineering **241** (2017) 012023 doi:10.1088/1757-899X/241/1/012023

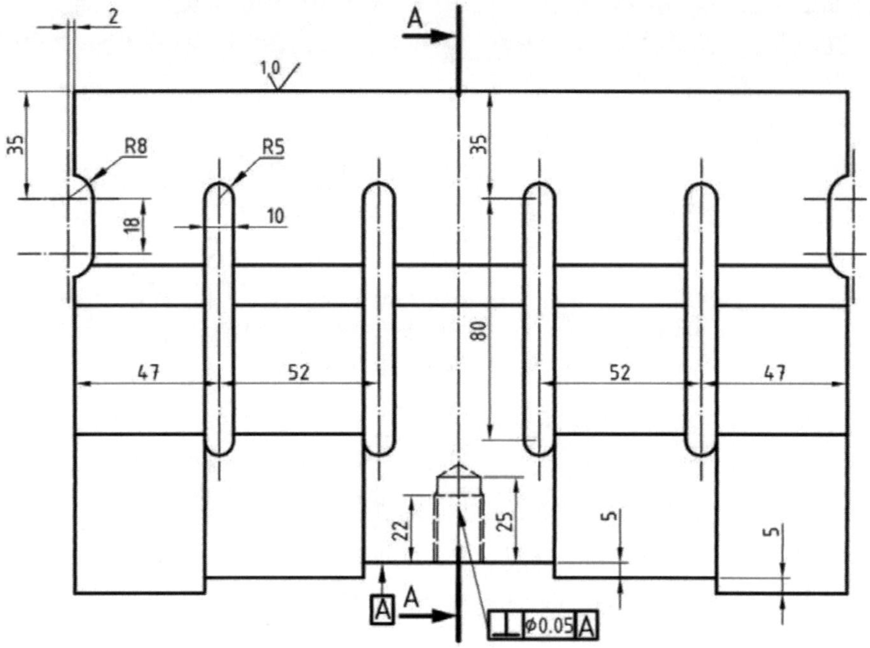

Fig. 4: Horn with necks and steps

Table 1: Material properties used in ultrasonic modelling

Component	Material	Elastic modulus [GPa]	Density [Kg/m3]	Poisson ratio
Front mass, back mass, booster, bolt	Steel	186	7800	0.29
Horn	Al, 7075	69	2700	0.33
PZT	P8	-	7700	-
Electrode	Copper	115	8942	0.31

In this model, materials are divided into two groups: alloys and piezoelectric ceramic. Front mass, back mass, booster, bolts and electrodes are defined by alloys with density, elastic modulus and Poisson ratio. Piezoelectric rings are defined with density and electrical permittivity. The detailed properties of materials can be seen in Table 1. Electrical potential is applied to two sides of the surface of the PZT rings. The flange of the booster is fixed while two ends of ultrasonic stack are free of constraints. In this model, no external forces are applied while internal forces come from fastening bolts with 33 kN. For comparison the amplitude distribution, another horn shape is intendedly designed by making steps at the input surface and necks at both sides of the horn, Fig. 4.

The simulated amplitude of the horn designed by CARD software is large at the center of the width and small at the two sides, Fig. 5(a). The change of the horn shape affects significantly the amplitude distribution on the output surface, Fig. 5(b). Therefore, the uniformity of the amplitude can be achieved by carefully designing the necks at the two sides and steps at the input surface. The 7075 aluminum horn is then manufactured by CNC machine. Impedance and working frequency were measured at 7.86 kΩ and 14.91 kHz, respectively, using ACTP impedance analyzer, Fig. 6.

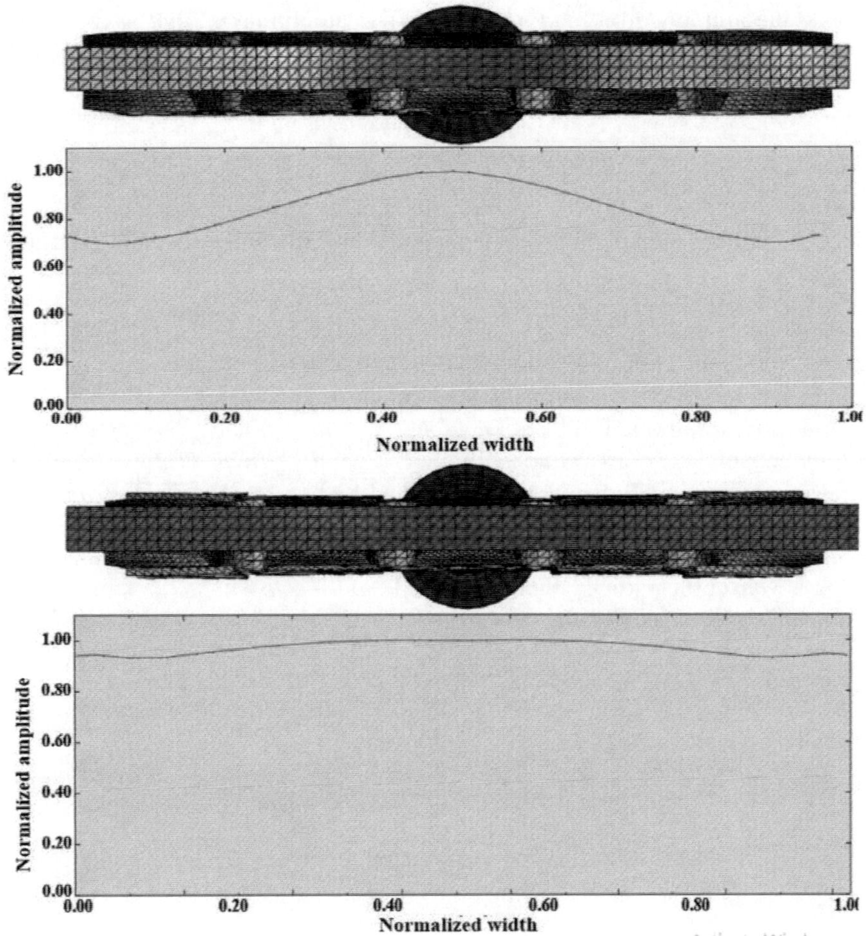

a) b)

Fig. 5: Normalized amplitude along width of the horn designed by: a) CARD software b) changing the shape

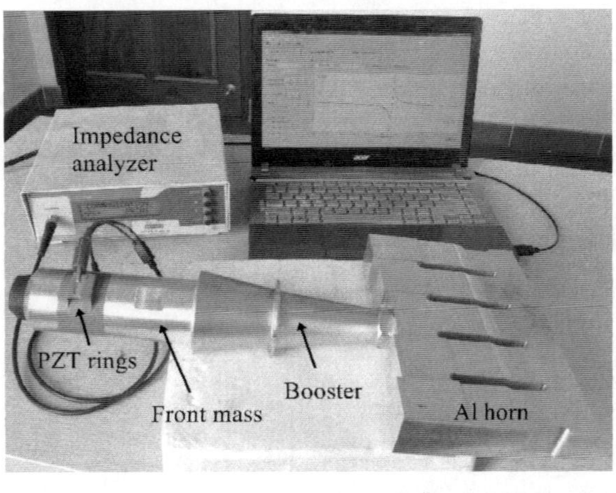

Fig. 6: Experiment set up for measurement impedance and working frequency

4. Conclusion

In this paper, the stepped wide-blade horn is roughly designed by CARD software with working frequency of 15 kHz, aluminum alloy and welding area of 250 x 34 mm^2. This horn is simulated by

FEM to investigate the amplitude on the top surface. The amplitude is high at the center and low at both two sides. Based on this amplitude distribution, the shape of the horn is corrected to achieve the uniformity of amplitude by making two necks at both sides and steps at the input surface. The impedance and working frequency are measured, however, the amplitude distribution will be done in the future.

Acknowledgements
This research is funded by Vietnam National University HoChiMinh City (VNU-HCM) under grant number C2016-20-02.

References

[1] M. Lucas and A. C. Smith Trans. 1997 *ASME*, Vol. **119**, p. 410
[2] I. Stanasel and F. Ardelean 2010 *Non. Tech. Rev* Vol. **4**, p. 54
[3] D. M. Patel and A. U. Rajurkar 2011 *Inst. Tech., Nirma Univ., Ahmedabad*, p. 1
[4] S. R. Kim, J. H. Lee, C. D. Yoo, J. Y. Song and S. S. Lee 2011 *IEEE Trans.*, Vol. **58**, p. 2194
[5] B. C. Behera, S. K. Sahoo, L. N. Patra, M. P. Rout and K. K. Kanaujia 2011 *Proc. of 5th ICAME, India*, p. 60
[6] J. C. Hung, Y. P. Tsai and C. H. Hung 2012 *Trans. Tech. Publication, Switzerland.*
[7] A. S. Nanu, N. I. Marinescu and D. Ghiculescu 2011 *Non. Tech. Rev*: Vol. **4**, p. 25
[8] K. R. Deibel and K. Wegener 2013 *Proc. of NAMRI/SME*, Vol. **41**

IOP Conf. Series: Materials Science and Engineering **241** (2017) 012024 doi:10.1088/1757-899X/241/1/012024

Operational Design of Magnetostrictive Inkjet PrintHead

Young-Woo Park

Department of Mechatronics Engineering, Chungnam National University, Daejeon 34134, Korea

ywpark@cnu.ac.kr

Abstract. This paper presents the redesign of the first magnetostrictive inkjet printhead (MagJet). The redesign includes the structural redesign and operational redesign. After completing the process successfully, the redesigned MagJet is subjected to testing. The effect of backpressure from a commercial syringe pump is evident. The operating frequency with backpressure reaches up to 1.5 kHz, although that without backpressure is 650 Hz. The droplet size at a frequency of 800 Hz is about 53 μm, less than half of the nozzle size. It can be concluded that various droplets can be made by combining frequency with a proper flow rate. Backpressure prevents the MagJet from air inclusion in the nozzle, so continuous ejection of the droplets are possible.

1. Introduction

Inkjet printing is demanding for various applications due to its technological and material flexibility [1-3]. One of the challenges is to design a proper inkjet printhead for demanding trend. Current inkjet printheads are mostly based on either piezoelectricity or heat. Both of them have a limited force generation, then are restricted to use very low-viscosity fluids. Some there exist to find out a novel actuation mechanism to overcome current limitation. Magnetostrictve actuation is one of them. Magnetostriction is the deformation of a body in response to a change in its magnetization. Yoo et. al. developed and characterized the first magnetostrictive inkjet printhead [4,5]. Droplet formation was successful, but not always. Also, fluid leakage was observed. Thus, this paper presents the redesign of the first magnetostrictive inkjet printhead (1st MagJet).

2. Redesign Process

The 1st MagJet was developed and characterized by one of the authors [4-5]. Even though it was possible to eject fluid droplets successfully, some problems arise such as fluid leakage and low frequency operation. The redesign process is focused on overcoming these problems: one with structural redesign of printhead, and the other with operational redesign.

3. Structural Redesign

As shown in Figure 1(a), the 1st MagJet consists of three components — one actuator (①) containing actuation mechanism, one receptacle (③) receiving a nozzle, and one center connector (②) interfacing the actuator and receptacle. Four screws are used to assembly the actuator and center connector, where most leakage occurs. The leakage is not likely to start up to a certain level of pressure, but is likely to start and get more as the pressure gets higher.

IOP Conf. Series: Materials Science and Engineering **241** (2017) 012024 doi:10.1088/1757-899X/241/1/012024

As shown in Figure 1(b), the redesigned MagJet (2nd Magjet) consists of two components — one actuator (①) and one receptacle (③). The center connector is incorporated into the design of a new recptacle. In addition, the inner surface of bottom end of the actuator and outer surface of top end of the receptacle are machined to have screw threads, which give more secure tightening between the components.

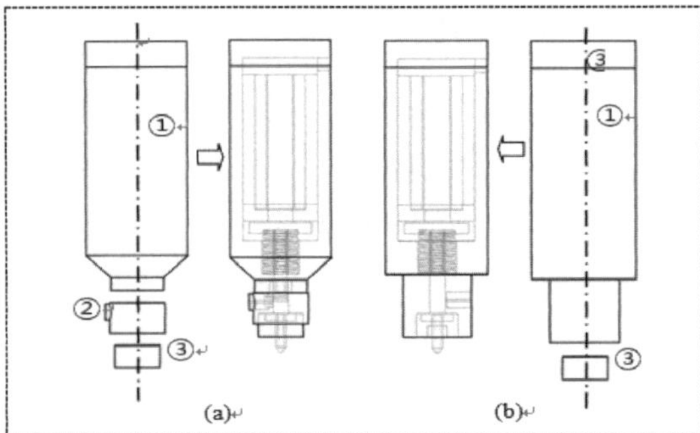

Figure 1. Structural redesign of MagJet: (a) 1st (b) 2nd.

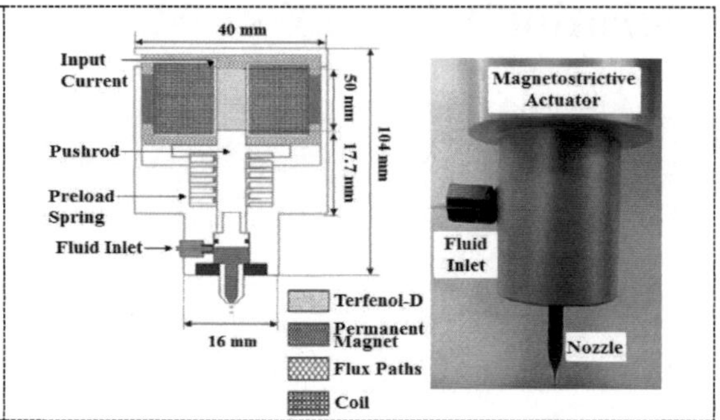

Figure 2. A schematic and photo of MagJet.

4. Operational Redesign

It is observed that the structural redesign helps improving operation as well as eliminating leakage. The maximum ejection frequency increases from 200 Hz at a current of 1.8 A to 650 Hz at a current of 1 A, and the minimum ejection current decreases from 0.35 A to 0.2 A. From practical point of view, a further operational improvement is still needed. So called operational redesign is conducted to achieve the goal. It is general that air comes in and remains in the nozzle during the operation. It is the reason why the 1st MagJet is operated at a lower frequency, and at a higher current. The core of the operational redesign is to add a commercial syringe pump to the fluid reservoir. The flow rate for droplet ejection is varied in accordance with the size, and must be determined experimentally. After applying the backpressure to the 2nd MagJet, the operation is improved significantly. It is ejected droplets continuously, and is operated at a frequency of up to 1.5 kHz.

5. Fabrication

Figure 2 shows a schematic cross-sectional view and a photo of the redesigned MagJet. A bobbin made of Al6061-T6 is wound by 1000 turns of coil (AWG25). The Terfenol-D rod (TD) with a diameter of 10 mm and a length of 50 mm is placed inside the bobbin. The top end of the TD is filled with the flux path made of SUS430, which acts blocked end with the top housing. The bottom end of the TD is connected to the push rod, which transmits the displacement and force generated by TD. The permanent magnet, an Alnico5 with a thickness of 5 mm, is used to provide DC bias to TD, thus allows it to move in two ways. A preloading spring of 6.8 MPa is selected for a better performance of the MagJet. The glass nozzle with a diameter of 130 μm is fit into the receptacle with teflon gasket. The final dimensions of the MagJet are 40 mm in diameter and 104 mm in length.

6. Experimental Procedure

Figure 3 shows a schematic diagram for the experimental setup. LabVIEW-based control program generates two signals: one is a driving signal, and the other a trigger signal. The driving signal is sent to the current amplifier, and used to drive the MagJet. The synchronized trigger signal is sent to CCD camera and strobe LED, and used to take droplet images from the MagJet. Syringe pump is used to supply the fluid with an adequate flow rate to MagJet through the fluid inlet. The experimental fluid is a water-based ink with a density of 958.5 kg/m^3, a viscosity of 1.31 cPs, and a surface tension of 33.3 dynes/cm. The drive signal in Figure 4 is a bipolar waveform. The operating principle of the MagJet with the waveform is as follows: During a time interval of t_c and t_i, the TD contracts with a reversal current, resulting in chamber filling of fluid. During a time interval of t_e, the Terfenol-D rod expands with a conventional current, causing the volume reduction of the chamber.

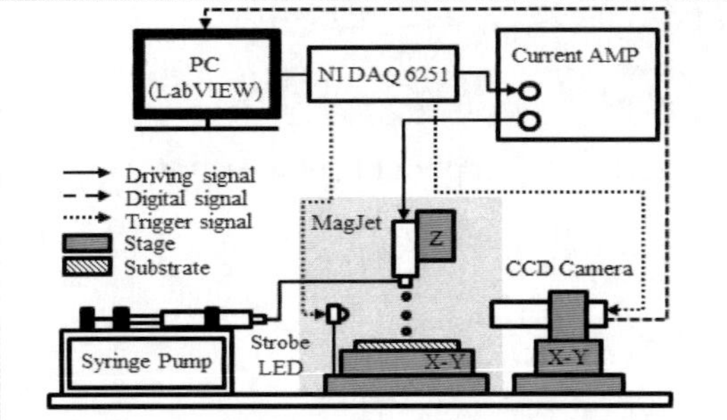

Figure 3. Schematic for experimental setup.

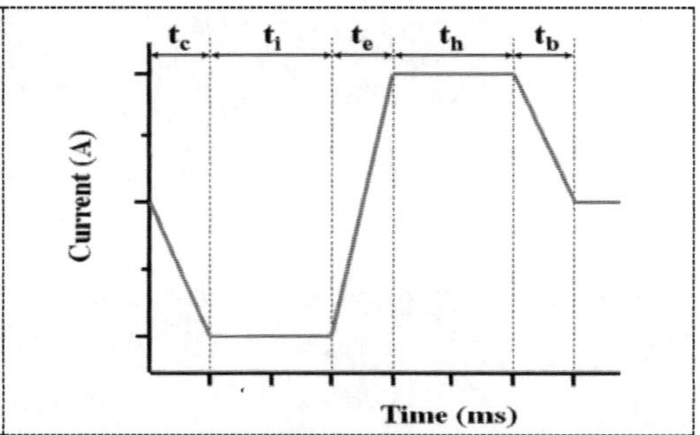

Figure 4. Driving signal.

7. Results and Discussion

Table 1 is the summary of the successful or failure ejection with or without a backpressure. The supplied current and applied frequency to the printhead range from 0.2 A to 1.0 A, and 100 Hz to 800 Hz, respectively. In the table, the respective S or F means "Successful" or "Failure", and the respective left or right side on the slash represents ejection with or without backpressure. It is observed that ejection with backpressure needs less current than one without backpressure, and that ejection frequency with backpressure is higher than one without backpressure. The maximum ejection frequency with backpressure reaches up to 1.5 kHz (although not listed in the table). It is also observed that the numbers of ejection with backpressure are countless, while them without backpressure are limited to 3 to 5 times. It is generally observed and is due to air inclusion into nozzle when the fluid is not supported by using backpressure.

Figure 5 shows the experimental comparison when ejections are performed with or without a backpressure. In Figure 5(a), it is observed that droplet is always formed at the tip of the nozzle. It makes the MagJet eject the sound droplets whenever needed. In Figure 5(b), it is observed that droplet is formed inconsistently, and that sometimes droplet is not even formed. It is mainly due to the air inclusion in the nozzle.

Table 1. Summary of successful or failure ejection.

Current / Freq.	0.2 A	0.3 A	0.4 A	0.5 A	1.0 A
100 Hz	S / S	S / S	S / S	S / S	S / S
200 Hz	S / S	S / S	S / S	S / S	S / S
300 Hz	S / S	S / S	S / S	S / S	S / S
400 Hz	S / F	S / S	S / S	S / S	S / S
500 Hz	S / F	S / S	S / S	S / S	S / S
600 Hz	F / F	S / F	S / S	S / S	S / S
650 Hz	F / F	S / F	S / F	S / F	S / S
700 Hz	F / F	F / F	F / F	F / F	S / F
800 Hz	F / F	F / F	F / F	F / F	S / F

Figure 6 shows the detailed procedure for one droplet ejection with a current of ±0.7 A. The ejection speed is estimated to be 4.5 m/s, and the droplet size to be 110 μm.

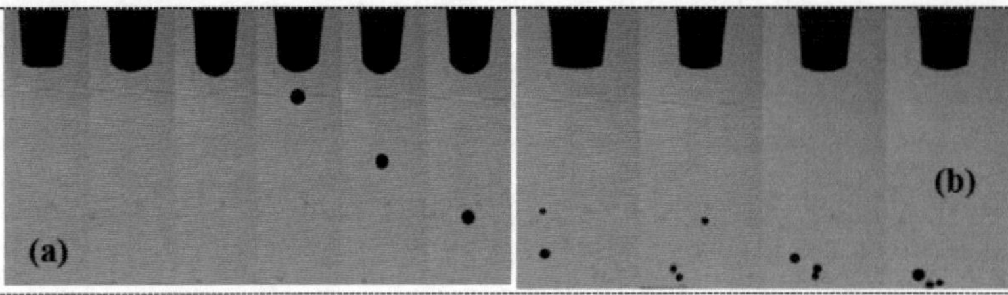

Figure 5. Droplet formation: (a) With backpressure (b) Without backpressure

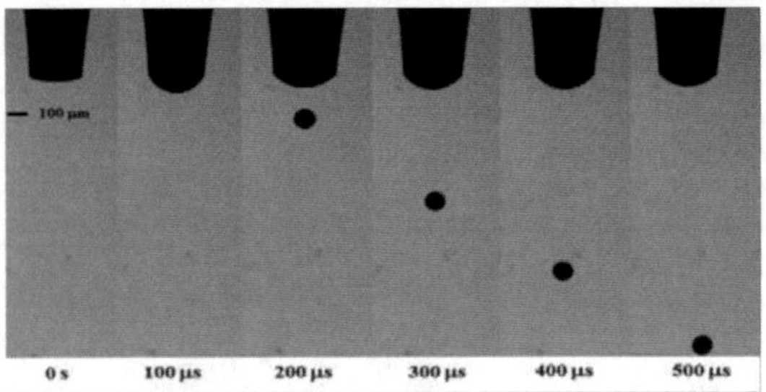

Figure 6. Droplet formation.

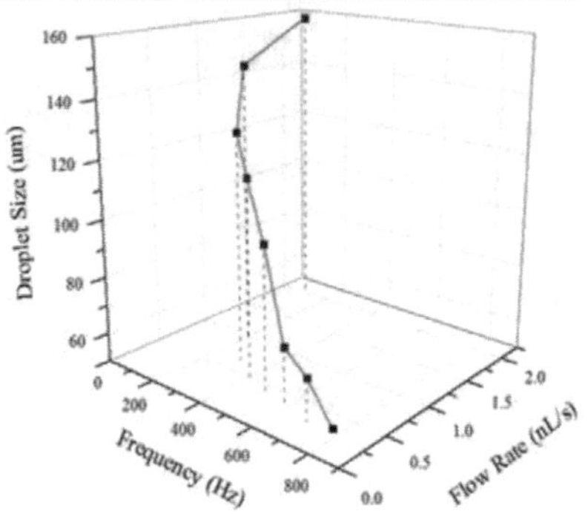

Figure 7. Droplet size vs. frequency and flow rate.

Figure 7 shows the droplet size versus frequency and flow rate. As frequency increases, the needed flow rate becomes lower, and consequently a smaller droplet. The dependency of a droplet size on frequency and flow rate is not linear with larger droplet (up to 120 μm), but is likely to be linear with smaller droplet (less than 120 μm). It is evident that various droplets can be made by combining frequency with a proper flow rate. It ranges from 53 μm to 160 μm. The droplet volume, calculated by assuming the droplet is sphere, ranges from 80 fL to 2 pL.

8. Conclusions

The 1st MagJet is redesigned in terms of structural redesign and operational redesign. The redesigned MagJet is subjected to characterization. Important findings are summarized as follows:

1) The structural redesign helps improving operation as well as eliminating leakage.
2) The operating frequency with a backpressure is much higher than that without a backpressure.
3) The droplet with backpressure is always formed at the nozzle tip, but the droplet without backpressure is formed inconsistently and sometimes droplet is not even formed.
4) As frequency increases, the needed flow rate becomes lower, and consequently a smaller droplet.
5) Various droplets can be made by combining frequency with a proper flow rate.

Acknowledgments

This research is supported by the National Research Foundation of Korea(NRF) grant funded by the Korean government(MSIP) (NRF-2016R1A2B4016482).

References

[1] Moon K, Choi J, Choi D, Kim S, Ha M, Nam H and Kim M 2008 *J MICROMECH MICROENG* 18, 125011
[2] Goldmann T and Gonzalez J 2000 *J BIOCHEM BIOPH METH* 42, 105
[3] Kim C, Park S, Sim W, Kim Y, and Yoo Y 2009 *COMPUT FLUIDS* 38, 602
[4] Yoo J and Park Y 2011 *CURR APPL PHYS* 11, p S353
[5] Yoo J and Park Y 2012 *J APPL PHYS* 111, p 07A936

IOP Conf. Series: Materials Science and Engineering **241** (2017) 012025 doi:10.1088/1757-899X/241/1/012025

Start-Stop Moment Optimization of Range Extender and Control Strategy Design for Extended -Range Electric Vehicle

ZHAO Jing-bo, HAN Bing-yuan, BEI Shao-yi

School of Automotive and Traffic Engineering, Jiangsu University of Technology, Changzhou, China
zhaojb1128@jsut.edu.cn

Abstract. Range extender is the core component of E-REV, its start-stop control determines the operation modes of vehicle. This paper based on a certain type of E-REV, researched constant power control strategy of range extender in extended-range model, to target range as constraint condition, combined with different driving cycle conditions, by correcting battery SOC for range extender start-stop moment, optimized the control strategy of range extender, and established the vehicle and range extender start-stop control simulation model. Selected NEDC and UDDS conditions simulation results show that: under certain target mileage, the range extender running time reduced by 37.2% and 28.2% in the NEDC condition, and running time UDDS conditions were reduced by 40.6% and 33.5% in the UDDS condition, reached the purpose of meeting the vehicle mileage and reducing consumption and emission.

1. Introduction

Electric vehicles which has a kind of energy-saving, environmentally friendly, independent of oil resources and smooth operation, quiet and other advantages of clean energy vehicles, is one vital way of solving the current environmental pollution and energy crisis which are the two major problems [1-2]. However, due to the lower specific power and power density of the current power battery, the pure electric vehicle (BEV) has a long charging time and a shorter driving range problem, which limits its large-scale commercial promotion [3]. The Extended-Range Electric Vehicle (E-REV) is equipped with a Range Extender (RE) which consisting of a small displacement engine and a generator on a pure electric vehicle, it can effectively extend the vehicle's driving mileage, as a smooth transition from a traditional car to a pure electric car, the extended range electric vehicle (Extended-Range Electric Vehicle, E-REV) has become an important branch of the current electric cars and drawn more and more attention.

For the E-REV power system, on the basis of meeting the whole vehicle dynamic performance, the coordination with the mechanical and electrical components affect the fuel economy. It is the key problem in E-REV development that how to coordinate the energy distribution between the power battery and the Range Extender, so that the vehicle system has the highest efficiency [4]. Range extender is the core component of E-REV, its start-stop control determines the operation modes of vehicle. Since the E-REV power systems use structure in series, the engine only used for power generation, and not directly output power to drive the vehicle, thus the engine can be adjusted in the vicinity of the optimum operating point stable operation, in addition, it is expected that on the basis of meeting the mileage of the vehicle, it can reduce the excess power of the extended range, save unnecessary fuel consumption, and achieve the best fuel economy and emission [5].

In order to minimize the use of extenders, this paper based on studying the constant power control strategy of the extended range, taking the target mileage as the constraint condition and combining the different driving cycle conditions of the vehicle, correcting the range extender power battery SOC (State of Charge) value, optimizing the start and stop control strategy to reduce the number of its start and stop and running time which to meet the car mileage while reducing fuel consumption and emissions purposes. Finally, based on ADVISOR platform and MATLAB software, the E-REV vehicle and RE start-stop control simulation model is established to verify the feasibility and effectiveness of control optimization.

2. E-REV Overall Designing

2.1. E-REV Structure and Vehicle Parameters

The basic structure of a typical E-REV is shown in Figure 1. Wherein, engines and generators connected coaxially to form extended-range system is essentially a generator set. When the battery is fully charged, the range extender is turned off and the car is running in pure electric mode. Only when the battery is low, the range extender starts and the vehicle runs in extended mode [6].

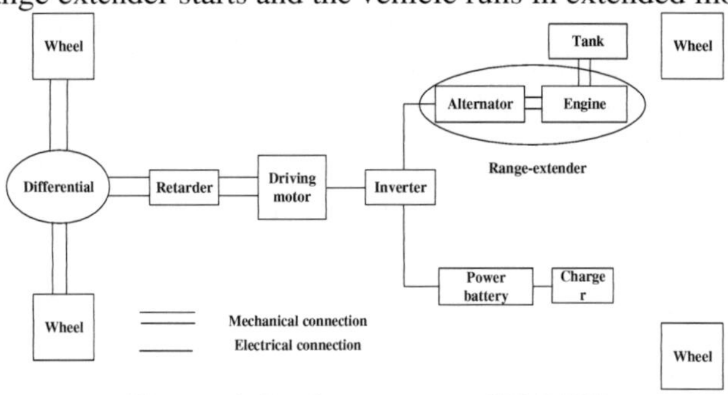

Figure 1. Basic structure of E-REV

In this paper, an example of a certain type of electric vehicle in the laboratory is selected, and the basic parameters of the selected vehicle are as shown in Table 1.

Table 1. Model parameters

Parameter name	Value
Vehicle length, width, height /mm	4498×1798×1430
Axis number	2
Driving form	4×2 FW
Curb weight m1/Kg	1715
Full load mass m/Kg	2015
Wheel radius r/m	0.314
Windward area A/m^2	2.0
Drag coefficient C_D	0.29

When the basic parameters of the vehicle are determined, performance index is shown in Table 2 .

Table 2. Performance indicators

Category	Parameter name	Value
Performance parameter	Maximum speed Vmax/km/h	161
	0-100Km/h Acceleration time t_a/s	9
	30Km/h climbing gradient α_{max}/%	≥30
	Pure electric driving range d_1/km	≥60
	Total mileage d/km	≥500

2.2. E-REV Power System

The E-REV powertrain consists of a power drive system (drive motor, gear unit, differential), energy storage system (power battery) and extended range system (engine, generator). According to the data of Table 1 and 2, the key components of the power system are selected. The main parameters are shown in Table 3.

Table 3. Parameters of power system component

Parameter type	Value
Motor type	Permanent magnet synchronous
Rated power (kw)/ Rated speed (r/min)	45/4000
peak power (kw)/ Maximum speed (r/min)	75/9000
Rated torque (N.m)	150
Peak torque (N.m)	240
Rated voltage (V)	300
Power battery type	Lithium iron phosphate
Capacity (A.h)	60
Rated voltage (V)	288
SOC range (%)	30-90
Engine type	Inline four cylinder
Displacement (L)	1.3
Rated power (kw)/ Rated speed (r/min)	38/3500
peak power (kw)/ Maximum speed (r/min)	55/6000
Maximum torque (N.m)/speed (r/min)	130/4250
Generator type	Permanent magnet synchronous
Rated power (kw)/ Rated speed (r/min)	35/4000
peak power (kw)/ Maximum speed (r/min)	60/9000

3. Analysis of Traditional E-REV Control Strategy

3.1. E-REV Energy Management Strategy

The energy management strategy of E-REV can be divided into Depleting Charge period and Charge Sustaining period. During the charge depleting period, the vehicle vehicles rely only on the power of battery, equivalent to pure electric driving, during charge sustaining period, the extender system began to start and power generation vehicle mainly rely on electricity to drive the car [7]. According to the upper and lower threshold SOC of power battery, based on rules of logic threshold switching control strategy to achieve a smooth transition of two period.

In order to ensure the drive motor within normal working voltage, while ensuring the performance of the power battery and prolong the life. In this paper, the SOC is set to 90%-30%.

3.2. E-REV Range-Extender Control Strategy

The E-REV range extender uses a constant power control strategy: According to the threshold of the battery SOC, it determines the opening and closing of the range extender .The control effect shown in Figure 2, first set the upper and lower limits of SOC, when the battery SOC over the upper limit SOCmax, the range extender is off, the power battery provides the energy requirements of the motor; when the battery SOC below the lower limit SOCmin, the range extender is on, and works in the optimal efficiency point, which driving the generator running power, to drive the motor providing energy to ensure that the vehicle continues to run, the excess energy will be used to recharge power battery; when the battery SOC is located between the upper and lower limits , Range extender maintain the working state before the last moment [8]. The advantage of this control strategy is that the engine is easy to be controlled, it can maintain low fuel consumption, high efficiency constant power output, effectively avoid the engine frequent start and stop and power fluctuations.

3.3. Selection of Engine Operating Points

Figure 3 is the curve of engine equivalent fuel consumption rate. E-REV uses tandem type power system, there is no mechanical connection between the engine and the transmission system, the engine output is disconnected with the ground load, and always working in low fuel consumption and low load, in the high fuel economic area. using direct start-stop technology to cancel the engine idle controlling, so that the engine starting work in the speed and torque respectively 3500r/min and 50N.m with constant speed and constant torque condition (Figure 3 point·), improving the efficiency of the engine, also reducing exhaust emissions. Engine power can be calculated at this time the engine output is constant power 18.3kW, It can meet the power demand of two driving conditions of the city streets and urban and highway combination.

3.4. Simulation Analysis of Traditional E-REV Control Strategy

Under the traditional E-REV control strategy, three typical driving cycle conditions NEDC, UDDS and CUDC were selected to simulate. Under different conditions, the battery SOC value with the car mileage changes shown in Figure 4.

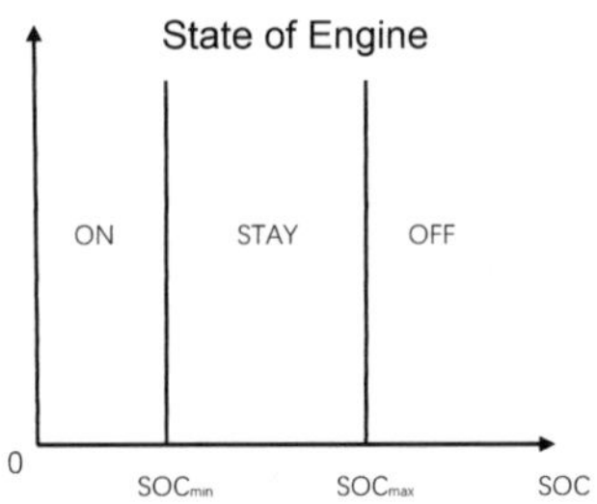

Figure 2. Constant power control strategy.

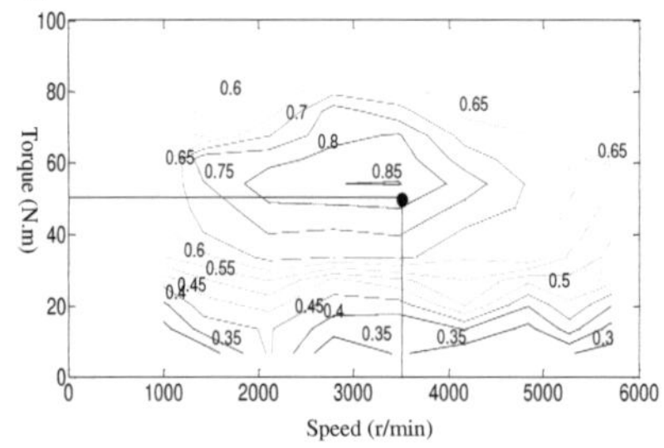

Figure 3. Engines equivalent fuel consumption curves.

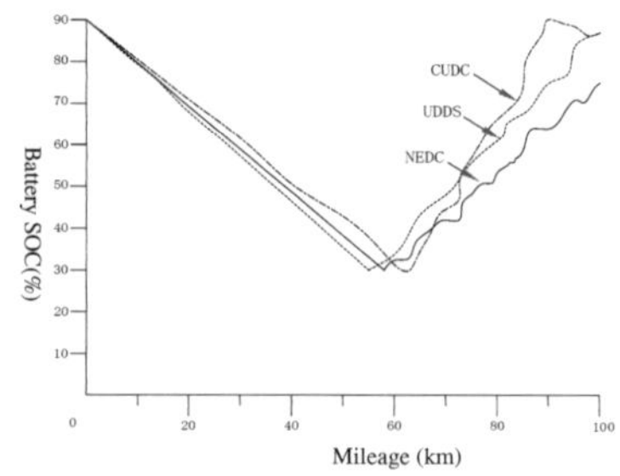

Figure 4. Relationship between SOC and mileage.

As is shown in Figure 4, in three different driving cycles, battery SOC value with the vehicle mileage increases continuously decline, before dropping to the 30% threshold, the range-extender is off, it indicated that the car meet various driving cycles of power demand in the period. Due to the different conditions of driving cycle, the power for vehicle's demand Preq are different, the battery SOC change rate is not the same, so the pure electric driving mileage Sare is different. but commonly in about 60km [9]. When the battery SOC value down to the lower limit, the range extender is on, extended range engine drive the generator power generation, on the one hand to drive the motor to provide energy to ensure that the vehicle continues to drive, on the other hand for the power battery charge, battery SOC is up until reach the upper limit, the range extender is off.

According to statistics from China's Ministry of Transportation, it shows that the average daily mileage is less than 60km. Therefore, E-REV basically rely on its own battery power energy to meet the daily needs of people to work and travel without starting the range extender system.

4. E-REV Control Strategy Optimization

4.1. Optimization of the Control Strategy of the Range- extender Starting and Stopping time

When the E-REV target distance D (the mileage adjacent to the two external charging) is greater than that of the pure electric vehicle mileage Sare, target distance will affect the range-extender start-stop times and running time, thus has impact on the vehicle's fuel economy, in order to reduce the start-stop times and running time, it should make full use of battery energy[10]. Therefore, when the E-REV is having long-distance travel, it should be reasonable for start-stop time for the range-extender.

According to the original control strategy, the relationship between the SOC value of the battery and the working state of the range-extander with the mileage is shown in Figure 5. In initial stage, the vehicle is maintaining pure electric mode, when battery state of charge (SOC) down to the lower threshold value, start the range-extender and enter the road charging mode until the battery state of charge (SOC) reaching the upper threshold, then the range-extender closed. To shorten the range extender running time, two methods can be used to optimize the original control strategy [11].

For starting the range-extender in advance during pure electric mode, in pure electric driving mode, when the SOC value falls to O point, start the system in advance, the simulation results shown in Figure 6. For stopping the range-extender in advance during the road charging mode, in the charging mode, when the SOC value is increased to point C, range-extender is stopped in advance and the simulation results as shown in Figure 7.

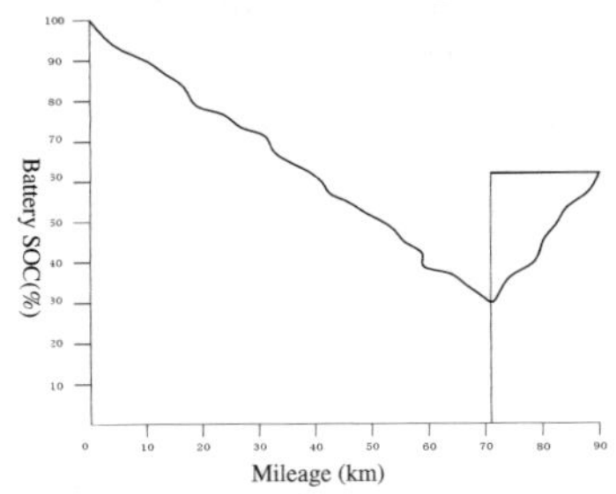

Figure 5. Original control strategy.

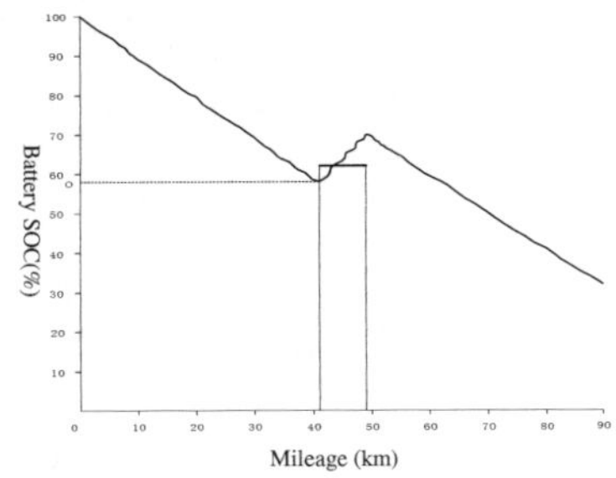

Figure 6. Open range extender advance.

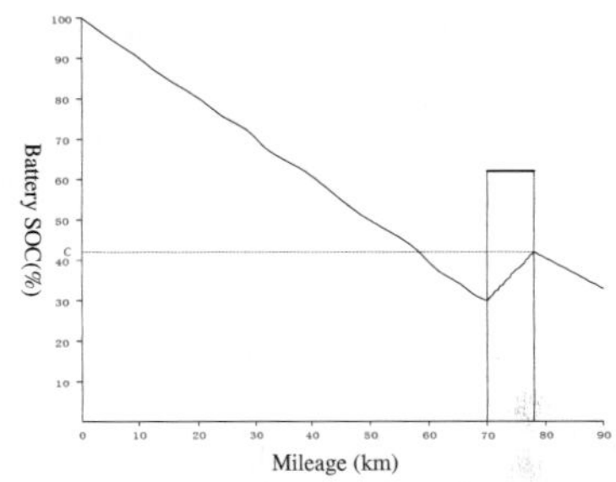

Figure 7. Close range extender advance.

In conjunction with Figure 8 and Figure 9, when the target distance is constant, these two optimization methods for start and stop time can be shortened range extender running time and the range extender power can just to run out before the next external charge [12]. The optimization strategy can make full use of battery energy, shortening range extender running time, reduce fuel consumption and cost savings.

When the engine is working in a particular point with a constant power output, the difference between the initial value of the battery SOC, the length of the target mileage, driving cycle conditions and other factors will affect the battery SOC decreasing speed, resulting in different start and stop times for range extender and different running time. Therefore, a combination of different driving cycle conditions is needed, to determine the range extender start and stop time.

The initial battery SOC value is set to 90%, with parameters k1, k2, k3 respectively SOCs is dropped to 30%; starting rage extenders SOC charged from 30% to 70%; Close range extenders after SOC decreased from 70% to 30% during these three mileage capacity ratios. According to results of simulation on three previous typical driving cycle conditions, the parameters k1, k2, k3 values are shown in Table 4.

Table 4. Mileage than capacity with different cycle conditions

Cycle condition	Mileage capacity ratio		
	k1	k2	k3
NEDC	0.87	0.75	0.78
UDDS	1.06	0.66	0.93
CUDC	1.07	0.43	1.05

In pure electric driving with starting the range-extender in advance, the target distance D is obtained when the range extender to start in advance. During the charging process, stop the range-extender in advance. Based on the numerical value of Table 4, the SOC of the starting and stopping time of the range-extender with different cycle conditions can be modified.

IOP Conf. Series: Materials Science and Engineering **241** (2017) 012025 doi:10.1088/1757-899X/241/1/012025

4.2. Comparison of simulation results before and after control strategy

To verify the effectiveness of the optimized control strategy, selecting the NEDC and UDDS two cycle conditions are simulated, battery SOCs value is set 90% and target D values of mileage is 100km. According to equation (3) and (5), drawn NEDC conditions SOC RE_on and SOC RE_off were 39.2% and 60.6%, and the simulation results shown in Figure 8.

By Figure 8, under the NEDC cycle condition target mileage is 100km, according to the original control strategy (a), the range-extender is running 2890s, after optimize control strategy, during pure electric vehicle driving (b), open the range-extender in advance, the total running time is 1806s, during driving charge period (c), shut down the range-extender in advance, accumulating the run time is 2068s, that is reduced by 37.2% and 28.2% respectively. The results of the specific comparison are shown in Table 5. According to above, UDDS conditions SOC RE_on and SOC RE_off were 46.2% and 51.2%, and the simulation results shown in Figure 9.

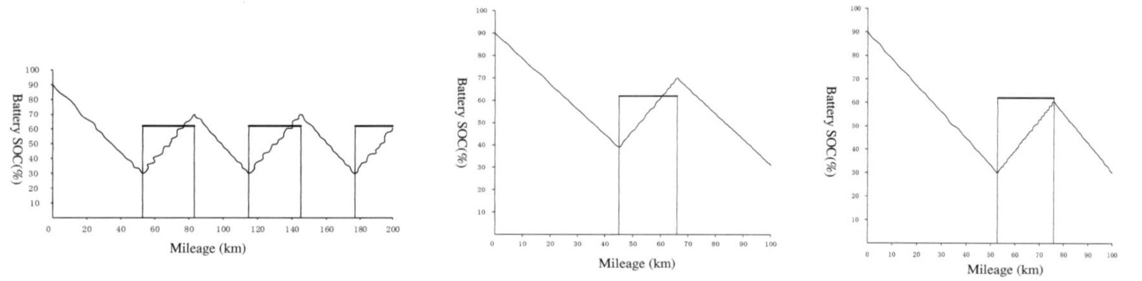

(a) The original control strategy (b) Start the range-extender (c) Stop the range-extender

Figure 8. Simulation results of the NEDC conditions

Table 5. Comparison of before and after optimization for NEDC condition

Control Strategy	SOC Control parameters/%	RE running time	Reduce RE running time /%
Traditional control strategy	30（SOC_{RE_on}） 70（SOC_{RE_off}）	2890	-
RE Start early	39.2（SOC_{RE_on}）	1806	37.2
RE close early	60.6（SOC_{RE_off}）	2068	28.2

By Figure 9, under the UDDC cycle condition target mileage is 100km, according to the original control strategy (a), the range-extender is running 2560s, after optimize control strategy, during pure electric vehicle driving (b), open the range-extender in advance, the total running time is 1516s, during driving charge period (c), shut down the range-extender in advance, accumulating the run time is 1708s, that's reduced by 40.6% and 33.5% respectively. The results of the specific comparison are shown in Table 6.

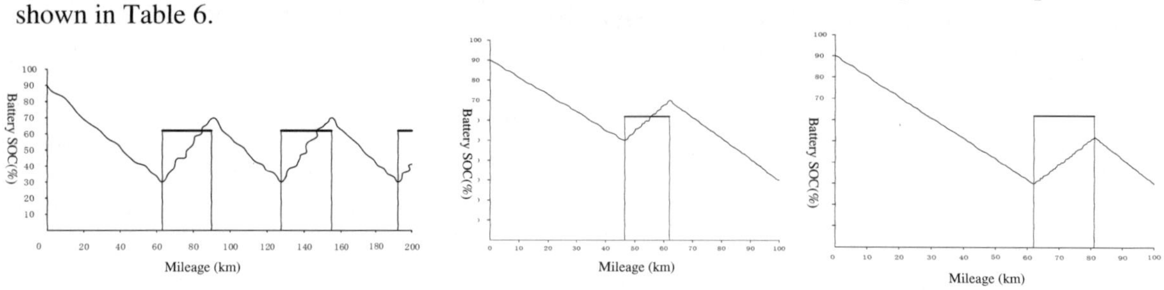

(a) The original control strategy (b) Start the range-extender (c) Stop the range-extender

Figure 9. Simulation results of the UDDS condition

According to Figure 8 and 9, after optimize the control strategy, the running time of two cycles conditions are significantly reduced while reaching the target mileage, besides the power that charged into the battery can basically run out, making full use of the energy of the battery at the same time also ensuring that the car charging power through the outer supply to prepare for the next driving.

Table 6. Comparison of before and after optimization for NEDC condition

Control Strategy	SOC Control parameters /%	RE running time	Reduce Rerunning time /%
Traditional control strategy	30 （SOC_{RE_on}） 70 （SOC_{RE_off}）	2560	-
RE Start early	46.2 （SOC_{RE_on}）	1516	40.6
RE close early	51.2 （SOC_{RE_off}）	1708	33.5

5. Summary

Footnotes should be avoided whenever possible. If required they should be used only for brief notes that do not fit conveniently into the text.

The range-extender electric vehicle technology has been generally recognized by the industry, the E-REV's start-stop control determines the mode of the car. Using the two methods that during pure electric driving start and stop the range extender in advance to reduce the running time when the target mileage is constant, and correcting the value of the battery SOC to precisely control the start and stop time.

The simulation results show that the control strategy is validated after the optimization of the control strategy. Under the condition of meeting the mileage of the vehicle, the running time of the extender is reduced by 37.24% and 28.2% respectively in the NEDC working condition, and in the UDDS operating conditions were reduced by 40.6% and 33.5% respectively, improving fuel economy and reducing emissions.

Acknowledgment

Funding from the National Natural Science Foundation of China (Grant No:61503163), the "333 project" of Jiangsu Province (Grant No: BRA2016440) and the six talent peaks project in Jiangsu Province (Grant No: ZBZZ-024) are gratefully acknowledged.

References

[1] Rahman K M, Jurkovic S and Stancu C 2015 IEEE Transactions on Industry Applications 3 2479
[2] Zhou Su, Li Feihong and Tiancai M A 2011 International Conference on Electric Information and Control Engineering(ICEICE) 4746
[3] Zhu X, Zhang H and Cao D 2015 Mechanical Systems & Signal Processing s58-59 15
[4] Patrick M Walsh SAE Paper 2011-01-0915
[5] Rahman K, Jurkovic S and Stancu C 2012 Energy Conversion Congress and Exposition 4152
[6] Sciarretta A, Serrao L and Dewangan P C 2014 Control Engineering Practice 6 287
[7] Chin P F, Miller J M and Lehmen A J 2015 US8956265
[8] Budde-Meiwes H, Drillkens J and Lunz B 2013 Proceedings of the Institution of Mechanical Engineers Part D Journal of Automobile Engineering 5 61
[9] Krenek T, Lauer T and Geringer B 2013 Proceedings of the FISITA 2012 World Automotive Congress 573
[10] Sah J J F, Heap A H and Kaminsky L A 2015 US 8989930B2
[11] Hu Jian-jun, Yang Zhi-peng and Zhao Jin-long 2013 Advances in Information Sciences and Service Sciences 5 1142
[12] Zhu X, Zhang H and Fang Z 2015 Mechanical Systems & Signal Processing s64-65 46

IOP Conf. Series: Materials Science and Engineering **241** (2017) 012026 doi:10.1088/1757-899X/241/1/012026

An Overview of the Thermal Calculation and the Cooling Technology for Active Magnetic Bearing

ZHANG Li[1],YU Meiyun[1], LUO Yanyan[1], LIU Jun[1], REN Yafeng[2]

[1] Qingan group co., LTD, 710077, Xian, China
[2] Shaanxi Engineering Laboratory For Transmissions and Controls, Northwestern Polytechnical University, 710072 , Xi'an, China

19871027zhangli@163.com

Abstract. The cooling process of AMB is that the energy loss is sent out to the outside world when the system is operating. The energy loss transfers to the surrounding medium in the form of heat, which leads to raise the temperature of system components and influences the performance of the system. So it is necessary to study the internal loss of the magnetic bearing system and thermal calculation method. Three kinds of thermal calculation methods are compared, which is important for the design and calculation of cooling. At the same time, the cooling way, the cooling method, and the cooling system is summarized on the basis of cooling technology of active magnetic bearing, and the design method of the cooling system is studied. But for the active magnetic bearing system, when designing the cooling system, heat dissipation of the motor can not be ignored. It is important not only for the performance of the active magnetic bearing system and stable operation, and but also for the improvement of the cooling technology.

1. Introduction

Active Magnetic Bearing (AMB) is a typical mechatronic system, which is no contact, wear and lubrication, controlling its dynamic behavior and other series of characteristics[1]. AMB can transform the electrical energy into mechanical energy, which levitates a rotor against gravity. In the process of the running, a certain energy losses will be generated. It leads to raise the temperature of system components and the loss diffused into the surrounding medium in the form of heat. If the cooling is not immediate, the performance of AMB system will be seriously affected. However, since AMBs are complex mechatronic systems, there are many potential power loss mechanisms. The minimizing of losses is the sum of various measures and depends very much on the requirements of the application. In fact, turbomachinery is the main application fields of magnetic bearing. It covers a small turbine vacuum molecular pump, large megawatt of turbine generator and compressor[2].

2. Losses and thermal calculation method in AMB

2.1. Losses

Similar to the motor, analysis of AMB loss and calculating the heat distribution are the important premise for the research of AMB temperature and cooling. Heat sources are related with the system losses. Seen from the Fig.1, the typical spatial layout of magnetic bearing system is represented, but the energy loss distribution of magnetic bearing system is also reflected. The losses mainly include electric loss (copper loss) of the stator, magnetic loss (hysteresis loss and eddy current loss), magnetic loss and aerodynamic losses of the rotor[3,4], as shown in figure 1. In switched amplifiers, switching losses are dominant. The switching losses are about proportional to the switching frequency and depend on the design of the electronic switches and the properties of the specific switching transistors employed. The ohmic losses in the cables may present a substantial part of the total losses. The losses in the cables depend on the length and the cross section of the cable wires. The losses in the cables have to be covered by the power amplifiers.

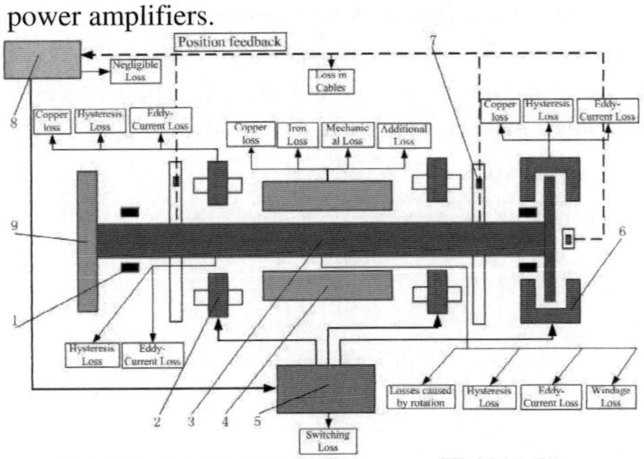

Figure 1. Distribution of energy loss and the spatial layout of AMB.

2.2. Thermal calculation method for AMB

Thermal calculation aims to understand the thermal flow of effective part and temperature distribution in the AMBs. The basic requirement of calculation is the accuracy of the highest temperature and average temperature for the coil. Average temperature is the most suitable operation index, because the highest temperature will limit power under the heat insulation level. Due to heat transfer, the thermal flow travels from the coil heads into the cooling medium, and due to heat conduction, through the insulation between the coil and the iron core. Finally, due to heat transfer, thermal flow travels from the iron core to the air.

*2.2.1 Thermal parameter method.*According to the principle of superposition, the rise of temperature of the AMB are identified. Assumptions: the calorific value of coil is determined by electric loss Pm1, the wastage of the coil is constant, which does not depend on other parts of the loss. Under the condition of the same cooling, the stator core and rotor loss P_2 and P_3 can be determined by using the heating coefficient k_2 and k_3, see the Eq. (1).

$$\begin{cases} \Delta T_{m1} = R_m P_{m1} \\ \Delta T_{m2} = k_2 R_m P_{m2} \\ \Delta T_{m3} = k_3 R_m P_{m3} \end{cases} \tag{1}$$

Where R_m is the equivalent thermal resistance. Under the stable thermal state, the total temperature rise calculation by Eq.(2). Thermal parameters method is relatively simple, but because the consideration is not comprehensive, it is suitable for rough calculation in the early design process.

$$\Delta T_m = \Delta T_{m1} + \Delta T_{m2} + \Delta T_{m3} = R_m (P_{m1} + k_2 P_{m2} + k_3 P_{m3}) \tag{2}$$

*2.2.2 Equivalent thermal network method.*The graph theory and circuit theory is introduced to combine with the heat transfer principle, and the equivalent thermal network method is developed. The bearing model has been realized as a thermal network. According to the unified form of the heat exchange equation (the Fourier law), which is similar to the ohm's law, Eq. (3) is derived.

$$P = \frac{\lambda S \Delta T}{\delta} = R_\lambda \Delta T \tag{3}$$

Where λ is the heat conduction coefficient, S is the average area of the heat transfer surface, ΔT is the temperature rise on the length δ, R_λ is the thermal resistance.

Assumptions: the real source of heat and thermal resistance is replaced by a finite number of concentrated heat source and the equivalent thermal resistance. the equivalent thermal resistance does not change with the size of the thermal flow. As a result, the equivalent thermal network is a linear thermal circuits and algebraic overlay. Thermal resistance is through the medium parameters (the thermal conductivity, the heat capacity, etc.) is identified. In the thermal circuit, temperature depends on the size of the heat source. Figure 2 shows the equivalent thermal network about the radial magnetic bearing. Each pair of coils is regarded as a heat source. Each heat source is represented with the mean surface temperature ϑ_{cu}. The bearing is air cooled with a temperature ϑ_0. Λ_{cu} is the thermal conductivity coefficient of the coil head. For the transfer from the iron core to the air Λ_{fe} is obtained. For the transfer from the insulation to the iron core Λ_i is obtained.

Figure 2. Equivalent thermal network of the radial magnetic bearing.

*2.2.3 Finite temperature field method.*It is based on the solution of the heat conduction differential equation, according to general thermal transfer law these equations are constructed. In differential equations and boundary conditions, using equivalent thermal network thermal contact will be expressed. In equation of the temperature field method the heating temperature of cooling medium and winding groove to make the transition to the end of the structure characteristics have been considered. Through this method the temperature field of magnetic bearing is analyzed. Therefore, the differential method is a good basis to analyze the effect of parameters of the effective part, and effective practical conclusions are concluded. Such as the diameter, the pole angle, the back-iron width on the temperature effect is different. But, in small or micro magnetic bearing, the thermal effect is more significant between different parts, such as the rotor and the stator, the core and the coil, etc. In these cases, therefore, the magnetic bearing of the equivalent thermal network is first recognized, and then each unit by using the method of temperature field is analyzed.

3. Cooling technology of AMB

Cooling medium can be a gas or liquid, the corresponding cooling way is a gas or liquid cooling. Gas may be air, compressed air, hydrogen or helium gas. In different gases, the thermal conductivity coefficients, heat capacity and density are different. Due to the calorific value is smaller for the tiny magnetic bearing, it can be air cooling or natural cooling. WANG Yan, XIE Heng[5,6] have analyzed the cooling system of magnetic bearing in high temperature gas-cooled test reactor. Helium cooling is instead of steam cooling. Because the helium gas has high thermal conductivity coefficient and heat capacity, and the small density, an effective cooling system can be established. Liquid can be distilled water or oil, it takes only a small amount of power to take away a lot of heat, this kind of cooling method has been applied in the large structure of AMBs.

Generally, the running environment is divided into totally enclosed and open for AMBs. When the surrounding environment is open form, the surrounding air is used to blow its heat source such as coil winding, eventually aiming at degrading the operating temperature. Yet this cooling way is not suitable for high humidity, dust, high temperature environment. Therefore, enclosed AMBs is separated by a closed cavity with the outside world. Because air cooling or cooling liquid is filled into a closed cavity in the body, the part of the heat is taken out to achieve the purpose of cooling. When designing cooling system needs to be fluid calculation, determine the flow resistance of the cooling system, pressure loss and flow velocity along the cooling channels, and make up for the loss of the cooling system, suitable coolant flow rate and power is selected. In the initial design, thermal parameter method may be used for the comparison of the different cooling scheme in order to eliminate unreasonable cooling scheme in preliminary heat calculation. If cooling groove has not been open or it can't satisfy cooling design, then the shaft directly exchange with cooling medium to cool magnetic bearing for heat. The Fig. 3 represents the flow charts of the cooling system design on active magnetic bearing.

Calculation of the ventilation system is to determine the flow resistance, pressure loss and the flow rate of the cooling medium on the cooling system. According to the law of conservation of energy, the flow of cooling medium is determined by thermal calculation. Determining the network features and setting up the equivalent circuit diagram of the cooling system, so it is simple and intuitive to calculate the ventilation system. But for the magnetic bearing system, when designing the cooling system, heat dissipation of the motor can be not ignored. Motor as a product, some manufacturers have to provide the cooling system. However, analysis and calculation of the cooling of the motor need to be for the integration of system design, because the heat coming from the motor is large.

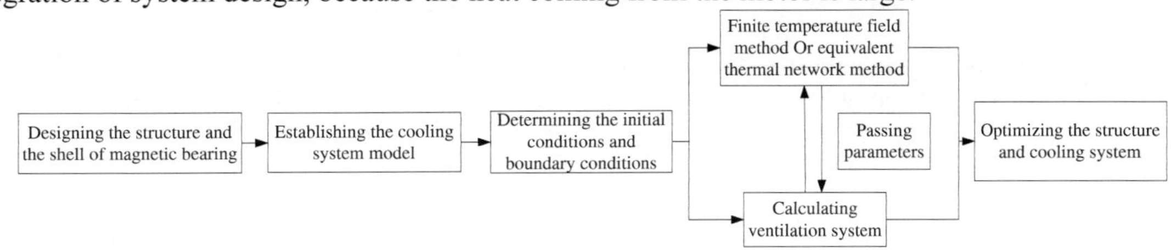

Figure 3. Flow charts of the cooling system design for AMB.

4. Conclusion

It is helpful for the optimization design of the AMB by studying the energy loss of AMB. Three kinds of thermal calculation methods are compared, which is important for the design and calculation of cooling. The cooling way, the cooling method, and the cooling system are summarized on the basis of cooling technology of active magnetic bearing. But for the active magnetic bearing system, when designing the cooling system, heat dissipation of the motor can not be ignored. It is important not only for the performance of the active magnetic bearing system and stable operation, and but also for the improvement of the cooling technology.

References

[1] R.Vuillemin, B.Aesehlimann,ete. Low Cost Active Magnetic Bearing for Hard Disk Drive Spindle Motors[J]. Proeeedings of the Sixth International SymPosiumon Magnetic Bearings, MIT, USA. Augusts5-7,1998.

[2] G.Schweitzer, H.Maslen. Magnetic bearing-Theory, design, and application to rational machinery [M], 2009.

[3] SUN Shouqun, GENG Haipeng, YU Lie. Thermoanalysis of Thrust Magnetic Bearing with Solid Magnet[J]. Transactions of China Electrotechnical Society, Oct. 2002.

[4] LI Xinsheng, YANG Zuoxing, ZHAO Lei, ZHAO Hongbin. Heating of Axial Active Magnetic Bearings[J]. Journal of Tsinghua University(Science and Technology), Aug.2002.

[5] WANG Yan, XIE Heng, ZHAO Lei. The Design of Electromagnetic Bearing Cooling Scheme in High Temperature Gas-Cooled Test Reactor with the Gas Turbine[J].Nuclear Power Engineering, June 2007.

[6] WANG Yan, XIE Heng, ZHAO Lei. The Analysis of the Electromagnetic Bearing Cooling System of Power Conversion Unit in High Temperature Gas-Cooled Test Reactor[J]. Journal of Tsinghua University(Science and Technology), Mar.2008.

IOP Conf. Series: Materials Science and Engineering **241** (2017) 012027 doi:10.1088/1757-899X/241/1/012027

The Influence of Heat Transfer Augmentor on the Performance of Window Air Conditioners

J.Y Li and T.P Teng

Department of Industrial Education, National Taiwan Normal University, No. 162, Sec. 1, He-ping E. Road, Da-an District, Taipei City 10610, Taiwan, ROC.

peter4351@hotmail.com

Abstract. This study explored the effects of different designs of a heat transfer augmentor (HTA) on the performance of window air conditioners (WACs). A HTA was fabricated from perforated aluminum plates with different pore diameters according to the surface temperature distribution of the evaporator. The WAC was tested under the T1 condition of CNS14464, and the performance before and after the installation of the HTA was compared. The experimental results showed that the pore diameters of the HTA in proportion to the surface temperature of the evaporator could effectively improve the performance of the WAC. The dehumidification capacity and the energy efficiency ratio (EER) were increased by 22.05% and 3.49%, respectively, compared with the original WAC.

1. Introduction

Air conditioners have almost become home necessities for the pursuit of a comfortable life. In developed countries, 20% of total electricity is consumed for air conditioning [1]. Atmospheric temperature exhibits a gradual upward trend because of environmental changes and the greenhouse effect. A 1 °C increase in summer temperatures has been correlated with a 3.8% increase in peak demand load for air conditioning [2]. An energy efficiency ratio (EER) enhancement of 0.1 for an air conditioning system can save approximately 4% of power consumption. Therefore, the performance of existing air conditioners should be studied and improved to enhance their EER.

Air-cooled heat exchangers (ACHEs) with non-uniform air distribution reduce the effective area of heat exchange and thus lower the efficiency of the heat exchanger [3-8], prevent the normal operation of the thermostatic expansion valve of the evaporator [9], increase the air pressure loss, increase the fan power consumption, and reduce the average air velocity [7, 8]. All these factors affect the overall performance of air conditioners. Improving the air distribution unevenness mainly involves changing the piping configuration of the heat exchanger [10-13] and installing air guide plates [6]. These methods adjust the temperature distribution of the heat exchanger to match the uneven air distribution or improve the uniformity of the air distribution to improve the efficiency of ACHEs.

These methods can effectively improve the performance of an ACHE or an air conditioner. However, they also have disadvantages such as the complexity of the modification procedure, high costs, and increased equipment volume. In this study, a heat transfer augmentor (HTA) with different pore diameters were installed directly on the return air (RA) side of the evaporator in a window air conditioner (WAC) to adjust the distribution relationship between the air flow and the surface temperature of the evaporator. This HTA improved the efficiency of the ACHE and the EER of the WAC without changing the original design to achieve the purpose of conserving energy.

2. Principle and Calculations

WACs use a vapor compression refrigeration system to achieve cooling. The enthalpies ($h_{ea,o}$ and $h_{ea,i}$) of the supply air (SA) and RA of the WAC can be obtained using the psychrometric chart from measurements such as the dry-bulb temperature (T_{DB}), wet-bulb temperature (T_{WB}), and relative humidity (ψ) of the SA and RA. The cooling capacity (Q_{WAC}) of the WAC is calculated as shown in Eq. 1. The mass flow rate (m_{eao}) of the SA can be converted using the volume flow rate ($G_{ea,o}$) with the air density ($\rho_{ea,o}$) of the SA in Eq. 2.

$$Q_{WAC} = m_{ea,o} \times (h_{ea,i} - h_{ea,o}) \quad (1)$$

$$m_{ea,o} = \rho_{ea,o} \times G_{ea,o} \quad (2)$$

The dehumidification capacity of the WAC (W_{WAC}) can be obtained from the humidity ratios ($w_{ea,o}$ and $w_{ea,i}$) of the SA and RA. W_{WAC} is calculated as shown in the Eq. 3.

$$W_{WAC} = m_{ea,o} \times (w_{ea,i} - w_{ea,o}) \quad (3)$$

The EER of the WAC can be expressed as:

$$EER = Q_{WAC} / P_{WAC} \quad (4)$$

Where P_{WAC} is the total power consumption of the WAC and usually includes the power consumptions of the compressor, evaporator fan, condenser fan, and related electronic control elements.

3. Experimental Design and Procedure

3.1. Heat Transfer Augmentor Design

Fig. 1 shows a thermal image of the surface temperature of the evaporator of a R410A WAC (SRA-22b, SANYO, Taiwan) measured using an infrared thermal imager (Testo 881, Testo, Germany). The analysis software of the infrared thermal imager was used to determine temperatures at each end of the pipe, which are presented in Fig. 2 The evaporator of the WAC comprised the upper and lower components of a sub-evaporator, assembled in parallel to form a double-loop configuration. Fig. 2 reveals that the surface temperatures of the upper (loop 1, M6–M12) and lower (loop 2, M1–M5) evaporators tended to gradually rise from bottom to top. However, the lower evaporator was affected by the temperature of the upper evaporator, and the temperature of M5 on loop 2 decreased. In theory, the refrigerant passes through the capillary into the bottom of the evaporator to absorb heat through vaporization, and the temperature of the refrigerant gradually increases. Thus, the temperature of each evaporator circuit (loop 1 and loop 2) should gradually increase from bottom to top.

In this study, two HTAs (S1 and S2) were designed with perforated aluminum plates with pore diameters of 5 and 8 mm, respectively, and the HTA was installed between the evaporator and the filter of WAC (Fig. 3). A larger pore size reduces the air resistance, increasing the volume flow rate of air (G), and vice versa. Therefore, the design principles of the two HTAs were that the lower surface temperature of an evaporator corresponds to a larger G (S1) and that the lower surface temperature of an evaporator corresponds to a smaller G (S2). Each HTA was fabricated from four perforated aluminum plates with the same size and different pore diameters. The HTA was installed between the evaporator and the filter to reduce the contamination of the HTA and not introduce additional space requirements.

3.2. Test Standard and Method

The performance test of the WAC was based on the CNS14464 standard with T1 environmental conditions [14]. The T_{DB}/T_{WB} was 35/24 ± 1 °C ($\psi \approx$ 40% RH) and 27/19 ± 1 °C ($\psi \approx$ 47% RH) for outdoors and indoors, respectively. In this study, the original WAC (control group) was denoted S0, and WACs with HTA (experimental group named St) were denoted S1 and S2. The thermocouples (T type: accuracy: ±0.75%) were installed on the SA and RA sides to measure the T_{DB} and T_{WB}, the pressure sensors (JPT-131S, Jetec, Taiwan; accuracy: ±0.25%) were installed on the refrigerant system to measure the high and low side pressure, and the temperature and pressure data were recorded continuously by using a data logger (TRM-20, TOHO, Japan). P_{WAC} was measured and recorded using a digital power meter (CW121, YOKOGAWA, Japan; accuracy: ±0.4%). The $G_{ea,o}$ of the WAC was measured and calculated using an anemometer (Testo 445, Testo, Germany) with an air hood and its cross-sectional area. The $G_{ea,o}$ for S0, S1, and S2 was 0.1145, 0.1093, and 0.1101 m^3/s, respectively. The $G_{ea,o}$ of the S1 and S2 was reduced by approximately 4.54% and 3.84% to compare with the S0, respectively.

3.3. ***Data Analysis***

The total time of each test was 2 hours, and the final hour of experimental data was averaged and analyzed. The data from the air side of the evaporator were analyzed using ASHRAE Psychrometric Analysis [15]. Then, the Q_{WAC} and W_{WAC} of the WAC were calculated using Eqs. 1 to 3 The EER of the WAC was then calculated using Eq. 4 with the Q_{WAC} and the P_{WAC} of the WAC. The test results of S1 and S2 in the experimental group (St) were compared with those of S0 to identify the percentage difference as shown in Eq. 5.

$$CS\% = [(St - S0)/S0] \times 100\% \quad (5)$$

4. Results and Discussion

Fig. 4 shows the Q_{WAC} values of S0, S1, and S2. The Q_{WAC} of S1 was lower than that of S0, whereas the Q_{WAC} of S2 was higher than that of S0. Compared with S0, the rate change in Q_{WAC} for S1 and S2 was −2.62% and 2.98%, respectively. Although the G values of S1 and S2 were lower than 4.54% and 3.84% of that of S0, the configuration of S2 provided a higher enthalpy difference (total heat) between the SA and RA, and the Q_{WAC} was improved. S1 greatly reduced G values, preventing improvements in the heat exchanger efficiency.

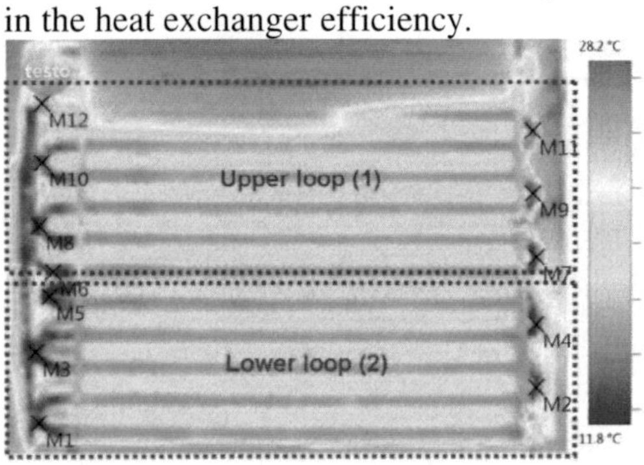

Figure 1. Infrared thermal image of the evaporator of an original WAC.

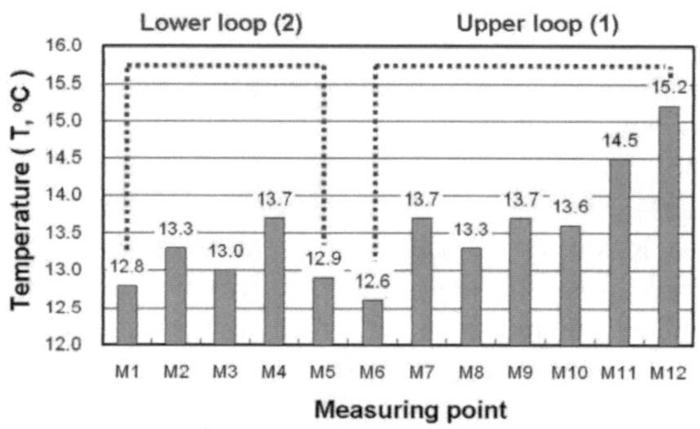

Figure 2. Temperature of the measurement point for the evaporator of an original WAC.

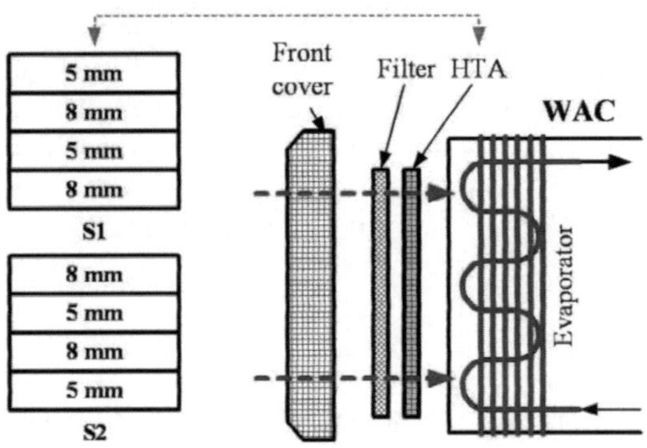

Figure 3. Design and installation of the HTA.

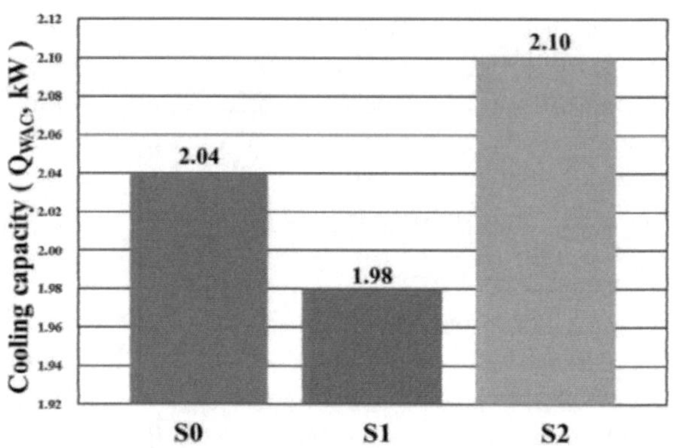

Figure 4. Cooling capacity of the WAC with various experimental parameter configurations.

Fig. 5 displays the W_{WAC} values of S0, S1, and S2. The W_{WAC} values of S1 and S2 were higher than that of S0 by 20.58% and 22.05%, respectively. This phenomenon is mainly due to the pores of the HTA causing local air turbulence to extend the contact time between the air and the fin of the evaporator, thus enhancing the heat and mass transfer of air and fins surface and the water vapor condensation efficiency. Experimental results showed that the W_{WAC} of S2 was better than that of S1; therefore, S2 should have an air distribution superior to that of S1.

Fig. 6 shows the EER values of S0, S1, and S2. The installation of an HTA for the WAC had no significant effect on the P_{WAC}. Compared with S0, the rate change of EER for S1 and S2 was −2.53% and 3.49%, respectively. The EER incorporates both the Q_{WAC} and the P_{WAC} of the WAC and is an important index of system performance. The experimental results indicated that the HTA of S2 improved the performance of the WAC and was the optimal configuration in this study.

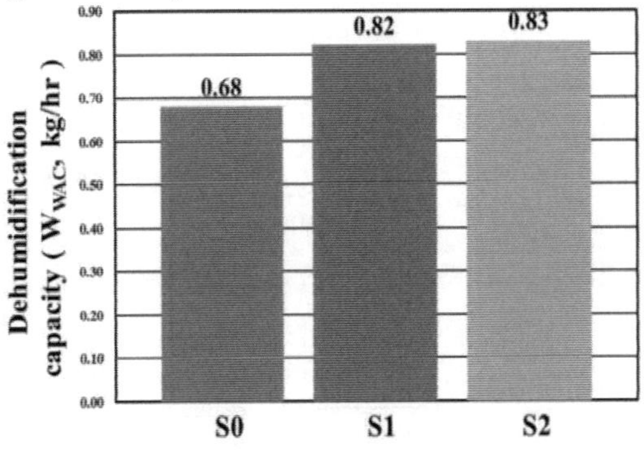

Figure 5. Dehumidification capacity of the WAC with various experimental parameter configurations.

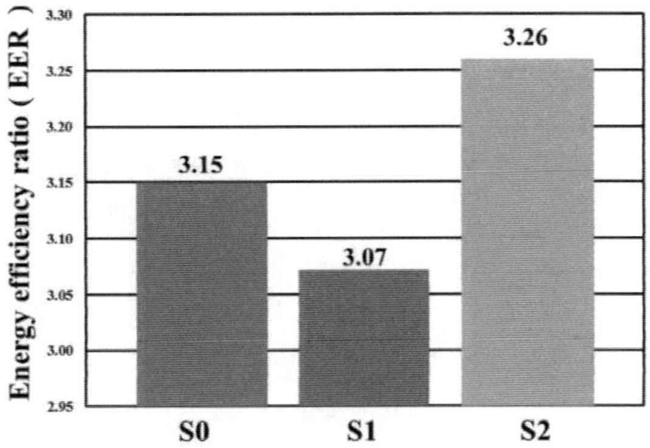

Figure 6. Energy efficiency ratio of the WAC with various experimental parameter configurations.

5. Conclusions

In this study, a HTA for the performance improvement of WACs was successfully developed using perforated aluminum plates with different pore diameters. The HTA did not require changes to the original design of the evaporator of the WAC, was introduced at a very low cost, and increased the EER to achieve the purpose of energy conservation. The results showed that the configuration of S2 exhibited the optimal performance, and the W_{WAC} and EER increased by 22.05% and 3.49%, respectively, compared with the original WAC. The main factor of performance enhancement of WAC was the high temperature difference between the heat transfer and the effective surface area of the evaporator caused by a more uniform air distribution.

References

[1] Park Y S, Jeong J H and B H 2014 *Ahn: Appl. Energy.* Vol. 132, p 99
[2] Green roofs and the urban heat island effect: http://www.buildings.com/DesktopModules/BB_ArticleMax/ArticleDetail/BBArticleDetailPrint.aspx?ArticleID=8620&Template=Standard_Print.ascx&siteID=0
[3] Aganda A A, Coney J E R and W C G 2000 *Sheppard: Appl. Therm. Eng.* Vol. 20, p 515
[4] Zhang L Z 2009 *Int. J. Heat Mass Transf.* Vol. 52, p 4500
[5] Kærn M R, Brix W, Elmegaard B, Larsen L F S 2011 *Int. J. Refrig.* Vol. 34, p 696
[6] Song X, Huang D, Liu X and Chen Q 2012 *Appl. Therm. Eng.* Vol. 40, p 284
[7] Mao J N, Chen H X, Jia H, Wang Y Z and Huc H M 2013 *Int. J. Therm. Sci.* Vol. 73, p 46
[8] Lee W J and Jeong J H 2016 *Int. J. Refrig.* Vol. 69, p 85
[9] Gong J, Gao T, Yuan X and Huang D 2008 *Energy Convers. Manage.* Vol. 49 p 1645
[10] Brix W, Kærn M R and Elmegaard B 2009 *Int. J. Refrig.* Vol. 32, p 1736
[11] Tuo H and Hrnjak P 2013 *Int. J. Refrig.* Vol. 36, p 2176
[12] Lee W J, Kim H J and Jeong J H 2016 *Int. J. Heat Mass Transf.* Vol. 98, p 462
[13] Kim J H, Braun J E. and Groll E A 2009 *Int. J. Refrig.* Vol. 32, p 1283
[14]CNS 14464 2003 *Chinese National Standard- Non-ducted air conditioners and heat pumps–testing and rating for performance*, National Bureau of Standards
[15]ASHRAE 2002 *Psychrometric Analysis*, version 6. ASHRAE, Inc.

IOP Conf. Series: Materials Science and Engineering **241** (2017) 012028 doi:10.1088/1757-899X/241/1/012028

Concurrent Monitoring of Chip Formation and Prediction of Roundness in CNC Turning Using Wavelet Transform

Somkiat Tangjitsitcharoen[1,a] and Mumin Sassantiwong[2]

[1,2] Department of Industrial Engineering, Faculty of Engineering, Chulalongkorn University Phayathai Road, Patumwan, Bangkok, 10330, Thailand

[a]somkiat.ta@eng.chula.ac.th

Abstract. The aim of this research is to investigate the chip formation, the roundness and the dynamic cutting forces in CNC turning process. The dynamic cutting forces are decomposed to classify the signals of the broken chip and the roundness. The Daubechies wavelet transform is utilized to identify those signals into different levels due to the different frequencies of itself. The experimentally obtained results showed that the decomposed cutting forces can be used to estimate the roundness under various cutting conditions.

1. Introduction

According to the previous researches, the dynamic cutting forces have been proposed to monitor the broken chip formation and the roundness in CNC turning [1,2]. The dynamic cutting forces have been used to be decomposed to monitor the straightness in order to avoid the broken chip and noise signals by using the wavelet transform [3].

Many methods have been proposed to monitor the roundness. Guo and Han [4] analysed the machining conditions to predict the diametral error of slender bar. Bugra et al. [5] utilized the cutting forces which is accurate enough to estimate the bar deflection. Bodi and Rongdi [6] proposed an artificial neural network model to investigate the influence of cutting parameters on the dimensional errors. It is proved that the diameter of the workpiece depends on the radial force and the feed force during the turning process. Xuefeng Bi et al. [7] presented that the dimensional error increases with the cutting velocity, the cutting time, the tool nose radius and the clearance angle. Shawky and Elbestawi [8] proposed an on-line measurement system in the bar turning to evaluate of workpiece size based on cutting forces.

The previous research has presented that the radial force (Fx) and the feed force (Fy) influence on the diameter error and the roundness error of the workpiece [9]. It is known that the peak-to-valley height depends on the feed rate and the tool nose radius [10] which causes the roundness error.

It is found that the decomposed cutting forces by using the Daubechies wavelet transform can be used to classify the roundness and straightness frequency effectively in both time and frequency domains, which appear on 8th level [1,3]. Hence, the aim of this research is to propose the Daubechies wavelet transform to decompose the dynamic cutting forces into 10 levels in order to differentiate the broken chip and roundness signals in CNC turning process.

2. Monitoring of chip formation and roundness

Figure 1 illustrates the dynamic cutting forces of broken chip and the corresponding time record of roundness profile. It is noticed that the amplitudes of dynamic cutting forces are larger than that of

IOP Conf. Series: Materials Science and Engineering **241** (2017) 012028 doi:10.1088/1757-899X/241/1/012028

continuous chip as shown in figure 2 [2]. It is implied that the chip breaking signals are included in the dynamic cutting forces. If those noise and chip breaking signals can be separated out from the dynamic cutting forces, the roundness will be estimated correctly from the use of dynamic cutting forces. The Daubechies wavelet transform is hence proposed to classify those signals from dynamic cutting forces into different levels due to the different frequency ranges. It is known that the higher frequency signal will occur at the lower level of wavelet transform of decomposed cutting force. On the other hand, the lower frequency signal will appear at the higher level of the wavelet transform [1,3].

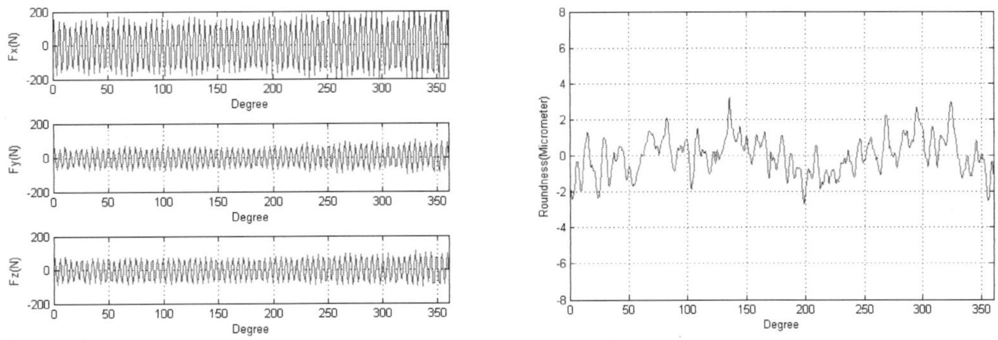

Figure 1. Illustration of dynamic cutting forces of broken chip and roundness profiles in time domain at cutting speed of 100 m/min, feed rate of 0.15 mm/rev, depth of cut of 0.8 mm, tool nose radius of 0.8 and rake angle of 11°.

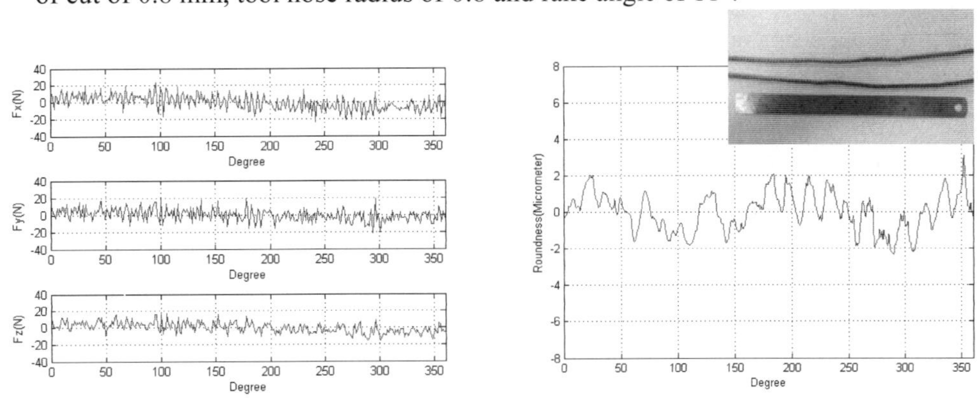

Figure 2. Illustration of dynamic cutting forces of continuous chip and roundness profile in the time domain at cutting speed of 150 m/min, feed rate of 0.15 mm/rev, depth of cut of 0.6 mm, tool nose radius of 0.8 and rake angle of -6°.

Since, the dynamic radial force (Fx) affects the roundness profile. While the dynamic feed force (Fy) may cause the roundness error, which is perpendicular to the roundness [2]. The dynamic radial force and the dynamic feed force are hence decomposed to monitor the roundness error during the cutting. The cutting forces are generalized by taking the ratio of the average variation of decomposed radial force to that of the decomposed feed force (AVFx/AVFy) as shown in figure 3. The relation between the decomposed cutting force ratio and the in-process roundness will be examined under various cutting conditions.

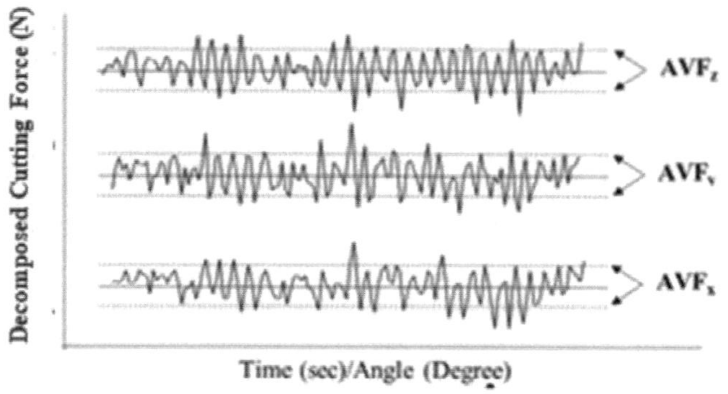

Figure 3. Illustration of average variation of cutting forces.

3. Prediction of roundness

The relation of the roundness error, the decomposed cutting force ratio and the cutting parameters are examined and proposed by employing the exponential function for the sake of it as shown in equation (1):

$$R_0 = C_1 \cdot (V)^{a_1} \cdot (f)^{a_2} \cdot (D)^{a_3} \cdot (R_n)^{a_4} \cdot (e)^{a_5 \gamma} \cdot \left(\frac{AVF_x}{AVF_y} \right)^{a_6} \quad (1)$$

Where R_0 is the roundness error in micrometer, V is the cutting speed in m/min, f is the feed rate in mm/rev, D is the depth of cut in mm, Rn is the tool nose radius in mm, γ is the rake angle in degrees, (AVFx)/(AVFy) is the decomposed cutting force ratio, a_1, a_2, a_3, a_4, a_5, a_6, and C_1 are the powers and the coefficients of the model, respectively.

4. Experimental setup and procedure

The cutting tests are conducted on CNC turning machine. The coated carbide cutting tools are used for the experiments with the different tool nose radiuses and the rake angles. The cutting conditions are summarized in the table 1.

The experimental setup is illustrated in figure 4. The dynamic cutting forces detected by dynamometer are sampling at 10 kHz and amplified through the low-pass filtered with cut-off frequency of 5 kHz prior to digitization and calculation within PC.

Table 1. Cutting conditions.

Cutting Tool	Coated carbide tool	
Tool geometry	DNMG150604FN	DCMT11T304HQ
	DNMG150608FN	DCMT11T308HQ
Rake angle	-6°	11°
Tool nose radius	0.4, 0.8 mm	
Workpiece	AISI 1045 Carbon Steel	
Cutting speed	100, 150, 200 m/min	
Feed rate	0.15, 0.20, 0.25 mm	
Depth of cut	0.4, 0.6, 0.8 mm	

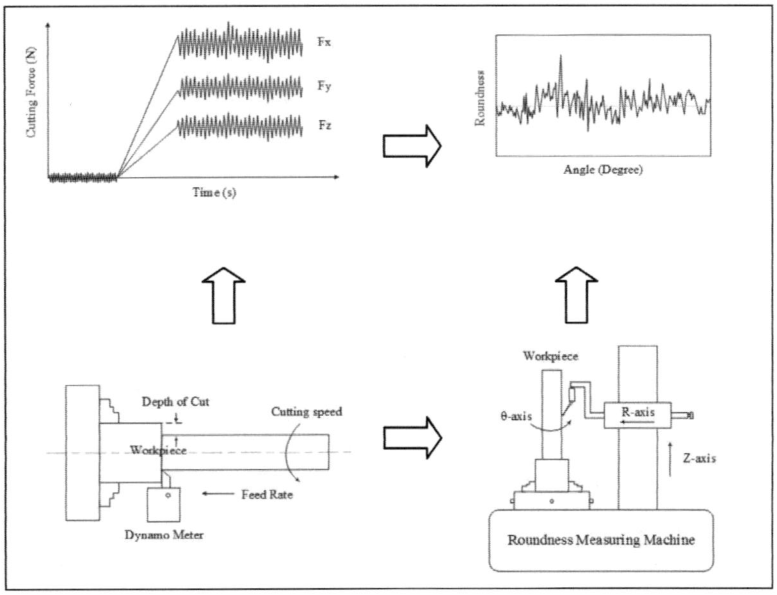

Figure 4. **Illustration of experimental setup.**

The following procedures are proposed to obtain the relation of the roundness, the cutting force ratio and the cutting conditions.

1. Start cutting and monitor the decomposed cutting forces referring to the cutting conditions.
2. Check the relations of the roundness and the decomposed cutting forces in time domain and frequency domain by using the minimum zone circle (MZC) method [11].
3. Take the ratio of the average variation of decomposed radial force to that of decomposed feed force (AVFx/AVFy).
4. Repeat the procedures (1) to (3) for other cutting conditions.
5. Model the in-process prediction of roundness and ratio of decomposed cutting forces under various cutting condition at 95% confident level.
6. Verify the obtained model with the new cutting conditions.

5. Experimental results and discussions

The examples of the cutting tests are illustrated in figure 5. The dynamic cutting forces are decomposed into 10 levels in both time domain and frequency domain by using the Daubechies wavelet transform. It is shown that the decomposed cutting force corresponds with the roundness profile at the 8th level, which are the same as previous research [3]. Hence, the decomposed cutting force from the 8th level can be monitored to estimate the in-process roundness as shown in figure 5.

As shown in figure 5, the broken chip and noise signals have been separated out from roundness signal to other levels due to the different frequency ranges of noise and chip breaking which appear in lower levels of wavelet transform. It means that the higher frequency signal will happen at the lower levels which correspond with the previous research [2,3].

6. Verification of roundness model and accuracy

The experimentally obtained in-process roundness model is expressed in equation (2) ;

$$R_0 = e^{2.3410} \cdot V^{-0.0937} \cdot f^{0.1589} \cdot D^{0.0568} \cdot R_n{}^{-0.0395} \cdot \gamma^{-0.004420} \cdot \left(\frac{AVF_x}{AVF_y}\right)^{0.3040} \quad (2)$$

IOP Conf. Series: Materials Science and Engineering **241** (2017) 012028 doi:10.1088/1757-899X/241/1/012028

The in-process roundness model is obtained at 95% confident level, which consists of the cutting speed, the feed rate, the depth of cut, the tool nose radius, the rake angle and the decomposed cutting force ratio. The new cutting tests are conducted as shown in table 2 for 32 experiments in order to verify the accuracy of the in-process roundness model by adopting the new cutting conditions as shown in figure 6.

The experimentally measured roundness versus the in-process predicted roundness obtained from the model is shown in figure 6. The prediction accuracy of the model within the ±10% error lines is about 95.51%. It indicates that the developed model can aid to predict the in-process roundness which is close to the measured ones during the cutting regardless of the cutting conditions.

7. Conclusion

The concurrent monitoring of chip formation and prediction of roundness has been proposed under various cutting conditions. The broken chip and noise signals can be detected and classified to other levels of decomposed cutting forces by using the Daubechies wavelet transform. The ratio of average variation of decomposed radial force to that of decomposed feed force is proposed to estimate the roundness. The exponential function is employed to represent the relation of roundness and decomposed cutting force ratio with cutting parameters. The multiple regression analysis has been utilized to calculate the powers of the in-process prediction of roundness model using the least square method at 95% confident level. It is proved by the cutting tests that the in-process roundness can be well predicted within ±10% of measured ones.

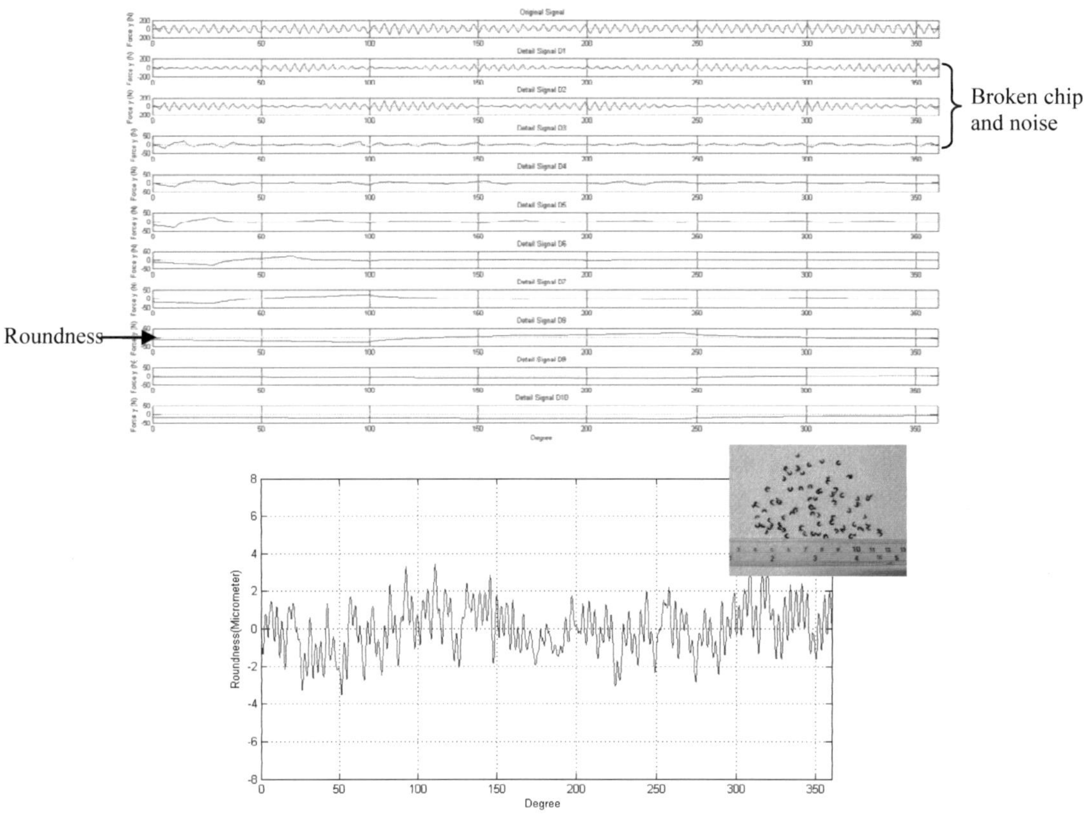

Figure 5. Illustration of decomposed feed forces of broken chip in time domain obtained at cutting speed of 150 m/min, feed rate of 0.25 mm/rev, depth of cut of 0.8 mm, tool nose radius of 0.4 and rake angle of 11°.

Table 2. New cutting conditions.

Cutting Tool	Coated carbide tool	
Tool geometry	DNMG150604FN	DCMT11T304HQ
	DNMG150608FN	DCMT11T308HQ
Rake angle	-6°	11°
Tool nose radius	0.4, 0.8 mm	
Workpiece	AISI 1045 Carbon Steel	
Cutting speed	120, 250 m/min	
Feed rate	0.18, 0.30 mm	
Depth of cut	0.2, 0.5 mm	

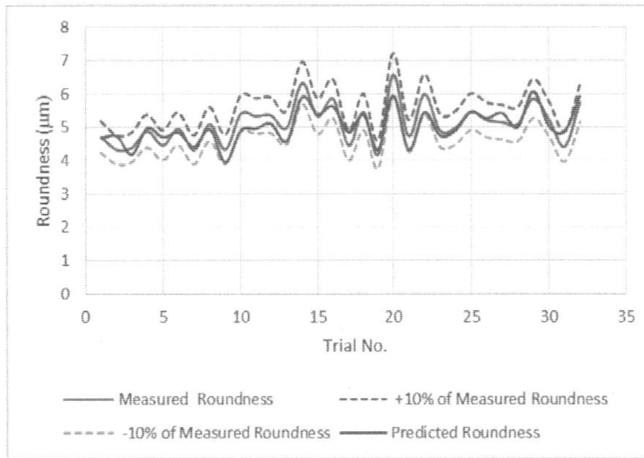

Figure 6. Verification of roundness model.

Acknowledgement
This work was performed by the partial funding of The Thailand Research Fund (TRF).

References

[1] S. Tangjitsitcharoen and K. Samanmit, "Monitoring of chip breaking and surface roughness in computer numerical control turning by utilizing wavelet transform of dynamic cutting forces", Journal of Engineering Manufacture, pp. 1-16, 2016
[2] D. Chanthana and S. Tangjitsitcharoen, "A Study of Relation between Roundness and Cutting Force in CNC Turning Process", Appl Mech Mater, vol. 799-800, pp. 366-371, 2015
[3] S. Tangjitsitcharoen and M. Sassantiwong, "In-Process Prediction of Straightness in CNC Turning by Using Wavelet Transform", 2nd International Conference on Green Materials and Environmental Engineering, pp. 199-203, 2015
[4] G. Jianliang, H. Rongdi, "A united model of diametral error in slender bar turning with a follower rest", 2006
[5] B. Kilic, J. Aguirr-Cruz, S. Raman, "Inspection of the cylindrical surface feature after turning using coordinate metrology", 1893-1903, 2007
[6] R. Han, B. Cui, and G. Jianliang, "In-Process Monitoring of Dimensional Errors in Turning Slender Bar" LNCS, 4402, pp. 277–286, 2007
[7] X. Bi, Yongxian Liu, Yang Liu, "Analysis and control of dimensional precision in turning process", Chinese Control and Decision Conference, 2009
[8] A.M. Shawky, M.A. Elbestawi, "In-process evaluation of workpiece geometrical tolerances in bar turning", Int. J. Mach. Tools Manufact, Vol. 36, pp. 33-46, 1996
[9] F. Sheng-bo, et al, "Prediction of diameter errors compensation in bars turning", Journal CSUT Vol.12, p.264-268, 2005
[10] Z. Liu1, P. Zhang, G. Peng and X. Ai1, "Surface Roughness in High Feed Turning with Wiper Insert", Key Engineering Materials, Vols. 375-376, pp. 406, 2008
[11] C.L.Du, C.X. Luo, Z.T. Han, and Y.S. Zhu, "Applying particle swarm optimization algorithm to roundness error evaluation based on minimum zone circle", Measurement 52, pp.12–21, 2014

Structure optimization of a micro drill bit with nonlinear constraints considering the effects of eccentricity, gyroscopic moments, lateral and torsional vibrations

Danh-Tuyen Nguyen[1,a], Tien-Dat Hoang[2,b] and An-Chen Lee[1,c]

[1]Department of Mechanical Engineering, National Chiao Tung University, Taiwan, R.O.C

[2]Faculty of International Training, Thai Nguyen University of Technology, Vietnam

[a]tuyen3586@gmail.com, [b]hoangdat@tnut.edu.vn, [c]aclee@mail.nctu.edu.tw

Abstract. A micro drill structure was optimized to give minimum lateral displacement at its drill tip, which plays an extremely important role on the quality of drilled holes. A drilling system includes a spindle, chuck and micro drill bit, which are modeled as rotating Timoshenko beam elements considering axial drilling force, torque, gyroscopic moments, eccentricity and bearing reaction force. Based on our previous work, the lateral vibration at the drill tip is evaluated. It is treated as an objective function in the optimization problem. Design variables are diameter and lengths of cylindrical and conical parts of the micro drill, along with nonlinear constraints on its mass and mass center location. Results showed that the lateral vibration was reduced by 15.83 % at a cutting speed of 70000 rpm as compared to that for a commercial UNION drill. Among the design variables, we found that the length of the conical part connecting to the drill shank plays the most important factor on the lateral vibration during cutting process.

1. Introduction

Nowadays, with the development of science and technology, there is a huge demand for micro drill bits to machine micro holes. Hence, many researchers have been making efforts to improve their cutting performance. The quality of micro drills is evaluated by many factors, for instance, material, burrs, chip removing, accuracy of drilled holes, length to diameter ratio, tool lifetime, etc. Several researchers investigated these factors in the past. To name a few, new material was created, e.g., the metal matrix composite of nickel ions and fine diamond powder was created to promote wear resistance of micro drills [1,2]. With coating technique, surfaces of micro drills were coated by special material to improve cutting performance. For example, the approach of coating Zr-C:H:N17% extended the tool lifetime by a factor of at least four compared to the uncoated micro drill, and obtained better cutting performance [3-5]. On the other hand, since drill point angle has a great effect on the tool lifetime of a micro drill, many researches have been conducted on this area [6]. The optimization of the geometry of micro drill point has been investigated theoretically and experimentally. In particular, helix angle and web thickness of a micro drill were optimized by Taguchi methods and response surface methodology to promote its cutting performance [7]. With the development mathematical formulation for the drill flute and geometrical features, the optimization was implemented by software Pro/Machanica to minimize the Von Mises stress of the micro drill below the compressive strength of the material property [8]. Generally, all these methods only focused on material, geometrical point angles, and flute of the drill to enhance its cutting ability. However, studies on the other portion of the drill bit, such as shank, and conical parts have been paid little attention in the past.

The structure of a micro drill is to be optimized to obtain the least lateral vibration at its drill tip for better performance in this research. Design variables are the diameter, and lengths of its cylindrical and conical parts. Constraints are its length, mass as well as the mass center location. The optimization problem is solved by function Fmincon in software Matlab. Results showed that the lateral vibration was reduced by 15.83% at the cutting speed of 70000 rpm as compared to that of a commercial UNION drill. From the sensitivity test, it is found that the length of the conical part connecting to the micro drill shank play the most important role on the lateral displacement during cutting process.

2. Optimization of the micro drilling system

The structure of the micro drilling system can be simplified as shown in figure 1(a). The micro drill E is grasped in the chuck C-D that is connected to the spindle A by a small shaft B. The system is supported by two identical linear bearings, and applied by a motor torque T_m for running the system at rotating speed Ω. It is assumed that the tip of the micro drill is applied by an axial force F_a, and a torque T at the steady-state during drilling machining. The whole system dimensions and parameters are referred from previous works [9-11]. Because the axial displacement is much smaller than the other displacements, it is neglected in this research [12-14]. The micro drill, chuck and spindle of micro drilling system are modeled as rotating Timoshenko beam elements to evaluate the lateral displacement at the drill tip. This vibration has a great impact to the accuracy of drilling process, so it is chosen as the objective function of optimization problem. Hence, the objective function is set as the common logarithm of lateral amplitude.

$$F = \log_{10}(\text{lateral amplitude}) \tag{1}$$

where lateral amplitude = maximum value of $\sqrt{x^2 + y^2}$, in which x and y are evaluated in our previous work [11]. The vibration at the micro drill tip is affected by many factors, such as micro drill structure, chuck, machine spindle, and cutting velocity, and so on. The micro drill structure is mainly concerned in this research. The design variables are lengths of cylindrical and conical parts of the micro drill (X_1, X_2, X_3, X_4), and the diameter of the cylindrical part (X_5) as shown in figure 1(b).

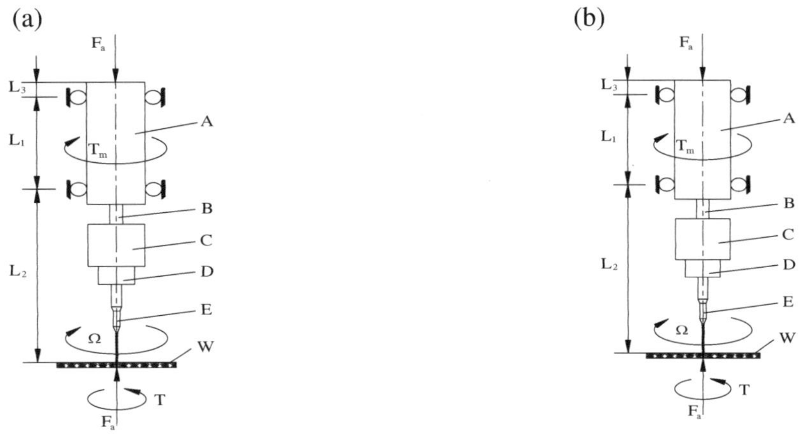

Figure 1. (a) Structure of micro drilling system, (b) design variables

The length, mass, and mass center location of the micro drill are constrained in the optimization problem. For reducing the vibration during cutting process, the drill mass location needs to be constrained inside the tool holder of the machine spindle. The drill mass is set as constant. The micro drill length L_d is set as variable within $[L_d^1, L_d^2]$, where, L_d^1 and L_d^2 are given parameters. The above nonlinear and linear constraints are described in the following:

$$L_d^1 \le L_d = L_g + L_f + X_1 + X_2 + X_3 + X_4 \le L_d^2 \ , \tag{2}$$

$$\begin{cases} f_m(X_1, X_2, X_3, X_4, X_5) = m_0 \\ L_{mass\_center}(X_1, X_2, X_3, X_4, X_5) \le L_g \end{cases} , \tag{3}$$

where $f_m(X_1, X_2, X_3, X_4, X_5)$ and $L_{mass\_center}(X_1, X_2, X_3, X_4, X_5)$ are the functions of mass and mass center location of the micro drill, respectively. The notation m_0 is the mass of the micro drill. The notations of L_d, L_g, and L_f represents lengths of the micro drill, the shank part grasped by the machine spindle, and flute part, respectively.

The micro drill mass function $f_m(X_1, X_2, X_3, X_4, X_5)$ in Eq. 3 is the sum of the drill element masses. A drill element mass m_i^e is evaluated by

$$m_i^e = \rho \cdot A \cdot L_{drill\_element} \ , \tag{4}$$

where ρ, A, and $L_{drill\_element}$ are the mass density, cross sectional area, and element length, respectively. The cross sectional area A is derived from our previous work [11].

The mass center location function $L_{mass\_center}(X_1, X_2, X_3, X_4, X_5)$ in Eq. 3 is calculated as follows:

$$L_{mass\_center}(X_1, X_2, X_3, X_4, X_5) = \frac{\sum_{i=1}^{N} m_i^e \times \left(L_i^e + \dfrac{L_{drill\_element}}{2} \right)}{m_0} \ , \tag{5}$$

where L_i^e is the position evaluated from the original B as shown in figure 1(b) of drill element i.

The optimization problem is resolved by Matlab function Fmincon to solve constrained linear and nonlinear optimization problems. The interior-point algorithm is chosen to solve the optimization problem [15-18].

A commercial Union micro drill is adopted for benchmarking and its parameters denoted as original variables are given in tables 1 and 2. The lower and upper bounds, as well as starting values of design variables are given in table 2. Applying the parameters in tables 1 and 2 and, objective function Eq. 1, and constraint functions Eqs. 2 and 3, one obtains the results as shown in figure 2(a). The optimum objective function is $F_{\text{Optimal}} = \text{-}5.4932$. The values of design variables are shown in table 1. The length of the optimal micro drill is

$$L_{d\_new} = L_g + X_1 + X_2 + X_3 + X_4 + L_f = 19.67 + 4.13 + 7 + 2.5 + 2 + 1.7 = 37 \text{ [mm]}. \tag{6}$$

Its mass and mass center location are $1.7g$ and 11.75 mm, respectively. The percentage reduction ΔX in table 1 is calculated by

$$\Delta X = \frac{X_i^{op} - X_i^{o}}{X_i^{o}} \times 100\% \ , \tag{7}$$

where X_i^o, and X_i^{op} are values of original and optimal variables, respectively. "+" and "-" signs denote that the increase and decrease, respectively, of the optimal value in comparison with the original value. The original objective function is $F_{Origianl} = \text{-}5.4183$, so the reduction percentage of lateral amplitude at the micro drill tip is

$$\Delta F = \frac{\text{Lateral amplitude(Optimal)-Lateral amplitude(Origional)}}{\text{Lateral amplitude(Origional)}} \times 100\% = -15.83\% \ . \tag{8}$$

IOP Conf. Series: Materials Science and Engineering **241** (2017) 012029 doi:10.1088/1757-899X/241/1/012029

Table 1. Values of design variables

Design variable	Lower bound (mm)	Upper bound (mm)	Starting value (mm)	Original variable value (mm)	Optimal variable value (mm)	Percentage deviation (ΔX) (%)
X_1	2	10	6	5	4.13	-17.40
X_2	2	7	4.5	4.43	7	+58.01
X_3	2.5	8	5.25	6.03	2.5	-58.54
X_4	1	2	1.5	1.27	2	+57.48
X_5	0.132	2.86	1.49	0.8	0.58	-27.50

Table 2. Parameters of the micro drilling system

Original Micro Drill	Radius R (mm)	Helical angle h_0	Semi point angle ρ_1	Length L_d (mm)	Flute length L_f (mm)	Drill mass m_0 (g)
	0.06	30^0	65^0	38.1	1.7	1.7
	Mass center location L_{mass_center} (mm)	Grasped length L_g (mm)	Cross section area of flute A (m^2)	Area moment of inertia of cross-section flute		
				I_u (m^4)	I_p (m^4)	I_v (m^4)
	11.9	19.67	4.929×10^{-9}	1.242×10^{-18}	8.783×10^{-18}	7.541×10^{-18}
	Young's modulus E (N/m^2)	Shear modulus G (N/m^2)	Mass density ρ (Kg/m^3)	Timoshenko's shear coefficient k_s	Eccentricity	
					e_u (mm)	e_v (mm)
Flute Part	650×10^9	280×10^9	15250	0.835	6×10^{-7}	-6×10^{-7}
Other Parts	200×10^9	90×10^9	7800	0.85	$e_u = 10^{-5}\cdot r,\ e_v = -10^{-5}\cdot r$, where r is radius of each cross section.	

Bearing stiffness	Bearing damping
$K_{xx}= K_{yy}=7\times10^7$, $K_{xy}= K_{yx}= K_{\theta xx}=0$ [N/m]	$C_{xx}= C_{yy}=1.7513 \times 10^3$ [Ns/m]
$K_{\theta xy} = K_{\theta yy} = K_{\theta yx} =0$, $K_{\varphi}=3 \times 10^4$ [Nm/rad]	$C_{xy}= C_{yx}= C_{\theta xx}= C_{\theta xy} = C_{\theta yy} = C_{\theta yx} =0$ [Nms/rad]
	$C_{\varphi}=1$ [Nms/rad]

Order of whirling at the drill tip n=2 , $\Omega = 7000$ [rpm], $F_a = 10^{-5}$[N], $T = 10^{-7}$[Nm],

$L_d^1 = 37$ [mm], $L_d^1 = 39$ [mm],

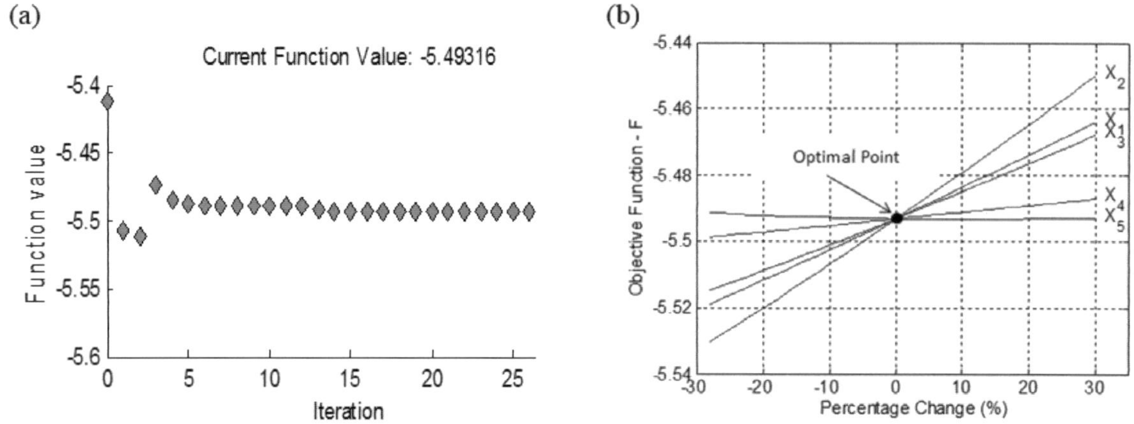

Figure 2. (a) Optimization results, and (b) sensitivity of design variables

The sensitivity of design variables to the objective function is carried out to examine their effect on the lateral displacement. As shown in figure 2(b), horizontal axis represents percentage change of design variables calculated by

$$X_i = X_i^O \pm \delta \cdot X_i^O \quad i = \overline{1,5} \, , \tag{9}$$

where X_i, X_i^O and δ are the design variable, optimal value of a design variable and percentage change, respectively, whereas vertical axis represents the objective function values corresponding to each design variable.

Since the lengths of cylindrical and conical parts of the micro drill affect the drill stiffness, i.e., the longer the lengths are, the smaller the stiffness is, increasing these lengths will enlarge the lateral vibration at the drill tip during cutting process. Consequently, the objective functions corresponding to design variable X_i, $i = \overline{1,4}$, the lengths of cylindrical and conical parts, increase when the percentage change increases as shown in figure 2(b). The increase of the cross sectional diameter makes the unbalance force and stiffness to increase. These two factors counteract with each other on the lateral displacement. Therefore, the objective function corresponding to the shank diameter (X_5) does not change much. One observes that the design variable X_2 has the highest sensitivity. In consequence, compared with other design variables, the length of conical part connecting to the shank has the greatest influence on the lateral vibration at the drill tip. This finding is very valuable in micro drill design.

3. Conclusions

The structure optimization of the micro drill bit was presented to give better cutting quality as compared to the original one. The results showed that the lateral vibration at the micro drill tip was decreased by 15.83% at the cutting speed of 70000 rpm. Compared with other design variables, the length of the conical part connecting to the drill shank is the most important factor that affects the lateral displacement at the drill tip. Furthermore, the usage of the transfer matrix approach to analyze the dynamics of the micro drilling system modeled as rotating Timoshenko beam elements with effects of axial force, torque, gyroscopic moments, and eccentricity provides an effective way in the design of micro drill.

Acknowledgments

This work was supported by the Ministry of Science and Technology of Republic of China, under Contract MOST 104-2221-E-009 -027 –MY2.

References

[1] H. Shi and H. Li 2013 Circuit World. **39** 75

[2] L. Fu, J. Qu, and H. Chen 2007 Circuit World. **33** 3
[3] W. H. Kao 2007 Surf. Coat. Tech. **201** 7392.
[4] C. W. Chu, J. S. C. Jang, H. W. Chen, and T. L. Chuang 2009 Thin Solid Films. **517** 5197.
[5] S. H. Yao, Y. L. Su, W. H. Kao, K. W. Cheng, and C. T. Su 2010 J. Mater. Process. Technol. **210** 660
[6] M. Irfan, S. N. Mehdi, N. Seetharamaiah, S. I. Sadaq, and M. A. Raheem 2014 Analysis **3**
[7] H.-S. Yoon, R. Wu, T.-M. Lee, and S.-H. Ahn 2011 Int. J. Precis. Eng. Manuf. **12** 871
[8] T. M. Bogale, F.-J. Shiou, and G.-R. Tang 2015 Arab. J. Sci. Eng. **40** 1497
[9] P. Yongchen, T. Qingchang, and Y. Zhaojun 2006 Int. J. Mach. Tool. Manu. **46** 1892
[10] S.-C. Hsieh, J.-H. Chen, and A.-C. Lee 2006 J. Sound Vib. **289** 294
[11] A.-C. Lee and D.-T. Nguyen 2016 MATEC Web of Conferences. **61** 01024
[12] H. P. Lee 1995 J. Sound Vib. **181** 169
[13] D.-M. Ku 1998 Mech. Syst. Signal Process. **12** 599
[14] Q. H. Qin and C. X. Mao 1996 Comput. Struct. **58** 835
[15] Global Optimization Toolbox User's Guide 2015
[16] Q. Wei, A. J. Flueck, and T. Feng 2005 IEEE Power Engineering Society General Meeting. **1** 447
[17] O. Stein and G. Still 2003 Siam J. Control Optim. **42** 769
[18] A. T. Azevedo, A. R. L. Oliveira, and S. Soares 2008 Ann. Oper. Res. **169** 55.

IOP Conf. Series: Materials Science and Engineering **241** (2017) 012030 doi:10.1088/1757-899X/241/1/012030

High speed tracking control of ball screw drives

Chao-Yi Liu[1,a], Ruei-Yu Huang[b] and An-Chen Lee[c]

[1]National Chiao Tung University, Department of Mechanical Engineering 30010 Hsinchu, Taiwan

[a]clotha16622@hotmail.com, [b]b2481b2481@gmail.com, [c]aclee@mail.nctu.edu.tw

Abstract. This paper presents a new method to achieve the requirement of high speed and high precision for ball screw drive. First, a PI controller is adopted to increase the equivalent structural damping in the velocity loop. Next, the design of the position controller is implemented by a two-stage method. The Doubly Coprime Factorization Disturbance Observer (DCFDOB) is developed to suppress disturbance and resist modelling error in the inner loop, while the outer loop is then designed based on method to extend the system bandwidth over first resonant frequency so that high speed and high accuracy can be achieved. Finally, a feedforward controller is implemented to improve tracking performance. The experiment results showed that the proposed method has smaller tracking error and better performance for suppressing disturbance when compared to the conventional cascaded P-PI control.

1. Introduction

Owing to the multiple transmission, the structure of the ball screw system is elastic, which makes actual movement has few differences from input command. These errors will also be magnified during fast motion. Moreover, the machine positioning accuracy will be affected by an external force when machining the workpiece. Therefore, how to fulfill the demand of high speed and high precision is very important in motion control technology.

A. Matsubara et al. [1] and M.S. Kim et al. [2] discussed a common ball screw system model by considering different coupling stiffness effects. S. Frey et al. [3] proposed a discrete model to derive the relationship of the table position and the first resonant frequency. External disturbance and measuring noise are important factors in the motion control. G. L. Luo et al. [4] used acceleration feedback in optimal control to compensate the effect of nonlinear term. K. Ohnishi et al. [5] proposed disturbance observer to suppress the effect from disturbance, noise and uncertainty to system. M. Zatarain et al. [6] predicted positon, velocity and acceleration based on state space observers and used these signals in feedback loop. As to other approaches, the interested reader may refer to [7-9].

The above-mentioned methods, however, have drawbacks to some degree. First, the designed disturbance observer is not optimum in sense of maximum bandwidth under modelling error in [5]. Second, some of them only work well in low frequency range in [7]. Thirdly, most of them did not consider the system change under different table positions except [9] and the others need additional sensor to set up controller [4]. To solve the above problems, this paper presents a new method to achieve the requirement of high speed and high precision for ball screw drive operating higher than the first resonant frequency.

2. System Model

The ball screw drive is actuated by applying voltage to a servomotor, which generates a torque to the screw connected to the servomotor by a coupling. Figure 1 shows the dynamic model of the ball screw system. From figure 1, we can obtain the transfer function from torque to table displacement:

$$\frac{x_t}{T} = \frac{RK}{JM_t s^3 + (M_t D + JC_t)s^2 + (JK + C_t D + R^2 KM_t)s + (R^2 KC_t + KD)} \cdot \frac{1}{s} \quad (1)$$

where T is the input driving torque, x_t is the table displacement, M_t is the table mass, J is the rotary inertia, K is the equivalent longitude stiffness, C_t is the damping coefficient of the guideways, D is the damping coefficient of the rotary motion, F_t is the force to the table, and R is the transform ratio from shaft angle to table position.

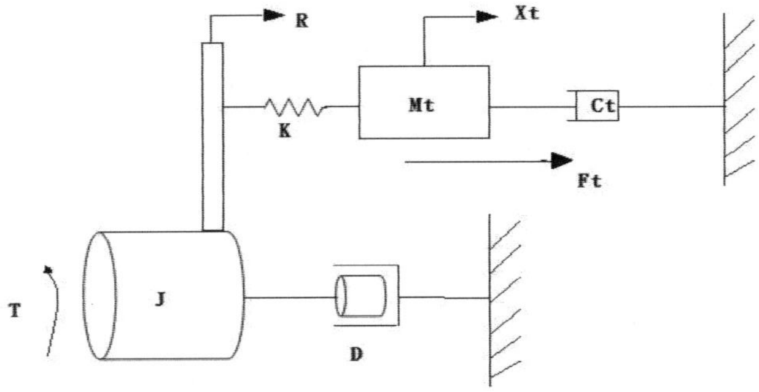

Figure 1. A dynamic model of ball screw system.

The equivalent stiffness K changes when table moves. This parameter is the major factor on machine vibration. However, as the second resonant frequency is much higher than the first one, we only discuss and design controller for the first resonant frequency. To design the controller easier, we add a PI controller to increase equivalent damping term. The parameters P and I are set 50 and 12, respectively, in underlying experiment. Because the system characteristics change with position, we divide the range of the ball screw into four regions and identify the model independently by using MATLAB system identification toolbox. The nominal plant is chosen by the minimum of the maximum of the gap metric between all models. The obtained nominal plant $G(s)$ is in equation (2) with its bode diagram shown in figure 2.

$$G(s) = \frac{10312s + 4.526 \times 10^6}{s^3 + 313.9s^2 + 5.446 \times 10^4 s + 4.642 \times 10^6} \quad (2)$$

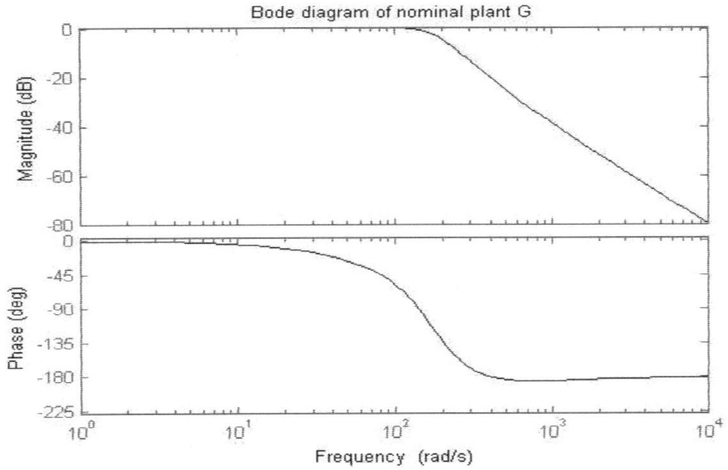

Figure 2. Bode diagram of the nominal plant.

3. Controller Design

3.1 DCFDOB Structure

Figure 3 shows the structure of DCFDOB, where the command reference u, the input disturbance d_i, the output disturbance d_o, the estimated input disturbance $\tilde{d}$, the system outputs y, the noise n, internal states e_r, e_d, e_n. The actual plant $P(s)$ is described as right coprime factorization as follow:

$$P=(N_n+\Delta_N)\cdot(M_n+\Delta_M)^{-1},(N_n,M_n,\Delta_N,\Delta_M\in RH_\infty) \tag{3}$$

Other factors of corresponding coprime factorization satisfy Eq. 4 as well.

$$\begin{bmatrix} X_r & Y_r \\ -\tilde{N}_n & \tilde{M}_n \end{bmatrix}\begin{bmatrix} M_n & -Y_l \\ N_n & X_l \end{bmatrix}=\begin{bmatrix} M_n & -Y_l \\ N_n & X_l \end{bmatrix}\begin{bmatrix} X_r & Y_r \\ -\tilde{N}_n & \tilde{M}_n \end{bmatrix}=I \tag{4}$$

where $\tilde{M}_n(s),\tilde{N}_n(s)\in RH_\infty$ are left coprime factorizations of nominal plant ($P_n=\tilde{M}_n^{-1}\tilde{N}_n$), and $X_r(s),Y_r(s),X_l(s),Y_l(s)\in RH_\infty$. The notation RH_∞ is the space of all real rational transfer function matrices without right-half-plane poles. From Fig. 3, the output y is expressed as:

$$\begin{aligned} y&=P_n\cdot r+P_n(I-Q(I-M_nX_r))\cdot(d_i-z_1)+(I-P_nQM_nY_t)\cdot(d_o+z_2)-P_nQM_nY_t\cdot\xi \\ &=P_n\cdot r+\tilde{N}_n^{-1}(I-\tilde{N}_nQY_l)\tilde{N}_n\cdot(d_i-z_1)+\tilde{M}_n^{-1}(I-\tilde{N}_nQY_l)\tilde{M}_n\cdot(d_o+z_2)-P_nQM_nY_r\cdot\xi \end{aligned} \tag{5}$$

For suppressing disturbance, we design the controller $Q(s)=\tilde{N}_n^{-1}\cdot J\cdot Y_l^{-1}$ (where J is a low-pass filter) to make $\tilde{N}_n^{-1}(I-\tilde{N}_nQ_DY_l)\tilde{N}_n\approx 0$ and $\tilde{M}_n^{-1}(I-\tilde{N}_nQ_DY_l)\tilde{M}_n\approx 0$ in an acceptable low-frequency range. Then, equation (5) can be rewritten as

$$y\approx P_n\cdot r-P_nQM_nY_r\cdot\xi\,. \tag{6}$$

Thus, the DCFDOB makes the perturbed plant behave like a nominal plant in the low-frequency range and the outer loop controller $C(s)$ can be designed easily.

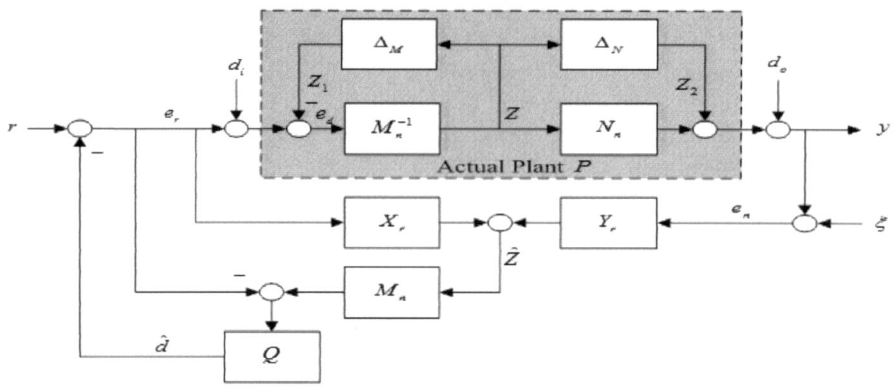

Figure 3. Block diagram of DCFDOB.

3.2 Cascade H_∞ Design Method

Figure 4 illustrates the control structure, including the inner-loop structure of DCFDOB and the outer-loop position controller. A design procedure which uses cascaded H_∞ design method is proposed [10]. The inner-loop and outer-loop controllers are then designed as

$$Q(s)=\frac{50.9s^3+8.568\times10^4s^2+2.509\times10^7s+2.041\times10^9}{s^4+797s^3+2.541\times10^5s^2+3.708\times10^7s+2.041\times10^9} \tag{7}$$

$$C(s)=\frac{2.07\times10^9s^4+5.59\times10^{12}s^3+1.66\times10^{15}s^2+2.80\times10^{17}s+2.29\times10^{19}}{s^5+1.45\times10^3s^4+7.65\times10^7s^3+2.16\times10^{11}s^2+3.02\times10^{14}s+9.69\times10^{16}} \tag{8}$$

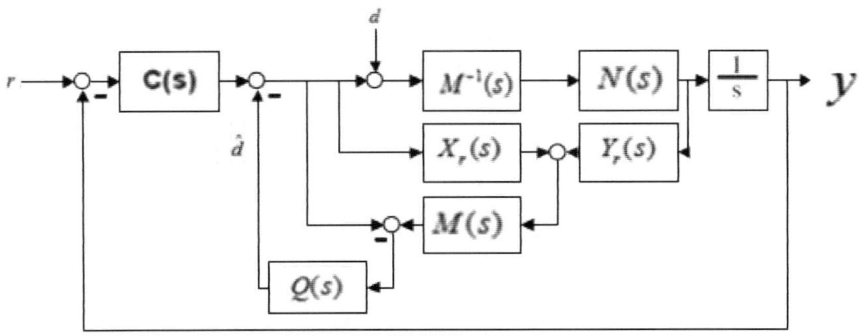

Figure 4. The structure of our controller.

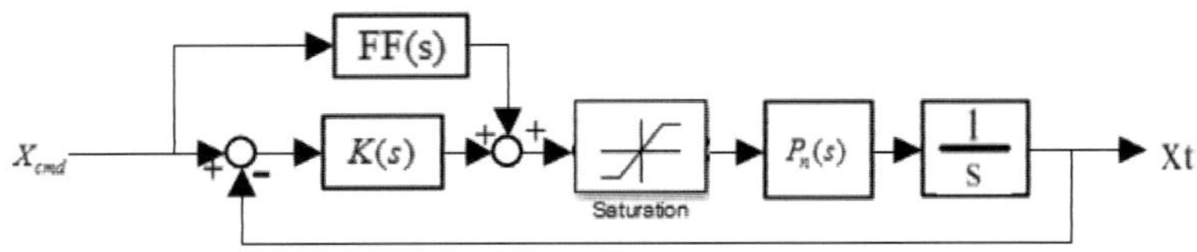

Figure 5. The structure of feedforward controller.

3.3 Velocity Feedforward Controller

Feedforward controller is used to improve tracking performance. The structure is shown as Figure 5. The idea of feedforward controller is to make input command and output response perfectly match. So, we design a feedforward controller $FF(\mathrm{s})$ to satisfy equation (9) in low-frequency range. The obtained $FF(\mathrm{s})$ is given in equation (10).

$$\frac{x_t}{x_{cmd}} = \frac{(C(s)+FF(s))P_n(s)}{s+K(s)\cdot P_n(s)} = 1 \tag{9}$$

$$FF(\mathrm{s}) = \frac{s}{P_n(\mathrm{s})} = \frac{s(s+246.7)}{240.5}\cdot\left(\frac{1000}{s+1000}\right)^2 \tag{10}$$

4. Experimental Results

A cascaded controller consisting of position P controller and velocity PI controller called cascaded P-PI is designed as benchmarking. The experiment set-up is shown in figure 6 and the parameters of two controllers are shown in table 1. Three experiments are conducted with results shown in figure 7(a-f). Case 1 used zero input command but added an artificial disturbance signal in the control effort through D/A converter and case 2 added an actual disturbance by applying 350g mass hitting on the moving table. Case 3 performed point-to-point positioning using an S-curve input command determined by the displacement $S = 0.2\mathrm{m}$, maximum velocity $V_{\max} = 0.4\mathrm{m/s}$, maximum acceleration $A_{\max} = 1\mathrm{m/s^2}$, and average acceleration $A_{avg} = 0.6\mathrm{m/s^2}$.

Table 1. The parameters of two controllers.

Parameters	Bandwidth (rad/s)		K_{vi}	K_{vp}	K_{pp}
	Velocity-loop	Position-loop			
Cascaded H_∞	180	500	12	60	N/A
Cascaded P-PI	180	110	12	60	50

IOP Conf. Series: Materials Science and Engineering **241** (2017) 012030 doi:10.1088/1757-899X/241/1/012030

Figure 6. Experiment set-up

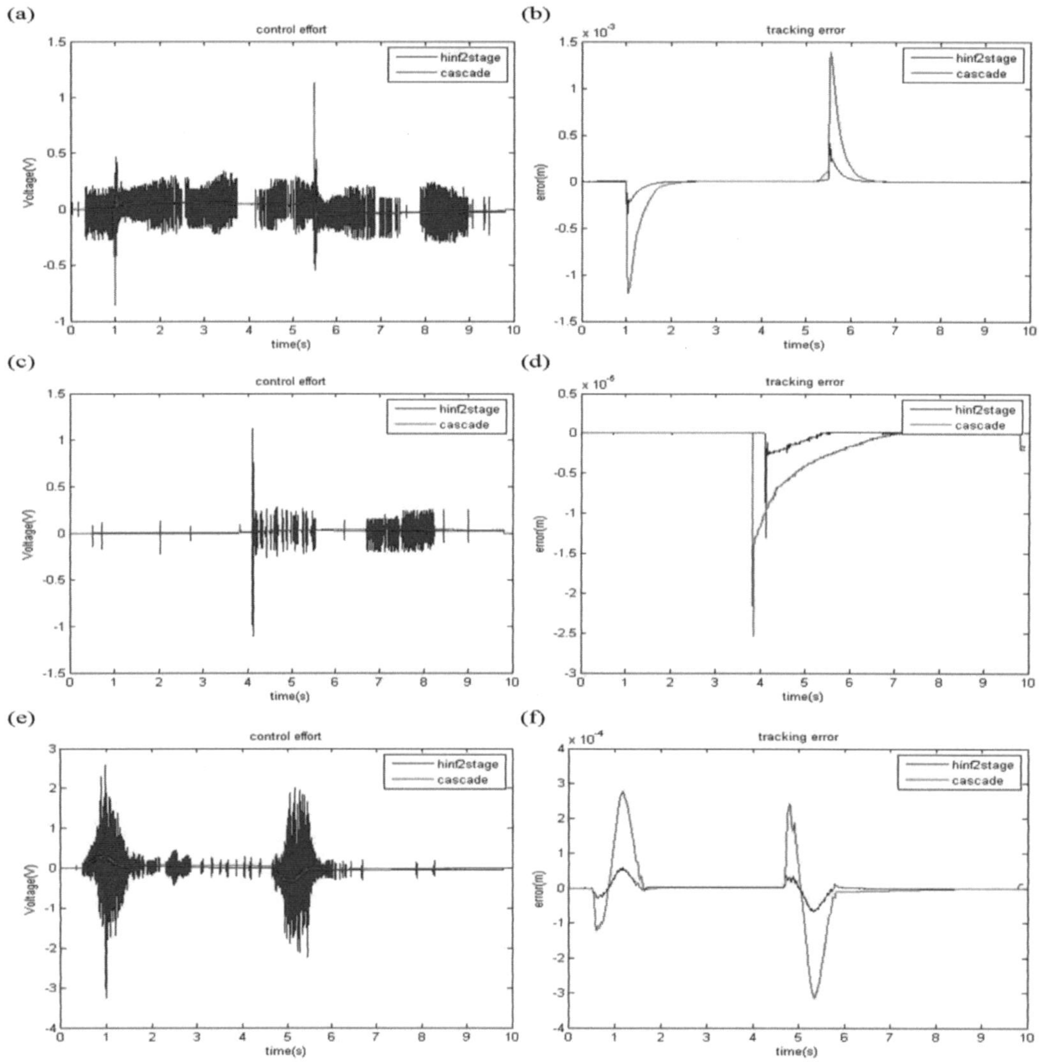

Figure 7. (a) Control effort of case 1, (b) tracking error of case 1, (c) control effort of case 2, (d) tracking error of case 2, (e) control effort of case 3, (f) tracking error of case 3.

Table 2. Result of each experiment

Case	Method	Control effort (V)	Position MSE (m^2)	Maximum tracking error (m)
1	Cascaded H_∞	6.2649×10^{-3}	2.3843×10^{-9}	4.120×10^{-4}
	Cascaded P-PI	4.0550×10^{-3}	5.2464×10^{-8}	1.386×10^{-3}
2	Cascaded H_∞	2.1790×10^{-3}	5.2837×10^{-13}	1.302×10^{-5}
	Cascaded P-PI	1.9790×10^{-3}	9.3056×10^{-12}	2.531×10^{-5}
3	Cascaded H_∞	9.7926×10^{-3}	2.5422×10^{-10}	6.675×10^{-5}
	Cascaded P-PI	6.2259×10^{-3}	6.4876×10^{-9}	3.147×10^{-4}

The position mean square error (MSE) and the maximum tracking error of the proposed method are smaller than those of the cascaded P-PI method. The position MSE is about 95.5% less than that of the cascaded P-PI method in case 1, about 94.3% in case 2 and about 96.1% in case 3. The maximum tracking error is about 70.6% less than that of cascaded P-PI method in case 1, about 48.6% in case 2 and about 78.3% in case 3. But the control effort of the method is larger; however, it is under saturation constraint, i.e., it is about 54.5% larger than that of cascaded P-PI method in case 1, about 10.1% in case 2 and about 57.3% in case 3. From the experiment results, our method demonstrates good performance at disturbance suppression and trajectory tracking. Because of no control effort constraint, our controller contains more high-frequency components which leads to faster response.

5. Conclusion

In this paper, a cascaded H_∞ design method for high speed tracking control of ball screw drives is presented. The experiment results show that this method has better performance in trajectory tracking and disturbance rejection than the cascaded P-PI design method. Though the control effort of our method is larger, it does not violate the driver constraint $\pm 10\text{V}$.

Acknowledgments

This work was supported by the Ministry of Science and Technology of Republic of China, under Contract MOST 104-2221-E-009 -027 –MY2.

Reference

[1] Matsubara, Y. Kakino and Y. Watanabe 2000 *Proc. of 2000 Japan-USA Flex. Auto. Conf*
[2] M.S. Kim and S.C. Chung 2006 *Mechatronics* **16** 491
[3] S. Frey, A. Dadalau and A. Verl 2012 *Product. Eng.* **6** 205
[4] G.L. Luo and G.N. Saridis 1985 *IEEE J. of Robotics and Automation* **RA-1** 152
[5] K. Ohishi, M. Nakao, K. Ohnishi, and K. Miyachi 1987 *IEEE Trans. on Ind. Elec.* **34** 44
[6] M. Zatarain, I. Ruiz de Argandoña, A. Illarramendi, J.L. Azpeitia and R. Bueno 2005 *Annals of CIRP* **54** 393
[7] A. Verl and S. Frey 2012 *Annals of CIRP* **61** 351
[8] K. Erkorkmaz and Y. Hosseinkhani 2013 *Annals of CIRP* **62** 387
[9] L. Dong, W.C. Tang and D.F. Bao 2015 *ISA Trans.* **55** 219
[10] C.T. Yen 2015 *Master thesis of National Chiao Tung University in Mech. Eng.*

IOP Conf. Series: Materials Science and Engineering **241** (2017) 012031 doi:10.1088/1757-899X/241/1/012031

Technical Feasible Study for Future Solar Thermal Steam Power Station in Malaysia

Z. H. Bohari[1], N. N. Atira[2], M. H. Jali[2], M. F Sulaima[2], T. A Izzuddin[2], M. F Baharom[2]

[1]Faculty of Engineering Technology, Universiti Teknikal Malaysia Melaka, HTJ, 76100 Malacca, Malaysia

[1]Faculty of Electrical Engineering, Universiti Teknikal Malaysia Melaka, HTJ, 76100 Malacca, Malaysia

E-mail: zulhasrizal@utem.edu.my

Abstract. This paper proposed renewable energy which is potential to be used in Malaysia in generating electricity to innovate and improve current operating systems. Thermal and water act as the resources to replace limited fossil fuels such as coal which is still widely used in energy production nowadays. Thermal is also known as the heat energy while the water absorbs energy from the thermal to produce steam energy. By combining both of the sources, it is known as thermal steam renewable energy. The targeted area to build this power station has constant high temperature and low humidity which can maximize the efficiency of generating power.

1. Introduction

Electricity is one of the main things that are needed in human's daily life [1, 2]. In the era of technologies, experts from related fields are working together to invent or to find out more ways to generate electricity instead of burning limited fossil fuel such as coal, natural gas and oil [3, 4, 5]. Renewable energy is in a form of source that continuously replenished when it is used for a long term such as the wind, hydro, biomass and solar power. These sources are required because a generator needs the energy to rotate the blade or turbine to generate electricity. In peninsular Malaysia, most of the power stations produce electricity from fossil fuels. Fossil fuel produces a huge amount of CO2 emission. Thermal steam power plant station is a power plant which converts heat energy to mechanical energy to electrical power. The research works covered the estimated cost of the project, type of fluid used to produce steam energy, the design of power plant, the design of turbine, total power that can be generated and location. The idea of developing potential renewable energy is required to overcome the problem of using limited natural resources such as burning fossil fuels to generate electricity because it is not continuously replenished once they are finished. To help in the growth of technology and economy of the country, a new idea of inventing a low cost and high efficiency of power plants are required as well [2, 5].

2. Methodology

Figure 1 shows a block diagram of the research process involved in this project.

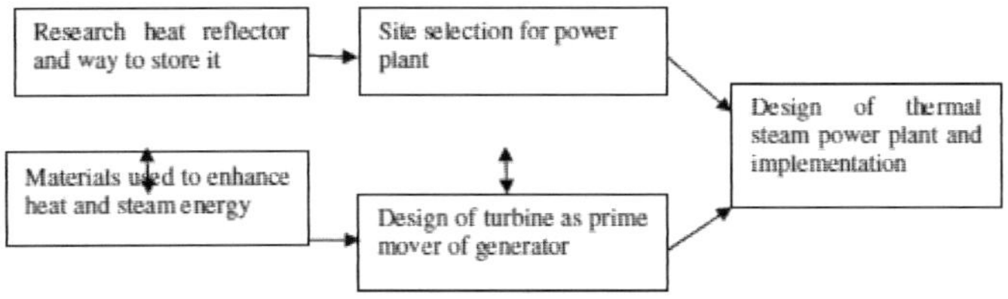

Figure 1. Research Processes

The parts that involved in the power plant project are heliostats, receiver, thermal energy storage tanks, air cooled condenser, steam generator, turbine and generator as Figure 2.

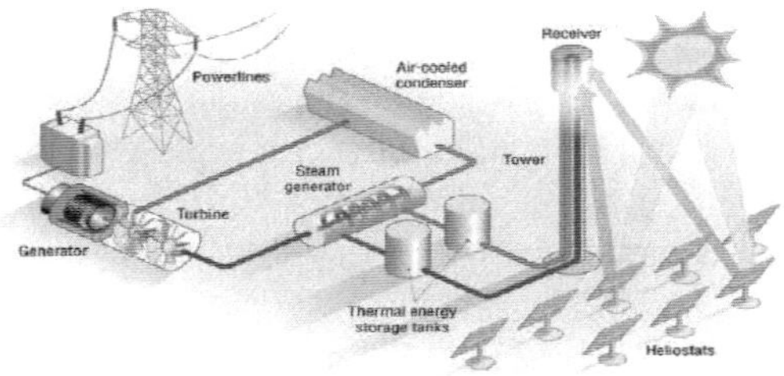

Figure 2. Design of thermal steam power plant

2.1 *Heliostat*

The heliostat is also known as power towers which are defined as an apparatus containing a movable or driven mirror which is used to reflect light to a fixed direction [6,7,8]. It shaped like a flat mirror which helps in sun tracking and focuses sunlight onto a receiver at the top of a tall tower. The heliostats are computer-controlled mirrors; it makes a small adjustment to position the mirror in order to keep the sunlight on the target as the sun moves across the sky. It is used for the production of concentrated solar power to generate electricity [4]. The design for heliostat's reflective components uses a second surface mirror which consists of a steel structural support, an adhesive layer, a protective copper layer, reflective silver and a top protective layer of thick glass to reflect sunlight fully.

2.2 Center Receiver Plant

The Center Receiver Plant is a type of solar furnace which is used a tower to receive the focused light from heliostats which are placed below it. The function of the receiver is to transform concentrated radiation into heat by transferring it to a fluid where water is chosen in this renewable energy proposal. Materials that will be used to build center receiver plant and heat pipe are by using copper [9]. The characteristic of copper is durable, corrosion-resistant, and strong and has a high melting point. Besides, copper also has a high thermal conductivity which is 394W/(mK). Heat transfers rapidly throughout the atoms in copper. Although silver has a higher thermal conductivity compared to copper in the aspect of price, silver cost much more which will then overall increased the cost of the project. Technology and absorber material of center receiver plant in concentrating solar power (CSP) system is capable of generating thermal power at a temperature greater than 650°C.

2.3 Thermal energy storage tank

The thermal energy storage acts like a battery for the power plant to overcome the problem of reduction in energy production when the weather is cloudy or rainy. In this project, two tank direct systems are applied. Solar thermal energy in this system is stored in the same fluid which is used to collect it [6]. The fluid is stored at one for high temperature and another one at low temperature. Fluid from low temperature flows through the solar receiver and it is heated, and then flows to the high temperature tank for storage [6, 8]. Fluid from the high temperature flows through a heat exchanger where steam is generated to produce electricity. Molten salt is used in the heat transfer. They are unique among solar electric technologies in the ability to efficiently store solar energy and deliver electricity to the grid when needed such as night or during cloudy weather.

IOP Conf. Series: Materials Science and Engineering **241** (2017) 012031 doi:10.1088/1757-899X/241/1/012031

2.4 Steam generator
The thermal energy storage tank stores thermal energy which is then transferred to water and converts to steam in the steam generator. The pressure and temperature keep on increasing till it becomes superheated steam energy which is strong enough to rotate the turbine of the power generator. Based on research, it may generate around 150 to 200 kg of steam in a day, depending on location and various other features that can save around 4500 liters of diesel in a year [9]. Meanwhile, to transfer steam energy, steam pipes are used.

2.5 Air cooled condenser
The air cooled condenser is needed in heat transferring system. It is a device used to condense a substance from gaseous to a liquid state. The steam condenser is installed on the exhaust steam from thermal power station where in this condenser; heat exchanger converts steam to the liquid state at a pressure below atmospheric pressure. The surface of the condenser which is normally used the water cooled shell and tube heat exchanger is installed on the exhaust steam. It is also acting as the separator of steam and water mixtures. Thus, steam is drawn out and distributed for further processes. Heating on saturated steam will make superheated steam to drive the steam turbine which is then used in generating electricity.

2.6 Turbine
Turbine, consists of a series of curved vanes or blades on a central spindle, is the device of a generating unit which is spun by the force of steam energy to drive an electric generator. In thermal steam power generation, steam is heated till it becomes superheated steam and causes the turbine to be spun at a very high pressure and temperature [7, 8]. A typical steam power plant steam turbine rotates at 1800-3600rpm. Non-condensing (back-pressure) and condensing steam turbine (extraction steam turbine) are the two types of steam turbine. Figure 3 below shows condensing steam turbine.

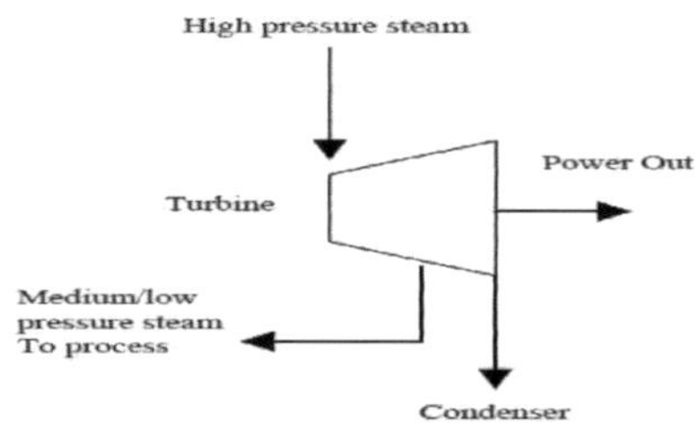

Figure 3. Diagram of flow of steam energy to turbine

It is most typically found in electrical power plants as these steam turbine exhaust steam in a partially condensed state. It has openings in its casing for extraction of a portion of steam at some intermediate pressure before condensing the remaining steam. The superheated vapor which contains high energy enters the turbine after it has exited from the boiler at high temperature and pressure. The high heated and pressured steam is converted into kinetic energy using the nozzle. Once the steam has exited the nozzle, it moves at high velocity and continues to flow to the turbine. A force is created on the blade due to the pressure of the vapor on the blades and causing them to rotate. Extraction steam turbine does not cool the steam so much, as it uses the remaining heat to make hot water in a system which known as combined heat and power (CHP).

2.7 Synchronous Generator
It is defined as a machine that converts mechanical power derived from steam, gas or hydraulic-turbine to alternating electric power. The chosen rotor design is non-salient pole which the utilization is for the high speed synchronous machine. As applied in the thermal steam power station, the prime mover of the generator will be rotated by steam energy which produced from the steam generator. However, the steam energy produces a very high pressure and temperature, thus direct-conductor cooling is needed by using water as coolant [9, 10].

Figure 4 shows the flow of diagram for three phase power emission. The turbine of the generator rotates and produces an alternating current which then flows to the rectifier to convert to direct current. It passed through a smoothing and filters circuit to produce a better characteristic of the waveform. Direct current continues to flow to an inverter which converts direct current to alternating current and finally the power is supplied to transmission tower to targeted areas.

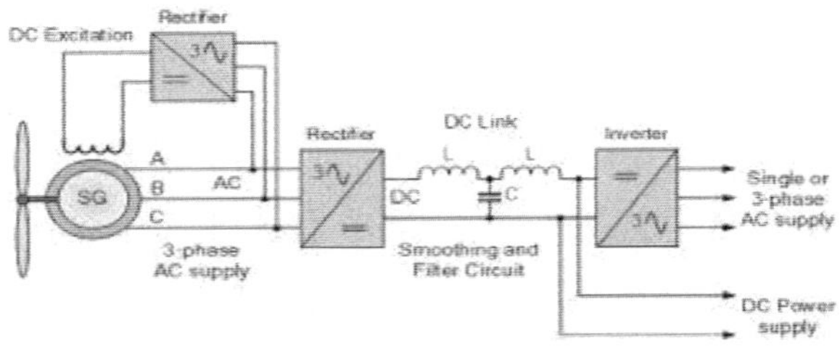

Figure 4. Flow of diagram for 3 phase power emission [10]

2.8 Location

The targeted location for this project must be a place with high temperature throughout the year [11]. The chosen location is Chuping, Perlis. Chuping is a small town of Perlis locating at the northeast of Kangar. According to Star News, Chuping has recorded the highest temperature of 40.1°C in the year 1998. On humidity research, Chuping received only 2000mm of rainfall annually which is then marked as one of the driest towns in Malaysia. Normally, the hottest and driest seasons occur from December to February.

3. Estimated cost

The cost estimation is based on the market research analysis on thermal steam power plants.

Parts	Cost (Ringgit Malaysia)
Heliostats	1.0mil
Receiver	600,000
Thermal energy storage tank	0.5mil
Steam generator	70,000
Air cooled condenser	400,000
Condensing turbine	0.5mil
Synchronous generator	30,000
Total	**3,000,000**

From the data analysis, the heliostats take up about 30% of the total investment of the project. Although the initial set up for the heliostat as the heat reflector to start up the power plant is high, but in the aspect of the long run, it is beneficial and profitable because it does not require any fossil fuels burning to produce the initial energy but only maintenance cost after it started to operate.

$$Payback\ Period\ (PP) = \frac{The\ Costs\ of\ the\ Project/\ Investment}{Annual\ Cash\ Inflows/\ Savings}$$

$$= \frac{RM\ 3\ 000\ 000}{RM\ 90\ 000}$$

$$= 33.33\ years$$

4. Results & Discussion

The total estimated power emission from the thermal steam power plant is targeted to be 1 MW which is able to supply the whole Chuping's street lights, homes and industry with enough electricity. One heliostat produces about 2000 watts, which by building 500 heliostats, the estimated power is 1 megawatts. The emission of power will be divided into 3 parts; one is the facilities on streets, housing area and industrial area. The electrical consumption for one LED Street light is 900mW, this project target is to build 2000 street lamps to upgrade the current's roads quality to enhance safeness and to promote a convenient journey for citizens and also tourists. The remaining power of 60% is supplied to the housing while 40% to the industrial area.

5. Conclusion

To conclude, the solar and steam energy is one of the most potential renewable energy sources. It is a backup power generation for the power station to work and to sustain balance and continuous electricity supply throughout the year. This project is aimed to be highly potential to work in Chuping, Perlis, Malaysia.

Acknowledgement

The authors like to acknowledge Universiti Teknikal Malaysia Melaka (UTeM) and Ministry of Higher Education, Malaysia for the financial support through RAGS/1/2014/TK06/FKE/B00052 for the research fund.

References

[1]Chapman, S. J., Electric Machinery Fundamentals, 5th Edition, New York: McGraw-Hill, 2012, pp. 192-224.
[2]Z. H. Bohari, H. S. Azemy, M.H. Jali., M.N.M. Nasir, M.F. Baharom, M.F. Sulaima, "Reliable short term load forecasting using self organizing map (SOM) in deregulated electricity market", Journal of Theoretical and Applied Information Technology, vol 79(3), pp 3898-394, 2015
[3]Bent Sorensen, Renewable Energy (Volume I), Earthscan 2011.
[4] Z.H. Bohari, M.F. Baharom, M.H. Jali, M.F. Sulaima, W.M. Bukhari, "Are Tidal Power Generation Suitable as the Future Generation for Malaysian Climate and Location: A Technical Review?", International Journal of Applied Engineering Research, Vol. 11(3), pp 7095-7099, 2016.
[5]MF Sulaima, SN Othman, MH Jali, MS Jamri, MNM Nasir, ZH Bohari, "A 33kV Distribution Network Feeder Reconfiguration by Using REPSO for Voltage Profile Improvement" , International Journal of Applied Engineering Research (IJAER), vol 9(18), pp 4569 - 4582, 2014
[6]Heinz P. Bloch, A Practical Guide to Steam Turbine Technology, New York: Mc-Graw-Hill, 1996.
[7]Rolf Kehlhofer, Combined-Cycle Gas & Steam Turbine Power Plants, PennWell Books, 2009.
[8]Rohr, B., The Promise of Small Heliostats, Northeast Sun, Spring 2009.
[9] Adrian Bejan, Allan D. Kraus, Heat Transfer Handbook, John Wiley & Sons, Inc, 2003.
[10] Matthew L. Wald, Costs Surge For Building Power Plants, The New York Times, July 2007.
[11] Mahlia, T.M.I., Emmision From Electricity Generation in Malaysia, Renewable Energy 2002.

IOP Conf. Series: Materials Science and Engineering **241** (2017) 012032 doi:10.1088/1757-899X/241/1/012032

Methods for automated semantic definition of manufacturing structures (mBOM) in mechanical engineering companies

Prof. Alexander Stekolschik[1, a]
TH Köln, University of Applied Sciences, Betzdorfer Straße 2, Cologne, Germany
[a]alexander.stekolschik@th-koeln.de

Abstract. The bill of materials (BOM), which involves all parts and assemblies of the product, is the core of any mechanical or electronic product. The flexible and integrated management of engineering (Engineering Bill of Materials [eBOM]) and manufacturing (Manufacturing Bill of Materials [mBOM]) structures is the key to the creation of modern products in mechanical engineering companies. This paper presents a method framework for the creation and control of e- and, especially, mBOM. The requirements, resulting from the process of differentiation between companies that produce serialized or engineered-to-order products, are considered in the analysis phase. The main part of the paper describes different approaches to fully or partly automated creation of mBOM. The first approach is the definition of part selection rules in the generic mBOM templates. The mBOM can be derived from the eBOM for partly standardized products by using this method. Another approach is the simultaneous use of semantic rules, options, and parameters in both structures. The implementation of the method framework (selection of use cases) in a standard product lifecycle management (PLM) system is part of the research.

1. Introduction

A BOM is an extensive list of parts (components), assemblies, raw materials, and linked documents required to describe a product from different points of view (e.g. engineering design, manufacturing, and service). The product structure (BOM) is usually represented in a hierarchical format, with the top level displaying the finished product and the bottom level showing individual parts and materials [1]. The engineering structure (eBOM) is created by the product development groups [2]. It represents the so-called ‘as designed’ structure and is mostly customer order-neutral. EBOM composition represents the multidisciplinary view of the functional definition of the product and is similar to the 3D CAD assembly structure. The eBOM is usually combined with related product specification documents, CAD models, non-CAD parts (e.g. oil), drawings, and annotations that should give a clear definition of the product. MBOM is a configuration of the product to show how it will be manufactured, assembled, serviced, and the like. An mBOM typically consists of a structured list of assemblies and sub-assemblies that describe the manufacturing process. This ‘as shipped’ structure is based on eBOM (it consumes eBOM items). But most of the time, it is not the same as the eBOM, because of the manufacturing process and localization. Some process-based reasons for the structural differences are (Fig. 1):

- Make or buy decision: Flat BOM structure for outsourced assemblies, including raw parts for insourced parts
- Site-related manufacturing processes: Existing machine fleet or tool sets, specific manufacturing procedures and different staff (NC programmers, operators, etc.)

- Local logistics chain: Deployment of locally procurable materials, raw parts, and suppliers
- Mechanical assembly procedures: Assembly structure based on easy accessibility and assembly order

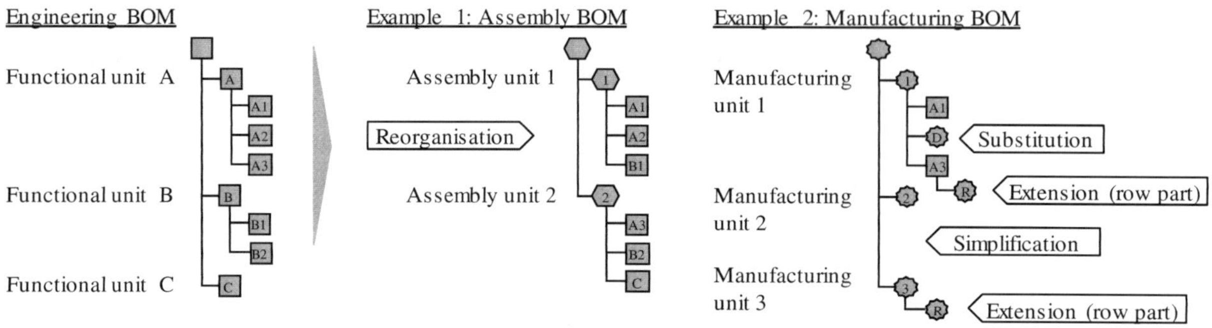

Figure 1: Examples of eBOM and manufacturing structures

The mBOM includes the equipment and tools required to build the product's components. It also includes details of packaging materials, user manuals, and brochures that make the product shippable. All these details help the manufacturing team better understand the design assembly and packaging procedures.

The differences between the engineering and the manufacturing structures often lead to a big effort in coordination along the creation process. The mBOM is mostly (logically) derived from the appropriate eBOM structure. But in most cases, it is created *manually* in the ERP or special BOM system. This results in time-consuming activities and errors. Moreover, the manual process makes efficient engineering change management difficult or even impossible [3].

2. MBOM creation process analysis

The mBOM definition process was analysed in the first step of the research. Different logics have been identified for the creation of the structure, depending on the process initiator.

2.1 Product structure (eBOM)-driven mBOM definition

This process is mostly used for mBOMs with structures similar to or even identical to those of eBOM. The definition of the eBOM or the eBOM configuration is done in the first step in engineering. Many companies copy their eBOMs and change the structure in the next step, i.e. manufacturing. The change types are (Fig. 1): reorganization of virtual assemblies (creation of new assembly nodes, movement of parts), structure simplification (flattening), structure extension (raw parts, blanks), and part substitution (global vs. local). In some cases, generic mBOM templates are used. These are reconfigured after the eBOM configuration. These processes are eBOM-driven, which means that the main part of the mBOM structure follows the eBOM composition, slightly adjusted for the local site purposes.

2.2 Manufacturing-driven mBOM definition

This process is applied if manufacturing requires structures that heavily differ from the initial product structure. An example is the assembly structure (Fig. 1), which consists of nodes that represent the local assembly process (e.g. assembly steps, parts in the near vicinity). The industrial engineer defines the necessary mBOM nodes in the first step (new mBOM or based on the mBOM template). For example, in the eBOM, the exhaust manifold is part of the engine compartment. During assembly, however, the exhaust manifold must be added after the body is built on the chassis, as the exhaust pipe must be put through the bumper. This type of regrouping and adding of manufacturing-specific parts is most efficiently performed in a separate assembly mBOM structure. EBOM parts (items) are used for the filling of these mBOM nodes in the second step until all parts of the product structure are consumed in the mBOM. The eBOM is not the driver but a 'part supplier' for the manufacturing structures so that

mainly the parts, and not the structures, are reused. This type of creation mechanism works well for industries in which the *manufacturing procedures are standardized*, such as the automotive or home appliances industries.
In the next step of the research, the requirements of the automated mBOM definition process were defined. The main requirements are common but different mBOM creation process patterns lead to different requirements of the automated mBOM definition process (Table 1).

Table 1: Selection of defined requirements

<table>
<tr><th>EBOM-driven mBOM definition</th><th>MBOM-driven definition</th></tr>
<tr><td colspan="2">MBOM structure handling (eBOM extension, etc.) without changing the eBOM, independent structure management (versioning, changes)</td></tr>
<tr><td colspan="2">Automated deriving of order-related mBOM structure and the possibility for manual actions in the mBOM</td></tr>
<tr><td colspan="2">Automated check-consumption usage completeness of eBOM items in the mBOM structure</td></tr>
<tr><td>A common configuration process for eBOM and mBOM based on the same configuration logics</td><td>MBOM-specific (eBOM structure-independent) configuration process</td></tr>
<tr><td>Common associative management of configuration rules and options e/mBOM</td><td>Management of generic mBOM templates and item selection rules</td></tr>
</table>

3. Approaches for automated semantic definition of manufacturing structures

In the next phase of the research, approaches based on the derived requirements were developed. These approaches are system-neutral so that they can be implemented after the adjustment in data and process-management-engineering IT tools.
Fig. 2 shows the approach for the *product-structure-driven process*. The main part of this process is the product structure—the so-called '150%' or 'maxi BOM' structure. This product configurator contains all possible variant configurations so that a shippable product cannot be built using this structure. Moreover, it contains rules or constraints that determine when the variant items exist in the structure, and also, the selections that may or may not be made in combination. This variant logic is combined in the so-called Configuration Set Engineering (Ce). The manufacturing process is based on the mBOM template, which is built similar to the eBOM and contains all possible configurations for the product parts and the required manufacturing items. The structure can be different from eBOM. The Ce is propagated to the mBOM template so that the order-related product configuration drives the option definition in the mBOM and the consumption of the eBOM items. MBOM-specific configuration tasks (e.g. blanks) are defined in the additional Configuration Set Manufacturing (Cm). Both sets (Ce and Cm) completely control the mBOM and allow the order-related configuration. Minor adjustments can be done in the final step of the mBOM creation process.
Fig. 3 shows the approach for the *manufacturing-driven process*. The main part is the generic manufacturing structure, consisting only of the main BOM nodes. The mBOM items can be included in the template, but this is not a condition. The generic nodes are extended for the configuration logic, part selection rules (C1, C2, etc.). These rules define which eBOM items should be included in the generic node in the order-related configuration process. The selection rules are based on the properties of the eBOM items (S in Fig. 3). Such properties could be meta-data attributes, part classification, etc. For example, an attribute, 'assembly unit', could be defined on an eBOM item so that the mBOM selection rule can identify the required part (e.g. values '1', '2'). Another example is the selection based on the part class: mounting parts (screws, bolts) can be identified by the selection rule based on their standard classification (e.g. eCl@ss). The definition of the selection property values in the eBOM structure is done by engineering in accordance with the business process. Some properties can be filled manually by the engineer and some can be a part of the eBOM template. The approach works effectively only if both structures are enriched for the rules and selection properties.

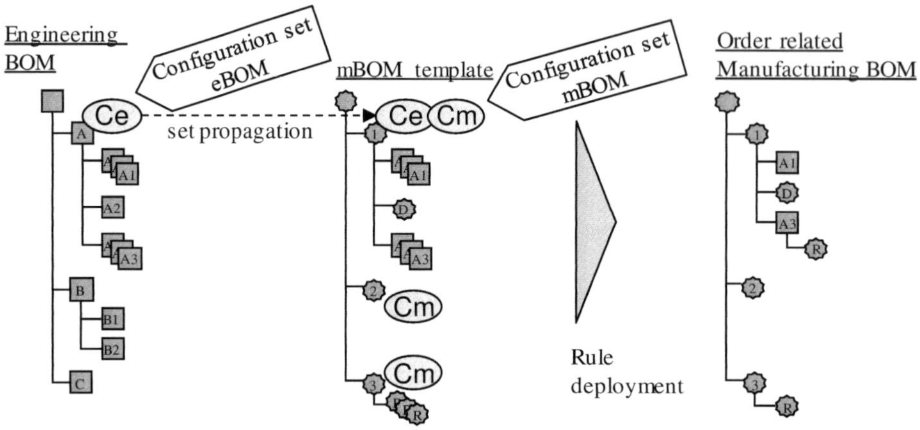

Figure 2: Approach for the product structure (eBOM)-driven mBOM definition

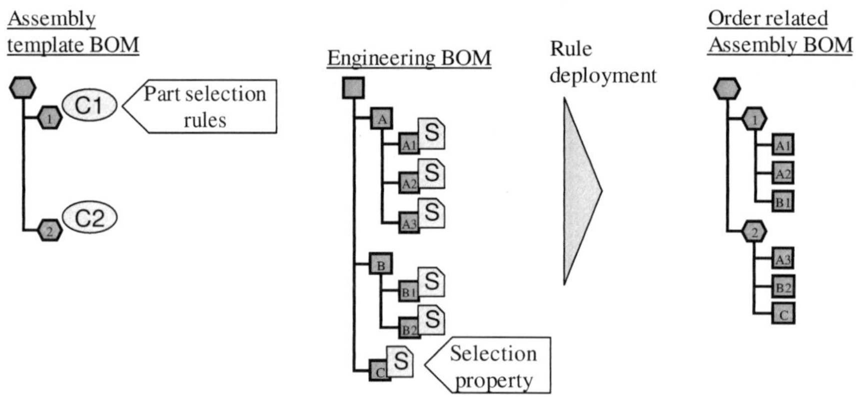

Figure 3: Approach for manufacturing-driven mBOM definition

4. Conclusions

The developed process framework was implemented as a prototype for different mechanical products. Engineering systems NX (for CAD modelling) and Teamcenter (product lifecycle management system [4,5] for engineering and manufacturing) were used. The BOM configuration logic, especially in the product-structure-driven approach, was implemented using the customized configuration of Teamcenter elements: options, variants, conditions, rule checks, constraints. The implementation proved the practical relevance and usability of the framework. The main conclusion of the research is that mBOM definition process can be (partly) automated, optimised and speeded up using modern engineering software systems and semantic configuration logic.

References

[1] A. Stekolschik and S. Schulte: *Approach for an integrated reference structure in the product lifecycle,* Tools and methods of competitive engineering 2014
[2] S. Vajna: *Integrated Design Engineering,* (Berlin: Springer)
[3] M. Malatesta, R. Raffaelo, M. Mengoni and M. Germani: *Supporting the modification process of products through a change management tool,* International Conference on Engineering Design - ICED 2013
[4] V. Arnold: *Product Lifecycle Management beherrschen,* (Berlin: Springer)
[5] S.-Y. Jung, B.-H. Kim, Bo-Hyun et. al.: *An integrated multi-BOM system for product data management,* Korean Journal of Computational Design and Engineering 2012 Vol 17/3

IOP Conf. Series: Materials Science and Engineering **241** (2017) 012033 doi:10.1088/1757-899X/241/1/012033

Investigation of Inrush Current in Nickel Microfuses

Ch Ketthanom[1] and R Phattanakun[2]

[1]Faculty of Engineering, Department of Electrical Engineering, Vongchavalitkul University, Nakhonratchasima, Thailand. Email: chitphong_ket@vu.ac.th
[2]BL6a: Deep x-ray lithography, Synchrotron Light Research Institute (Public Organization), Nakhonratchasima, Thailand. Email: rungrueang@slri.or.th

Abstract. This paper presents investigation of inrush current in nickel microfuses using microelectromechanical systems technology with high-aspect-ratio micromachining technology by x-ray lithography technique. The nickel microfuses were designed by the tether and saw-tooth pattern and studied the melting time, maximum melting current, maximum voltage and maximum energy consumption of the nickel microfuses in a blowing circuit. A microcontroller is used as a device to measure and collect the values of current, voltage, and energy consumption. The experimental results showed that tether fuses used melting time faster than saw-tooth fuses. But saw-tooth fuses can withstand melting current greater than tether fuses. By increasing the length of the two type fuses are ability to withstand current, power, energy consumption, and voltage are similar for the two type fuses. The I^2t of inrush current is increased by the length of nickel microfuses. The length of tether 500 μm has the I^2t value of inrush current a minimum at 0.0014 A^2-s and the length of saw-tooth 2000 μm has the I^2t value of inrush current a maximum at 0.0164 A^2-s.

1. Introduction

The power consumption of electronic appliances and electric system of automotive is increasing because of the short circuit in this equipment. Therefore, the fuses in such equipment are becoming large. As current increases, the voltage fluctuations become large when a fuse is blown due to a shorting accident or an overcurrent. Components in the equipment or in other equipment connected to the DC distribution might be damaged by the voltage fluctuations [1,2]. The fuse should be used to prevent fire and overheating. It should be chosen to limit the current in the circuit below the point where components could dangerously overheat and initiate ignition. Fuses won't even protect circuits against all faults. They typically will only help prevent overheating from a significant change in current. Fuses may protect against additional types of faults, but the fuse will not typically be the primary method of protection for those faults [3,4]. In this paper, We have studied I^2t of inrush current, melting time, melting current, blowing voltage, blowing power, energy consumption of nickel microfuses prototype. The design of structural fuses are two types of tether and a saw-tooth as shown in figure 1. The fuses are fabricated of the electrodes size 1000×1000 μm^2 and a line width of 100 μm fuses all the same. The difference structure is the length of the fuse, such as 500, 1000, 1500 and 2000 μm respectively.

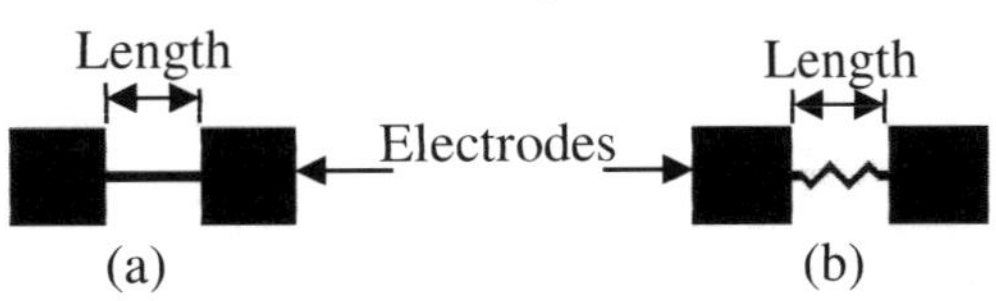

Figure 1. Fuse pattern (a) Tether design (b) Saw-tooth design

2. Principle and Theory

Inrush current refers to the maximum, instantaneous input current drawn by an electrical device when first turned on. When electrical equipment is first turned on, a large current flows that exceeds the steady-state current value. Figure 2 shows the current waveform when the power is turned on. When the power is turned on, current begins to flow, and the initial current flow reaches the peak current value that is larger than the steady-state current value. Following this, the current value gradually decreases until it stabilizes at the steady-state current. The part during which a large current flows before reaching the steady-state current is the inrush current. If the size of the inrush current exceeds that allowed by the part in use, depending on the magnitude of the inrush current (difference between the peak current value and the steady-state current value) and length of its duration (the length of time until the peak current value converges with the steady-state current value, hereafter called the pulse width), the part used in the circuit may overheat, potentially causing the electrical device to malfunction or break down [5].

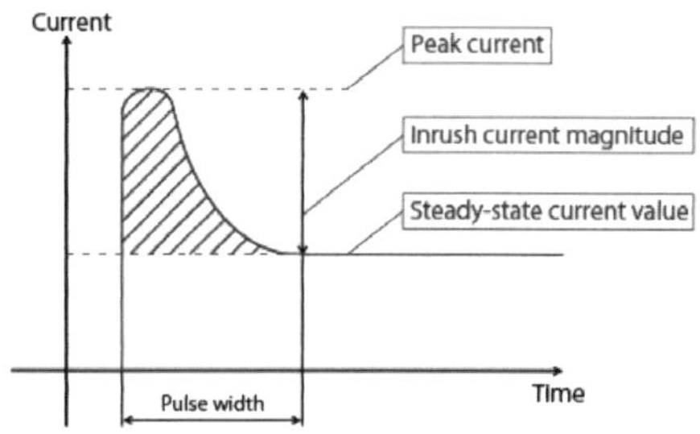

Figure 2. Current waveform when the device is powered up [5].

3. Experiments

DC power supply in the experiment were used for testing the melting current of nickel microfuses. A microcontroller is used as a device to measure and collect the values of current, voltage, and power consumption. If nickel microfuses get current and voltage over the coordinates of electricity, electric stress and heat. As a result, the melting fuse and disconnect the circuit in figure 3.

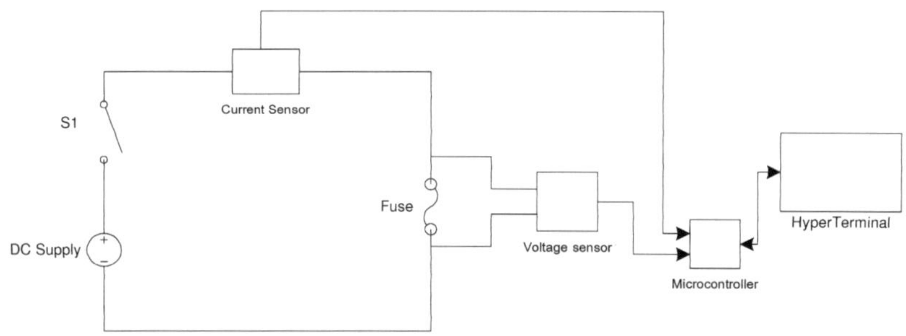

Figure 3. Testing fuse circuit block diagram [6].

Figure 3 shows voltage sensor, which uses to create a simple test circuit for the principle of voltage divider. Then forwarded to the microcontroller, which allows the resistors to divide the voltage does not exceed 5 Vdc so that the microcontroller can read an analogue input. Then processed for export to store data. The detector current using Ohm's law, since the microcontroller can be measured only in an analog voltage that does not exceed 5 Vdc only needed this principle. By measuring the voltage drop across the resistor relative to ground, then divided by the resistance to the current flowing through the fuse. Because the resistor used to measure voltage connected in series with the fuse. The current flowing through the resistor equals the current flowing through the fuse. The experiment consisted of nickel microfuses with circuit shown in figure 2. Microcontroller is used as a tool to measure the voltage drop across the fuse and the current flowing through the test. Then use the function of microcontroller to calculate the voltage and current to the power according equation 1.

$$P = IV \tag{1}$$

Where P is the electric power (W). I is current (A), and V is voltage (V). Then, the measured data is exported to the hyper terminal program and then plot a graph of the response of the fuse, Which can be calculated from the inrush current waveforms of the experiment. Then the waveform of the current surge in the experiment is shown in figure 4.

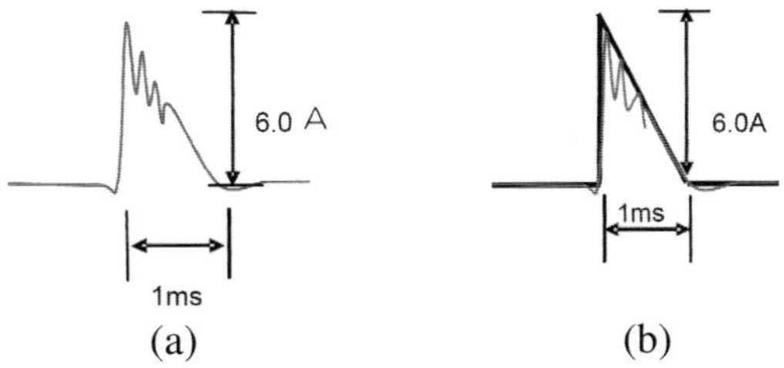

Figure 4. (a) Inrush current waveform (b) Inrush current waveforms. Grey Line is a waveform derived from the measurement. Black line is the waveform of the estimation [7].

Calculating the I^2t (Joule Integral) inrush currents estimated from the waveform. This formula is based on an estimated waveforms. For example, Figure 4 is waveforms with a width of 1 ms and a maximum current of 6 A, when an estimated waveforms is a triangle. The formula used to calculate the equation 2.

$$I^2t \text{ of inrush current} = \frac{1}{3} I^2{}_m t \tag{2}$$

Where $I^2{}_m$ is the peak current (A), t is the pulse applying time (s), and I^2t of inrush current is the peak current squared times time (A^2-s). Equation 2 calculates the I^2t of inrush current. Results of the I^2t of inrush current is $(1/3)\times 6^2 \times 0.001 = 0.012$ A^2-s. This method will be used to determine the I^2t of inrush current the following range of nickel microfuse sizes.

4. Results

In the experiments, the nickel microfuses are measured such as length of tether 500, 1000, 1500, and 2000 μm and length of saw-tooth 500, 1000, 1500, 2000 μm, respectively. We used a blowing circuit in figure 3, which can measure the voltage and the current flowing through the fuse and storage data of current and voltage. These data were calculated the power consumption that occurs, then the experiments and analyzed the results below. Table 1 shows the characteristic of nickel microfuses rated testing and important parameters, such as melting time, peak current, peak voltage, peak power and peak energy consumption, etc. The initial resistance was measured by a digital LCR Meter (Digicon Model DM-845) measured at a frequency of 1 kHz.

Table 1. Concluded other important parameters of the nickel microfuses.

Parameters	Tether length (μm)				Saw-tooth length (μm)				Units
	500	1000	1500	2000	500	1000	1500	2000	
I^2t of inrush current	0.0023	0.0045	0.0075	0.0130	0.0014	0.0059	0.0087	0.0164	A^2-s
Peak voltage	13.23	13.19	13.19	13.19	13.17	13.12	13.21	13.31	V
Average voltage	13.14	13.17	13.11	13.12	13.09	13.09	13.16	13.18	V
Peak current	0.37	0.44	0.50	0.57	0.29	0.47	0.54	0.64	A
Average current	0.21	0.24	0.22	0.24	0.16	0.22	0.22	0.24	A
Melting time	0.05	0.07	0.09	0.12	0.05	0.08	0.09	0.12	s
Peak power	4.88	5.80	6.60	7.52	3.82	6.15	7.12	8.42	W
Average power	2.70	3.22	2.83	3.13	2.15	2.92	2.95	3.21	W
Peak energy Consumption	0.24	0.41	0.59	0.90	0.19	0.49	0.64	1.01	J
Average energy consumption	0.14	0.23	0.25	0.38	0.11	0.23	0.27	0.39	J
Initial resistance	0.250	0.279	0.319	0.381	0.262	0.295	0.339	0.396	Ω
Final resistance	-	-	-	-	-	-	-	-	Ω

5. Conclusions

The experiments showed that the duration of current, while the melting time of the nickel microfuses tether types with a length of 500, 1000, 1500, 2000 μm and a saw-tooth types with a length of 500, 1000, 1500, 2000 μm and thickness all microfuses of 90 μm, which showed that the microfuses can withstand peak current following the in Table 1. The tether types take a melting time faster than saw-tooth types but saw-tooth types can withstand current and melting time over tether types. However, if the microfuses is increased the length, which increases the ability to withstand current, power, energy consumption is increased as well and the microfuses voltage are the same for all. The I^2t of inrush current is increased by the length of nickel microfuses. The length of tether 500 μm has the I^2t value of inrush current a minimum at 0.0014 A^2-s and the length of saw-tooth 2000 μm has the I^2t value of inrush current a maximum at 0.0164 A^2-s.

Acknowledgments

The authors would like to thank Vongchavalitkul University, Thailand for their support during the work being of this research and Dr.Rungrueang Phattanakun for their support at BL6a: Deep x-ray lithography Synchrotron Light Research Institute (Public Organization) or SLRI. We would also like to thank the editor and anonymous reviewers for their construction comments and suggestions.

References

[1] Tanaka T and Yamasaki M 2004 *Telecommunications Energy Conference 2004 (INTELEC 2004) 26th Annual International* Modeling of Fuses for Melting Time and Fusing Current Analysis pp 671–675

[2] Tanaka T Kawaguchi H Terao T Babasaki T and Yamasaki M 2007 *Telecommunications Energy Conference 2007 (INTELEC 2007) 29th International* Modeling of fuses for DC power supply systems including arcing time analysis pp 135–141

[3] Eckert T J 2009 Understanding Fuse Ratings *Product Compliance Engineering (PSES)* pp 1-6

[4] Huang R K and Nilsson S 2012 *Product Compliance Engineering (ISPCE) 2012 IEEE Symposium* Fuse selection criteria for safety applications pp 1–8

[5] "*What is inrush current?*" Murata Manufacturing Co Ltd Accessed from : http://www.murata.com/en-global/products/emiconfun/emc/2012/10/29/en-20121029-p1

[6] Ketthanom Ch 2015 *The Asian Conference on Engineering and Natural Sciences (ACENS 2015)* Investigation of Inrush Current in Glass Tube Fuses pp 248–256

[7] *Micro Fuse Product Data Sheet* 2007 Matsuo Electric Co Ltd

Multi-sensor information fusion method for vibration fault diagnosis of rolling bearing

Jing Jiao, Jianhai Yue and Di Pei

School of Mechanical, Electronic and Control Engineering, Beijing Jiaotong University, Beijng, China

E-mail: 15121285@bjtu.edu.cn

Abstract. Bearing is a key element in high-speed electric multiple unit (EMU) and any defect of it can cause huge malfunctioning of EMU under high operation speed. This paper presents a new method for bearing fault diagnosis based on least square support vector machine (LS-SVM) in feature-level fusion and Dempster-Shafer (D-S) evidence theory in decision-level fusion which were used to solve the problems about low detection accuracy, difficulty in extracting sensitive characteristics and unstable diagnosis system of single-sensor in rolling bearing fault diagnosis. Wavelet de-nosing technique was used for removing the signal noises. LS-SVM was used to make pattern recognition of the bearing vibration signal, and then fusion process was made according to the D-S evidence theory, so as to realize recognition of bearing fault. The results indicated that the data fusion method improved the performance of the intelligent approach in rolling bearing fault detection significantly. Moreover, the results showed that this method can efficiently improve the accuracy of fault diagnosis.

1. Introduction

As an important part of rotating machinery, rolling bearing plays an important role in the accuracy, performance, status and life of the equipment. With the wide application of the mechanical equipment, the fault caused by damage of mechanical equipment bearing occupy a large proportion, so the significance of rolling bearing fault diagnosis is obvious. Kun Yu et al. [1] employs a combined EEMD and SVD technique to extract useful fault features from the condition monitoring data of rolling element bearings. Ye Tian et al. [2] proposed an intelligent fault diagnosis method based on LMD, SVD and ELM.

However, the rolling bearing vibration signals are often non-stationary and represent non-linear processes. Information fusion technology can make full use of the information resources of multi sensors to make the best decision. Yan Tingai et al. [3] proposed a fusion method based on *n*-DCPD for rolling bearing fault diagnosis with two types of signals. Xiaoyun Gong et al. [4] suggests a new feature frequency extraction approach for gear malfunctions diagnosis using the FCS and EEMD method. Chuan Li et al. [5] addresses the use of DRFF technique to improve fault diagnosis performance for gearboxes by using acoustic emission sensor and accelerometer.

In this research, the vibration signal is collected from the high-speed EMU transmission test bench. And eight parameters in both time domain and frequency domain were used to obtain the feature sets. LS-SVM is employed in feature fusion. In order to increase the classification accuracy, data fusion method using D-S evidence theory is employed. The advantage of the proposed procedure is to detect fault in rolling bearings.

2. Multi-sensor Information Algorithm

2.1. Least Square Support Vector Machine (LS-SVM)
Proper kernel function and optimum kernel parameters are of importance in LS-SVM classifier. Given the training sample set $\{(x_1, y_1), \dots, (x_i, y_i), \dots, (x_n, y_n)\}$, $x_i \in R_n$, $y_i \in \{-1,1\}$. The regressive function is shown as

$$y(X)=\omega TX+b \tag{1}$$

where $X=(x_1, x_2, \dots, x_n)$ is a nonlinear transform function, which maps the input space to higher dimension space, $\omega=(\omega_1, \omega_2, \dots, \omega_n)$ is the weight coefficient, and b is threshold value.

Based on structural risk minimization (SRM), the regressive optimization problem is summarized to minimize the function:

$$min \frac{1}{2}\|\omega\|^2 + \frac{1}{2}c\sum_{n=1}^{n}\xi_i^2 \tag{2}$$

which is subjected to the constraint:

$$y_i=<\omega \cdot X>+b+\xi_i \tag{3}$$

where c is penalty factor, ξ_i is slack variable.

The optimization classification function can be obtained by introducing Lagrange function in:

$$f(X) = \sum_{n=1}^{n} \alpha_i K(X, X_l) + b \tag{4}$$

where α_i is lagrange multiplier and $K(X,X_l)$ is kernel function.

2.2. Dempster-Shafer (D-S) Evidence Theory
Evidence theory is the main means to deal with uncertain, incomplete and inaccurate information [6]. Assuming a finite nonempty set of mutually exclusive alternatives Θ is the frame discernment of θ, if the function m: $2^{\Theta} \rightarrow [0,1]$ satisfies $m(\Phi) = 0, \sum_{A\subseteq\Theta} m(A) = 1, \forall A \subseteq \Theta$, m is called the basic probability assignment (*BPA*). *Bel*(*A*) is used to express the belief levels of hypothesis in D-S evidence theory, and the expression formula is as follows:

$$\forall A \subseteq \Theta, Bel(A) = \sum_{B\subseteq A} m(B) \tag{5}$$

However, it is not comprehensive of using the *Bel*(*A*) to describe the belief levels of hypothesis. Therefore, *Pl*(*A*), which represents the plausibility belief level, is introduced to complement the description and can be defined as follows:

$$\forall A \subseteq \Theta, Pl(A) = 1 - Bel(\bar{A}) = \sum_{B\subseteq A} m(B) \tag{6}$$

In order to facilitate the judgment and decision making, the algorithm of synthesizing and combining two or more different *BPAs* were proposed, called as Dempster's combination rule [7]. Assuming that E_1 and E_2 belong to the same recognition framework Θ, and the corresponding basic probability assignment is m_1 and m_2. The combination rule can be described as follows:

$$m_{1\oplus 2}(A) = \begin{cases} \dfrac{\sum_{A_i \cap B_j = A} m_1(A_i) \bullet m_2(B_j)}{1 - K} & A \neq \varnothing \\ 0 & A = \varnothing \end{cases} \tag{7}$$

where $A_i, B_j, A \in \Theta, K = \sum_{A_i \cap B_j = \varnothing} m_1(A_i) \bullet m_2$.

3. Fault Diagnosis Based on Information Fusion

3.1. Experimental Setup
The research of this paper is based on the high-speed EMU transmission test bench, as is shown in figure 1. It simulates the transmission mode of a certain type of EMU running gear, which is mainly composed of motor, gear box, driving wheel, track wheel, bearings, drive axle and generator. The operation box has the function of start/stop, emergency stop, speed control and real-time parameters

(including given speed, actual frequency, output speed, actual pressure and actual current). The test rolling bearing is 351306 which are normal bearing (N), out- ring fault bearing (O) and inner-ring fault bearing (I).

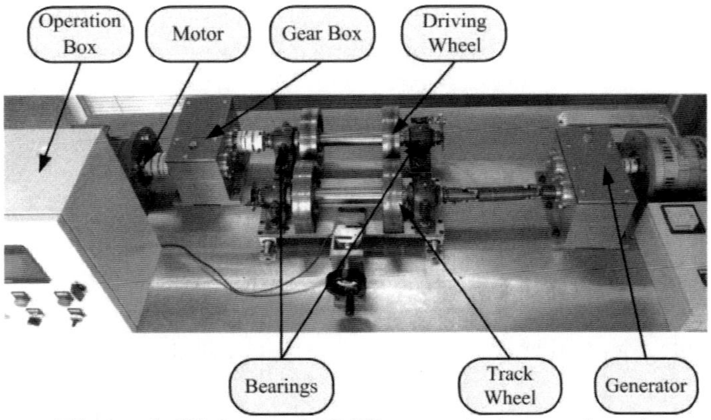

Figure 1. High-speed EMU transmission test bench

3.2. Structure of Fusion Prediction Model

The idea of fault diagnosis based on LS-SVM and D-S evidence theory can be summarized as follows. Firstly, wavelet de-noising was applied in processing stage to remove the signal noise. Afterward, some statistical feature parameters were used for extracting the useful diagnostic information from the signals. LS-SVM classifiers were employed for fault detection and classification. Data fusion method performed by D-S theory of evidence shows the flowchart of the proposed procedure. The used techniques are briefly explained in the following. The flowchart of the signal processing is shown as figure2.

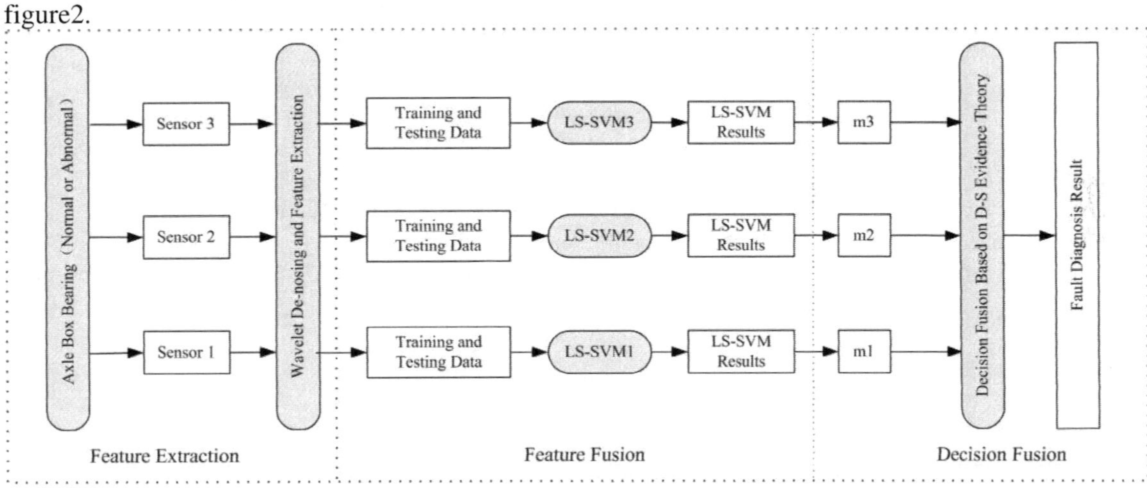

Figure 2. Flowchart of the signal processing

3.3. Bearing Fault Diagnosis

There are three kinds of rolling bearings near the motor side of the test bench, which are normal, outer-ring fault and inner-ring fault. The other bearings are normal. The speed of test bench is 3000 rpm, which is close to the speed of 250 km/h in real EMU running. All signals were simultaneously sampled with a 10k Hz sampling frequency. And the length of each original signal captured was 40,960. The processed signals contain a large number of data points and could not be used as inputs of the classifier, because high-dimensional data increases computational complexity so, training of the classifier becomes very difficult [8]. Therefore, some statistical features should be applied to reduce the dimensionality of data.

The time domain parameters are peak index K_{CF}, impulsion index K_{IF}, tolerance index K_{CLF} and kurtosis index K_u; the frequency domain parameters are mean frequency u_{mf1}, frequency center ufc_1, frequency root mean square (*RMS*) u_{rmsf1} and frequency standard deviation of frequency u_{stdf1}. Because of the different definitions of these parameters, it is necessary to normalize these feature sets [9]. The results of the fault identification of the LS-SVM in time domain and frequency domain are tabulated in table 1.

Table 1. Classification results of LS-SVM in time domain and frequency domain

	Bearing condition	N	I	O	Average
	N	0.83	0.10	0.07	
time domain	I	0.09	0.71	0.20	0.75
	O	0.05	0.23	0.72	
	N	0.89	0.03	0.08	
frequency domain	I	0.02	0.91	0.07	0.89
	O	0.05	0.08	0.87	

It can be seen from table 1 that frequency domain characteristic parameters are better than time domain characteristic parameters in classification. Therefore, in order to obtain higher fault recognition accuracy, the fault recognition result obtained by frequency domain characteristic parameters is used as the characteristic input of D-S evidence theory. The accuracy of decision-making layer fusion are shown in table 2.

Table 2. Classification accuracy of decision-making layer fusion

Bearing condition	N	I	O	Average
N	0.98	0.02	0	
I	0.01	0.96	0.03	0.97
O	0.01	0.01	0.98	

4. Conclusion

In this paper, LS-SVM and D-S evidence theory was proposed to solve the problem of bearing fault diagnosis. On the basis of the high-speed EMU transmission test bench, the multi-source vibration signals of the bearing were analysed by information fusing method. The experimental results show that the LS-SVM fault identification based on the frequency domain feature parameters is more accurate. The decision-making layer fusion result shows that the accuracy of the fault diagnosis is improved greatly through the combination of the feature layer and the decision layer information fusion, which is of great significance for the bearing fault diagnosis.

Acknowledgments

The study is supported by Technology Division of China Railway (M15D00230).

References

[1] Yu K, Lin T Rn and Tan J W 2017 *Applied Acoustics* **121** 33-45
[2] Tian Y, Ma J, Lu C and Wang Z 2015 *Mechanism and Machine Theory* **90** 175-186
[3] Ai Y, Guan J, Fei C, Tian J and Zhang F 2017 *Mechanical Systems and Signal Processing* **88** 123-136
[4] Gong X, Ding L, Du W and Wang H 2017 *Procedia Engineering* **174** 918-926
[5] Li C, Sanchez R, Zurita G, Cerrada M, Cabrera D and Vásquez R E 2016 *Mechanical Systems and Signal Processing* **76–77** 283-293
[6] Shafer G 1976 *A mathematical theory of evidence* Princeton University Press, Princeton 10-15
[7] Si L, Wang Z, Tan C and Liu X 2014 *Measuremen* **54** 140-151
[8] Zhang L, Xiong G, Liu H, Zou H and Guo W 2010 *Expert Systems with Applications* **37** 6077-6085
[9] Jing J and Jianhai Y 2016 *1st International Workshop on Structural Health Monitoring for Railway System* 296-303

IOP Conf. Series: Materials Science and Engineering **241** (2017) 012035 doi:10.1088/1757-899X/241/1/012035

Rolling bearing fault diagnosis based on information fusion using Dempster-Shafer evidence theory

Di Pei, Jianhai Yue and Jing Jiao

School of Mechanical, Electronic and Control Engineering, Beijing Jiaotong University, No.3 Shangyuancun,Haidian District, Beijng, China

E-mail: 15121291@bjtu.edu.cn

Abstract. This paper presents a fault diagnosis method for rolling bearing based on information fusion. Acceleration sensors are arranged at different position to get bearing vibration data as diagnostic evidence. The Dempster-Shafer (D-S) evidence theory is used to fuse multi-sensor data to improve diagnostic accuracy. The efficiency of the proposed method is demonstrated by the high speed train transmission test bench. The results of experiment show that the proposed method in this paper improves the rolling bearing fault diagnosis accuracy compared with traditional signal analysis methods.

1. Introduction

In recent years, many methods have been used for bearing fault detection. The methods are broadly classified as acoustic measurement [1], current and temperature monitoring [2], wear debris detection [3] and vibration analysis. Compare all diagnosis methods, vibration analysis is widely used. However, vibration analysis also has its disadvantage. For example, using a single sensor has the limitation in the amount and type of data. Once the sensor failed, bearing vibration signal cannot be obtained. So, using multiple sensors to detect the working state of bearing can get more diagnostic information and improve diagnostic reliability.

Multi-sensor information fusion is a combination of data from multiple sensors and related information to obtain more detailed and more accurate reasoning results than a single sensor. Evidence theory is initially based on Dempster's work and extended later by Shafer. As an extension to Bayes theory, D-S evidence theory uses belief and plausibility functions to quantify evidence and uncertainty. D-S evidence theory models how the uncertainty of a given hypothesis or discourse diminishes as pieces of evidence accumulate during the reasoning process [4].

The applications of D-S evidence theory in mechanical fault detection and diagnosis are reported in references [4-7]. In reference [4], D-S evidence theory is used for engine valve fault diagnosis. Reference [5] proposed a method that combine SVM with D-S evidence theory to resolve conflicting results generated from each SVM model. In reference [6], neural network and D-S evidence theory are combined to fuse the data from vibration signal and stator current signal for motor fault diagnosis. Reference [7] describes how evidence theory and fuzzy set theory are combined to improve the detection quality of weld defects.

2. Fault diagnosis based on D-S evidence theory

2.1. Basic of D-S evidence theory

Let Θ be a finite non-empty set of mutually n exhaustive and exclusive hypotheses about some problem domain. Evidence theory first assumes the definition of a set of hypotheses $\Theta = \{A_1, A_2, \ldots, A_n\}$, called the frame of discernment. The set composed with the proposition A of Θ is denoted as 2^Θ.

$$2^\theta = \{\varnothing, \{A_1\}, \{A_2\}, \ldots, \{A_n\}, \{A_1, A_2\}, \{A_1, A_3\}, \ldots, \theta\} \tag{1}$$

Assuming Θ is the frame discernment, if the function m: $2^\Theta \rightarrow [0,1]$ satisfies

$$m(\varnothing) = 0, \ \sum_{A \subseteq \Theta} m(A) = 1, \ \ \forall A \subseteq \Theta, \tag{2}$$

m is called the basic probability assignment (*BPA*), also called the mass function. $m(A)$ expresses the A's value of basic probability assignment obtained by analysis.

The belief function (*Bel*) is defined as:

$$\forall A \subseteq \Theta, \ \ Bel(A) = \sum_{B \subseteq A} m(B), \tag{3}$$

Bel is used to express the belief degree of A's corresponding proposition.

Assuming that the function Pl: $2^\Theta \rightarrow [0,1]$ satisfies

$$\forall A \subseteq \Theta, \ \ Pl(A) = 1 - Bel(\bar{A}) = \sum_{A \cap B} m(B), \tag{4}$$

Pl is called the plausibility function, reflecting the confidence degree of A's corresponding proposition.

Assuming that $m_1, m_2, \ldots, m_n$ are n *BPA*s in the same Θ. According to the Dempster's orthogonal rule, we have

$$m_{1\oplus\cdots\oplus n}(A) = \begin{cases} \dfrac{1}{1-K} \sum\limits_{\bigcap_{i=1}^{n} E_i = A} m_1(E_1) \cdot m_2(E_2) \cdots m_n(E_n), & A \neq \varnothing \\ 0 & , A = \varnothing \end{cases}, \tag{5}$$

$$K = \sum_{\bigcap_{i=1}^{n} E_i = \varnothing} m_1(E_1) \cdot m_2(E_2) \cdots m_n(E_n) > 0, \tag{6}$$

the value of K reflects the conflict degree of evidence. The higher value of K, the higher degree of evidence confliction, the less the information obtained after evidence combination.

2.2. Calculation of mass functions

In this paper, the mass functions are calculated by feature vector method introduced in Ref. [4]. Assuming that the number of bearing fault is N, and these faults are independent of each other.

Let

$$X = [x_1 \ x_2 \ \cdots \ x_n] \tag{7}$$

X is a feature vector that is composed of the feature parameters of vibration signal of known fault bearings, x_i represents the ith feature, n is the number of feature parameters. Thus, for N types of bearing faults (including normal state), the working state can be described by a matrix H:

$$H = \begin{bmatrix} X_1 \\ X_2 \\ \vdots \\ X_N \end{bmatrix} = \begin{bmatrix} x_{11} & x_{12} & \cdots & x_{1n} \\ x_{21} & x_{22} & \cdots & x_{2n} \\ \vdots & \vdots & \vdots & \vdots \\ x_{N1} & x_{N2} & \cdots & x_{Nn} \end{bmatrix} \tag{8}$$

$X_j=[x_{j1} \ x_{j2} \ \ldots \ x_{jn}]$, and x_{ji} is the ith feature parameter of the jth fault, $j=1,2,\ldots,N$, $i=1,2,\ldots,n$. The matrix H can be regarded as the fault prototype matrix. Similarly, the measurement feature vector S_k can be defined as

$$S_k = [s_{k1} \ s_{k2} \ \cdots \ s_{km_k}], \ \ k = 1, 2, \ldots, M \tag{9}$$

s_{ki} is the ith element of S_k, m_k is the number of feature parameters obtained by the kth sensor, $\sum m_k=n$. The distance between S_k and X_j is denoted as

$$d_{kj} = \begin{cases} \left[\sum_{i=1}^{m_k} \left(s_{ki} - x_{ji}\right)^{\alpha}\right]^{1/\alpha} & k = 1 \\ \left[\sum_{i=1}^{m_k} \left(s_{ki} - x_{j(i+m_k-1)}\right)^{\alpha}\right]^{1/\alpha} & k > 1 \end{cases} \quad k = 1, 2, \ldots, M \quad j = 1, 2, \ldots, N \tag{10}$$

Then, defining $p_{kj}=1/d_{kj}$, the value of p_{kj} can be regarded as the *BPA* of the *j*th fault, as far as the *k*th sensor is concerned, that is,

$$m_k(A_j) = p_{kj} \tag{11}$$

3. Experiment and analysis

3.1. Experiment Setup

The experiment of this paper is based on the transmission test bench of high speed train. The test bench is mainly composed of motor, gear box, rolling bearing, bearing seat, wheelset, transmission shaft and generator. The model of bearing in this experiment is 351306, and the bearing can be changed. Acceleration sensors are set on bearing seats, as shown in figure 1.

Figure 1. The transmission test bench of high speed train

In experiment, we set bearing 1 as normal (N), inner ring fault (I) and outer ring fault (O) respectively. Bearing 2, bearing 3 and bearing 4 are set as normal. In this paper, in order to calculate the fault prototype matrix *H*, three groups of experiments (denote as experiment A) were set up. Each group of experiment A has 60 times measurement. In order to calculate the measurement feature vector S_k (k=1,2,3,4) , 60 groups of experiments (denote as experiment B) were set up. Each group of experiment B has 3 times measurement in normal, inner ring fault and outer ring fault condition of bearing 1 individually. The settings of experiment A and B are shown in Table 1.

Table 1. The settings of experiment A and B

Experiment settings	Group	Fault types		
		Bearing 1	Bearing 2	Bearing 3	Bearing 4
A	1	N	N	N	N
	2	I	N	N	N
	3	O	N	N	N
B	1	N I O	N	N	N
	2	N I O	N	N	N
	⋮	⋮ ⋮ ⋮	⋮	⋮	⋮
	60	N I O	N	N	N

3.2. Fault diagnosis and analysis

Before using D-S evidence theory for fault diagnosis, experiment A was done to collect 180 groups of data for diagnosis based on traditional vibration signal analysis methods: time domain analysis and frequency domain analysis. Then, we carry on the fault diagnosis by using D-S evidence theory. The peak-to-peak value and the peak frequency of bearing 1 vibration signal from experiment A are selected as the feature parameters.

Then, we carry on the fault diagnosis by using D-S evidence theory. The fault prototype matrix H can be calculated by feature parameters. In order to calculate the measurement feature vector S_k and fuse sensors data, experiment B was designed. The mass function can be calculated by equation (9) to (11). Data of sensor 1 and sensor 2, sensor 1 and sensor 3, sensor 1, sensor 2, sensor 3 and sensor 4 were fused respectively according to equation (5) and (6). Meanwhile, we set up a contrast test that only judge the bearing 1 condition by sensor 1. The fault diagnosis results of bearing 1 from different diagnostic methods are shown in table 2.

Table 2. Comparison of different diagnostic methods in diagnostic accuracy

		Fault diagnosis methods					
		Time domain analysis	frequency domain analysis	S_1	S_1+ S_2	S_1+ S_3	S_1+ S_2+ S_3+ S_4
Correct number of diagnosis	N	45	48	42	47	48	57
	I	42	45	39	49	51	54
	O	45	45	45	51	54	57
Accuracy (%)		73.33	76.67	70.00	81.67	85.00	93.33

The diagnosis accuracy based on time domain analysis and frequency domain analysis are 73.33% and 76.67% respectively. In the case of non-fusion, the basic probability assignment of the fault obtained by the sensor 1 is used only, and the corresponding condition of maximum basic reliability distribution value is taken as the diagnosis result, the diagnosis accuracy is 70%, lower than time domain analysis and frequency domain analysis. Fuse the data from sensor 1 and sensor 2, the diagnosis accuracy reaches to 81.67%; Fuse the data from sensor 1 and sensor 3, the diagnosis accuracy reaches to 85.0%; Fuse the data from the sensor 1, sensor 2, sensor 3 and sensor 4, the diagnosis accuracy reaches to 93.33 %.

4. Conclusion

The experiment result based on transmission test bench of high speed train allow to draw the following conclusions:

1. Compared to traditional vibration signal analysis methods for rolling bearing fault diagnosis, using D-S evidence theory to fuse the sensors data can improve the diagnosis accuracy remarkly.
2. The diagnosis accuracy cannot be increased only by taking mass function of D-S evidence theory as judgement without evidence fusion.
3. When fusing the data from sensors, the location and number of sensors have an impact on the accuracy of the diagnosis to some extent, in general, the more the number of evidence, the higher the diagnostic accuracy.

Acknowledgments

The study is supported by Technology Division of China Railway (M15D00230).

References

[1] Hecke B V, J Yoon and D He 2016 *Applied Acoustics* **105** 35-44
[2] Seo J J, Yoon H, Ha H, Hong D P and Kim W 2011 *Advanced Materials Research* **31** 295-297 1544-47
[3] Zmitrowicz A 2005 *Journal of Theoretical & Applied Mechanics* **43** 3-35
[4] Basir O and Yuan X 2007 *Information Fusion* **8** 379-386
[5] Hui K H, Meng H L, Leong M S and Al-Obaidi S M 2017 *Engineering Applications of Artificial Intelligence* **57** 160-170
[6] Yang B S, Kim K J 2006 *Mechanical Systems & Signal Processing* **20** 403-420
[7] Kaftandjian V, Dupuis O, Babot D and Zhu Y M 2003 *Pattern Recognition Letters* **24** 547-564

IOP Conf. Series: Materials Science and Engineering **241** (2017) 012036 doi:10.1088/1757-899X/241/1/012036

Fire hazard analysis of alcohol aqueous solution and Chinese liquor based on flash point

Qinpei Chen[1], Guoting Kang[2], Tiannian Zhou[1], Jian Wang[1]

[1]State Key Laboratory of Fire Science, University of Science and Technology of China, Hefei, Anhui 230026, China
[2]Puyang Green Foam Co., Ltd, Puyang, Henan, 457000, China

E-mail: wangj@ustc.edu.cn

Abstract. In this paper, a series of experiments were conducted to study the flash point of alcohol aqueous solution and Chinese liquor. The fire hazard of the experimental results was analysed based on the standard GB50160-2008 of China. The result shows open-cup method doesn't suit to alcohol aqueous solution. On the other hand, the closed-cup method shows good applicability. There is a non-linear relationship between closed-cup flash point and alcohol volume concentration. And the prediction equation established in this paper shows good fitting to the flash point and fire hazard classification of Chinese liquor.

1. Introduction

China has a long history of liquor culture since Du Kang has made it four thousand years ago [1]. Chinese liquor can be divided into different flavor: Maotai type, Luzhou type, fragrance type, rice type and others. The decisive effect on the flavor style is not to account for 99% of the alcohol and water, but those of the trace materials, their different proportions and levels lead to different flavor style and different tastes. At the same time, fires caused by Chinese liquor happened very frequently. As for the fire hazard of Chinese liquor, it may be different with alcohol aqueous solution. The flash point is one of the major physical properties used to determine the fire hazard of fuels, and the lower the flash point, the greater the fire risk [2].

GB50694-2011 [3] has given an empirical equation to predict the flash point of Chinese liquor, but in the actual use of the empirical equation, it is found that the prediction is not accurate, especially when the alcohol volume concentration is low. To rebuild the prediction model, and reveal the relationship between Chinese liquor and alcohol aqueous solution in flash point and fire hazard, a series of flash point experiments were conducted in this paper.

2. Experimental approaches

2.1. Experimental Apparatus and Verification

Figure 1. The automatic open-cup and closed-cup flash point tester.

Figure 1 shows the automatic open-cup and closed-cup flash point tester [4] (SCKB3000, Zibo Shengkang Dianqi CO., LTD, China), which can perform the standard test methods of GB/T3536-2008 [5] and GB/T261-2008 [6]. Before the experiments, some standard substances closed-cup flash point are tested and compared with the Open Chemistry Database as listed in table 1 to verify the tester, and it can be seen that the test values of *n*-tetradecane, phthalate and *n*-hexadecane meet the requirements.

Table 1. Calibration results using the standard substance.

Test mode	Closed-cup test		Open-cup	
Test substance	*n*-Tetradecane	*n*-Hexadecane	*n*-Hexadecane	Phthalate
Standard values	372	408	373	491
Test values	371.5	407.7	372.6	491.7

2.2 Specimen

The specimens of the study were of the alcohol aqueous solutions with alcohol volume concentration (AVC) of 5%, 10%, 20%, 30%, 35, 40%, 50%, 60%, 70% and 100%. In addition, alcohol has a very low boiling point of 13°C compared with the environment temperature. So all the specimens were cooled to 5°C using a fridge before the test. Every test at the same condition was repeated several times to ensure reproducible results within permitted error ranges.

3. Results and discussion

3.1 Open-cup flash point

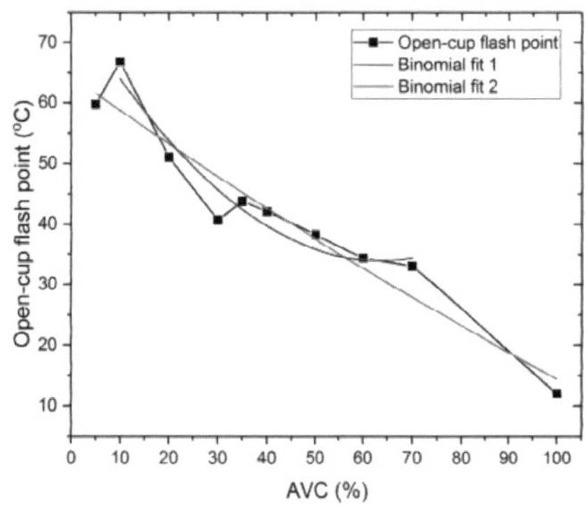

Figure 2. The open-cup flash point curves of alcohol aqueous solution.

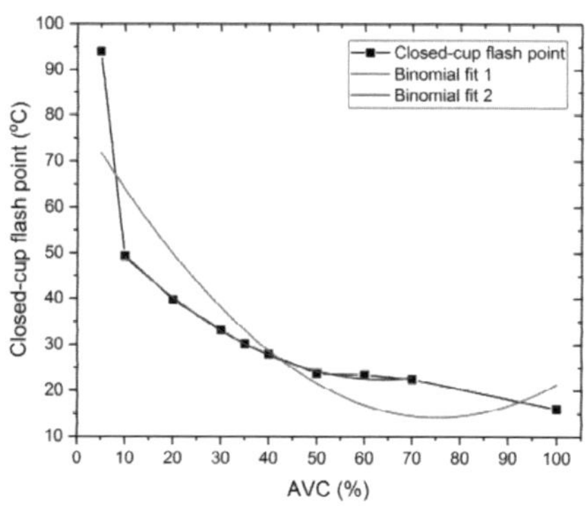

Figure 3. The closed-cup flash point curves of alcohol aqueous solution.

Figure 2 shows the open-cup flash point experimental result, and it can be found that the flash point value of alcohol aqueous solution drops with AVC increases in general. But there are two inflection point when AVC is 5% and 30%. In theory, when the alcohol volume concentration is lower, the lower the alcohol content in the vapour under the same heating conditions. The difference between theory and experimental result is because alcohol has a low boiling point of 13°C, alcohol in the solution has been volatile before reaching the open-cup flash point. For liquid fuels with a large vapourability, open-cup method shouldn't be recommended, closed-cup method is the right choice. Despite this, we use binomial fitting method for the experimental data, the fitting curve is approximately straight. In fact, the AVC of Chinese liquor is almost all between 10% and 70%, so only need to fit the data in this interval, shown as Figure 2, to predict the open-cup flash point of Chinese liquor and alcohol aqueous solution.

3.2 Closed-cup flash point

Figure 3 shows the closed-cup flash point experimental result, and it can be found that with AVC decreases, the closed-cup flash point grows. When AVC is less than 40%, the flash point value growth rate is gradually getting bigger. When AVC is equal to 5%, the growth rate become biggest, and the difference between 5% and 10% is 44.6°C. The alcohol content of the vapour in the closed-cup is gradually thinner, and it needs more energy to ignite the vapour. It can be concluded as once AVC is lower than 10%, the closed-cup flash point will become a sharp increase as figure 3. Using binomial fitting method for the experimental data, but the fitted curve is less consistent with the data when AVC is between 5% and 100%. So we just fitted the data in the interval from 10% to 70% as shown in figure 3, and R^2 of the curve is nearly 0.998, which means it can well predict closed-cup flash point of Chinese liquor and alcohol aqueous solution, shown as equation (1).

$$Y = 59.94315 - 1.17166X + 0.00915X^2 \quad (1)$$

3.3 Fire hazard analysis

The flash point is one of the major physical properties used to determine the fire hazard of fuels, and the lower the flash point, the greater the fire risk. Measurement of the closed-cup flashpoint provides a method of classifying flammable liquids fire hazard [2]. Many countries have developed their standards to evaluate the liquid fire hazard. In China, GB50160-2008 [7] is used to guide the classification. In this standard class A is the highest fire hazard risk level when the closed-cup flash point is lower than 28°C. When flash point is between 28°C and 60°C,it's class B. when the flash point is higher than 60°C, it's class C. According to the study of Lu [8], the common Chinese liquor closed-cup flash point is listed as table 2.

Table 2. The common Chinese liquor closed-cup flash point.

AVC (%)	Closed-cup test (°C)	AVC (%)	Closed-cup test (°C)
38	29.1	52	24.5
45	26.5	54	23.7
50	24.9	57	23.2

In the standard GB50694-2011 [3], a linear regression equation is established as equation (2). Plotted the Chinese liquor data of table 2, the fitted results of equation (1) and equation (2) in figure 4. First of all, figure 4 also shows the well consistent of equation (1) with the experimental closed-cup flash point results. The flash point development of Chinese liquor with AVC is similar to that of alcohol aqueous solution, and the closed-cup flash point of Chinese liquor grows as AVC decreases. Compared with equation (1), the prediction of flash point of equation (2) is not accurate. To a certain extent, it can be said to be serious deviation. Unlike national standard, the result shows it is a non-linear relationship between Chinese liquor flash point and AVC. And this paper recommends the use of equation (1) to predict the flash point instead equation (2) in the standard GB50694-2011.

$$Y = 36.6619 - 0.2430X \quad (2)$$

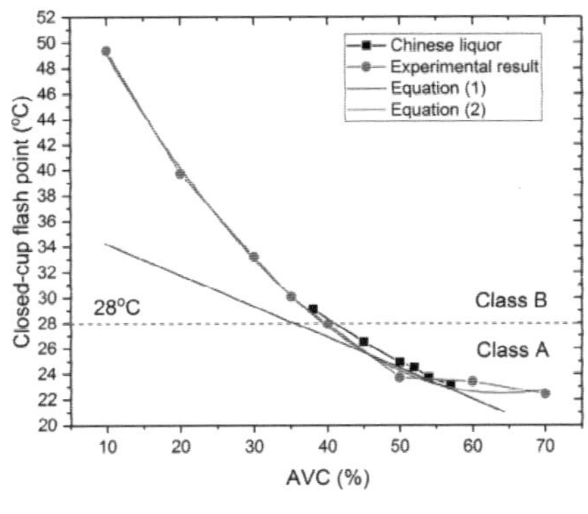

Figure 4. The fire hazard of Chinese liquor and alcohol aqueous solution.

Comparing the two sets data of equation (1) and Chinese liquor, it can be found that there are still differences. The differences result means the trace materials in Chinese liquor have some effect on closed-cup flash point to a certain extent. Because the Chinese liquor used in reference [8] is Maotai type, so it can be inferred that the flash point of different flavour style Chinese liquor in the same AVC may has slight difference.
As to fire hazard classification, it can be seen the critical AVC between class A and class B is 39% based on the experimental result and equation (1). The critical AVC of class A and class B is 41 based on the common Chinese liquor closed-cup flash point data. The relative error between the two is less than 5%, which means good agreement. Therefore, once AVC is higher than 40%, its fire hazard will be the highest level class A, which means the highest fire protection measures should be provided. At the same time, the prediction result is 35.6% based on equation (2), which has no reference value compared with equation (1) and the common Chinese liquor flash point data.

4. Conclusions

In this paper, after a series of experimental study, the following conclusions can be got.
(1) Closed-cup method is more suitable to test flash point of alcohol aqueous solution and Chinese liquor, compared with open-cup method, because alcohol is volatile.
(2) There is a non-linear relationship between alcohol aqueous solution / Chinese liquor flash point and alcohol volume concentration. And the prediction equation shows good agreement with Chinese liquor closed-cup flash point data.
(3) It can be inferred that the flash point of different flavour style Chinese liquor in the same AVC may has slight difference
(4) The critical alcohol volume concentration between fire hazard class A and class B is 40%, Once the alcohol volume concentration is higher than 40%, its fire hazard will be reached to the highest level class A, which means the highest fire protection measures should be matched to the alcoholic beverages factory.

Acknowledgments
This research was supported by National Natural Science Foundation of China (No: 51376172). The authors deeply appreciate the supports.

References

[1] Gong W Y 1993 *J. Pop. Cult.* **27**(2) 57-74.
[2] Hurley M J 2016 *SFPE Handbook of Fire Protection Engineering* (New York: Springer).
[3] Ministry of Construction of China 2011 *Code for design of fire protection of alcoholic beverages factory, GB 5050694-2011* (Beijing: China Standards Press).
[4] Ding C, Yao W, Zhou D, Rong J, Zhang Y, Zhang W and Wang J 2013 *J. Fire Sci.* **31** 469-477.
[5] General Administration of Quality Supervision, Inspection and Quarantine of the People's Republic of China 2008 *Determination of flash point-Cleveland open cup method, GB/T3536-2008* (Beijing: China Standards Press).
[6] General Administration of Quality Supervision, Inspection and Quarantine of the People's Republic of China 2008 *Determination of flash point–Pensky–Martens closed cup method, GB/T 261-2008* (Beijing: China Standards Press).
[7] Ministry of Construction of China 2008 *Fire Prevention Code of Petrochemical Enterprise Design, GB 50160-2008* (Beijing: China Standards Press).
[8] Lu D F 1999 *Fire Science and Technology* **3** 15-16+3

Flammability properties of typical aviation functional oils

Jianlong Zhong[1,2], **Tiannian Zhou**[1] **and Jian Wang**[1,*]

[1]State Key Laboratory of Fire Science, University of Science and Technology of China, Hefei, Anhui, 230026, China
[2]The First Aircraft Institute of AVIC, Yanliang, 710089, China

[*]Email: wangj@ustc.edu.cn

Abstract. A sequence of experiments were conducted in a cone calorimeter to measure the flammability properties of 4050# lubricating oil and 15# hydraulic oil. Meanwhile, the heat release rate, mass loss rate and ignition time were used to investigate the combustion characteristics of the two kinds of oil with varying external radiant heat fluxes. The results show that the 15# oil is easier to be ignited than the 4050# oil and -0.55 power of the ignition time is linear with the external radiant heat flux, which agrees with the previous studies.

1. Introduction

Lubricating and hydraulic fluids are two typical functional oils, which are widely used in transportations: vehicles, vessels and aircraft. The lubricating oil is commonly employed as a liquid lubricant to protect machines and components, comprising base oil, additives and other chemical components [1]. The hydraulic oil is used as a medium for power transfer and control, heat transfer or sealing. However, the flammability of the two kinds of oil could lead to a potential fire hazard. Meanwhile, a main hazard is the accidental leakage near to a source of ignition [2]. In particular, the fire risk in aviation safety is paid more attention. Hence, it is significant to investigate the flammability properties of aviation lubricating and hydraulic fluids.

The cone calorimeter based on oxygen consumption principle is a standard apparatus to measure the flammability properties of materials in fire tests [3]. Iwata et al. [4] conducted a sequence of pool fire tests in a cone calorimeter to study the combustion characteristics of crude oils and the results showed that heat release rate (HRR), mass loss rate (MLR) and smoke yield are dependent on its density. Chen et al. [1] employed the same instrument to investigate the influence of radiant heat flux on ignition and combustion behaviour of lubricating oils. Generally, a combustible fluid is easier to be ignited in the presence of an external radiation by a pilot such as spark, flame and hot surface. Thus, although the heat release rate is the single most important variable to determine the flammability of materials in fire hazard [5], the ignition time is also an essential parameter to understand the combustion characteristics of liquid fuels because it can be used to predict fire initiation.

The objective of this study is to measure the flammability properties of two typical aviation lubricating and hydraulic oils including heat release rate, mass loss rate and ignition time using a cone calorimeter. On this basis, a quantitative analysis for HRR and ignition time with varying radiant heat fluxes was carried out.

2. Experimental set-up

The experiments were conducted in a cone calorimeter, which consists mainly of a combustion chamber, load cell and gas analyser [3]. Two typical aviation functional oils: 4050# lubricating oil and

IOP Conf. Series: Materials Science and Engineering **241** (2017) 012037 doi:10.1088/1757-899X/241/1/012037

15# hydraulic oil were chose as the test samples. The fluid samples of 50 mL for each test were loaded by a square pan with the length of 100 mm. The liquid fuel was exposed to the radiant cone with a high temperature and the distance of the liquid-level was 60 mm away from the bottom of the cone radiant source. Considering there existing a high flash point for the lubricating oil [6], 3 radiant heat fluxes: 25, 35 and 50 kW/m^2 were used to heat the fluid sample so that it was faster to be ignited by an electric spark. It should be noted that the flash point of 15# is appropriate 82 ℃. Some typical specifications of 4050# lubricating oil and 15# hydraulic oil are summarized in Table 1.

Table 1. Specifications of 4050# and 15# oils.

Specifications	4050#	15#
Appearance	clear and colorless	red
Viscosity at 100 ℃ (mm^2/s)	4.9 – 5.4	4.9
Flash point (℃, open cup)	246	82
Work Temperature (℃)	–40 – 200	–54 – 135

In total, 16 experiments were conducted. The experimental conditions are shown in Table 2. Except the minor variations in ambient temperature and humidity, every experiment was repeated at least twice at the same radiant heat flux. Meanwhile, several parameters including HRR, MLR, ignition time and flame temperature were measured in real time during the experiments. Note that the ignition time was determined by the flame temperature histories combined with observing the experimental phenomena, for the reason that there was an abrupt change in temperature at occurring a flame.

Table 2. Summary of the experimental conditions.

Experiment No.	Sample	Heat flux (kW/m^2)	Ambient temperature (℃)	Humidity (%)
1-2	4050#	25	24	39.5
3-5	4050#	35	23.5	42
6-8	4050#	50	23.7	44
9-11	15#	25	24.5	51
12-13	15#	35	24.8	52
14-16	15#	50	24.8	51

3. Results and discussion

3.1. Heat release rate

HRR is the single most important parameter because it can indicate the size of fire and the growth rate of fire [7]. Consequently, these predicted values contribute to knowing the time available for safe evacuation or suppression in a real fire scenario. In the current study, the HRR was measured in a cone calorimeter based on the oxygen consumption principle, which assumes the heat released per unit of oxygen consumed is approximately the same value of 13.1 kJ/g with ±5% accuracy [8]. The heat release rates of the samples with varying radiant heat flux are shown in figure 1 and it are used to analyse the influence of external heat flux on HRR. Note that the initiation point is the ignition instant. This means the horizontal axis is a relative time but not beginning with heating fuels in figure 1.

Figure 1(a) shows the HRR histories of 4050# lubricating oil exposed to 3 different radiant heat fluxes. An obvious feature is the maximum HRR increases with the increase of the external heat flux. In addition, the variation of HRR under the condition of 25 kW/m^2 is more stable than others, which can be divided into three typical stages: (1) ascending stage; (2) steady stage and (3) descending stage. However, the steady stage gradually fades away with increasing of the heat flux. This indicates a higher radiant flux can enhance the burning intensity of oils. Likewise, this phenomenon can be observed in figure 1(b). However, a distinct steady burning for 15# hydraulic oil is not to occur. Furthermore, the average maximum HRR of 15# is 3 – 4 times greater than that of 4050# at the same radiant flux. This result is consistent with a lower flash point of 15#, which means the ignition time of 15# is earlier than 4050#.

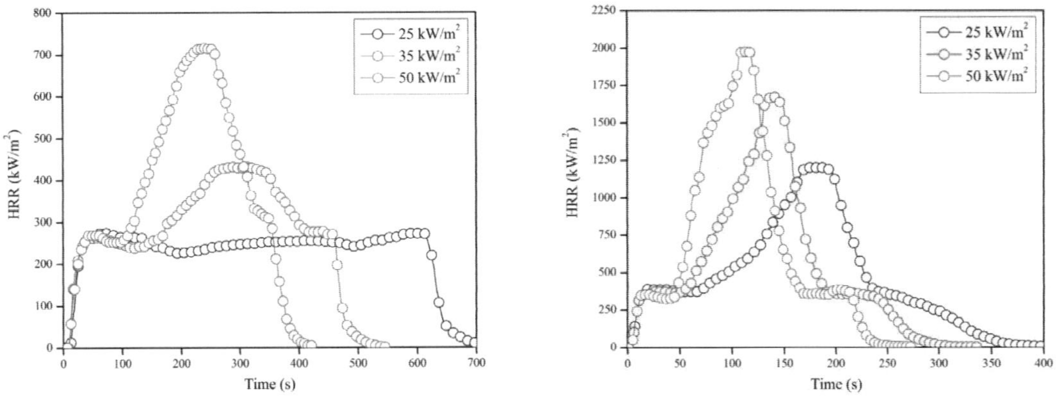

(a) (b)

Figure 1. (a) Heat release rate of 4050#; (b) heat release rate of 15# with varying radiant heat fluxes.

3.2 Mass loss rate

The average depth of oils in the current study is 5 mm that belongs to the range of the thin-layer fuel. However, the burning process for a thin-layer fuel is more complicated in comparison with that of pool fire with infinite depth due to the involvement of the boiling phenomenon [9]. According to previous study [10], this process can be divided into four stage: (1) pre-burning stage; (2) quasi-steady burning stage; (3) boiling stage and (4) decay stage. Compared with the simplified partition based on HRR, above four stages are more significant to understand the burning process of fluid fuels in the presence of external radiant heat flux.

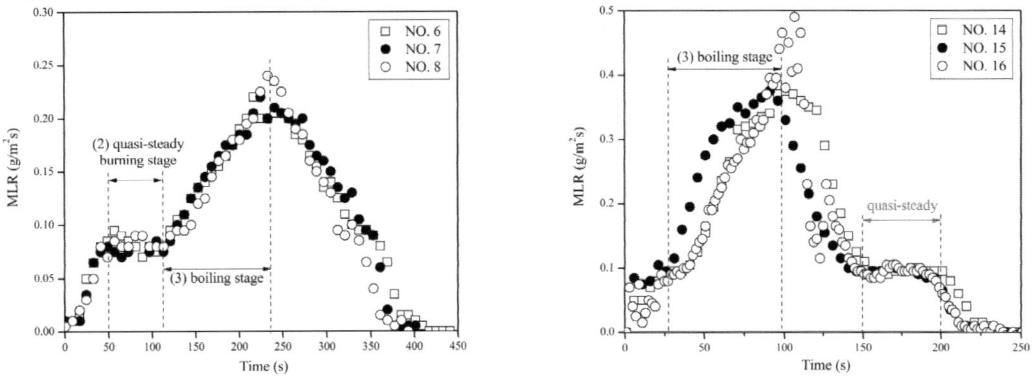

(a) (b)

Figure 2. (a) Mass loss rate of 4050#; (b) mass loss rate of 15# at 50 kW/m^2.

The mass loss rates of the samples at 50 kW/m^2 are shown in figure 2, where three repeated experiments for the same condition are all plotted. The 4050# lubricating oil and 15# hydraulic oil are corresponding to figures. 2(a) and 2(b), respectively. As figure 2 (a) shows, four typical stages can be observed. For 4050# samples, the quasi-steady burning stage occurs after 50 s of the pre-burning stage and lasts about 60 s, but the first two stages are not presented distinctly in figure 2(b). During this stage, the heat feedback from flame to fuel surface is approximately equivalent to the convective heat loss and the heat loss from gasification of the fuel. Subsequently, an abrupt increase in burning rate occurs with the consumption of fuel until reaching to peak, which is called the boiling stage. This is because the residual fuel has already been heated to the boiling temperature by the flame and hot oil pan.

Meanwhile, as figure 2(b) shows, an interesting phenomenon is noticed that a longer quasi-steady burning happens in the decay stage, which corresponds to a relatively stable flame. The reason for this is unknown, but one could speculate that the low boiling point for 15# hydraulic oil, resulting in a fast boiling before reaching heat equilibrium.

3.3. Ignition time

In the current work, the temperature histories were used to determine the ignition time associated with occurring flame. There exists a abrupt increase in temperature at a certain instant, which corresponds to the ignition time. For typical oils including diesel and aviation kerosene, Chen et al. [1] verified that the -0.55 power of the ignition time is linear with the external radiant heat flux. figure 3 shows the ignition time of 4050# and 15# with varying heat fluxes. As a result, the experimental data agrees well with the correlation $t^{-0.55} \sim \dot{q}_e''$.

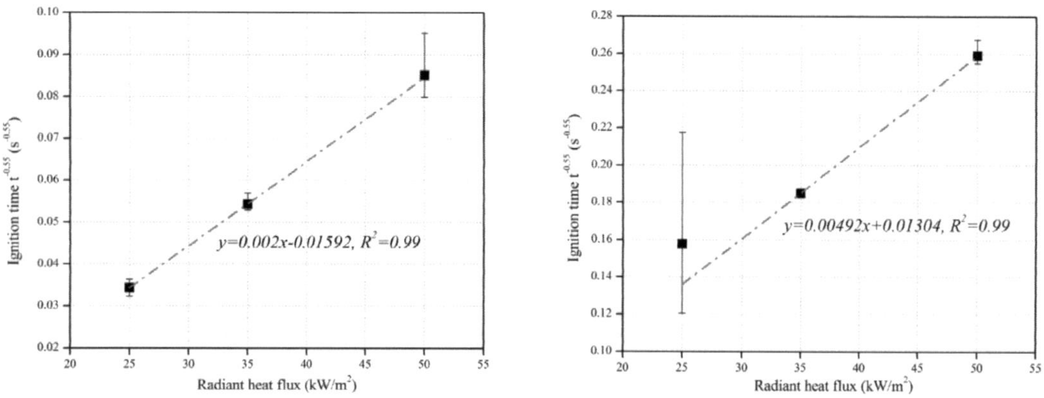

(a) (b)

Figure 3. (a) Ignition time of 4050#; (b) ignition time of 15# with varying radiant heat fluxes.

4. Conclusions

The major results and observations are summarized as follows:

1. It is easier to ignite 15# hydraulic oil at the same radiant heat flux than 4050# lubricating oil. In addition, although there exist four typical burning stages for the two kinds of oil, a longer quasi-steady burning happens in the decay stage for 15# but not following after pre-burning stage.
2. The relationship between the ignition time of the two kinds of oil and the radiant heat flux is consistent with the linear correlation $t^{-0.55} \sim \dot{q}_e''$.

Acknowledgements

This work was supported by National Natural Science Foundation of China under Grant No. 51376172.

References

[1] Chen X, Lu S, Li C, Zhang J and Liew K M 2014 *Fire Mater.* **38** 409-17
[2] Roberts A and Brookes F 1981 *Fire Mater.* **5** 87-92
[3] ISO 5660-1 2015
[4] Iwata Y, Koseki H, Janssens M and Takahashi T 2001 *Fire Mater.* **25** 1-7
[5] Babrauskas V and Peacock R D 1992 *Fire Safety J.* **18** 255-72
[6] Siregar M A M and Nugroho Y S 2014 *Appl. Mech. and Mater.* **493** 161-6
[7] Huggett C 1980 *Fire Mater.* **4** 61-5
[8] Chow W K and Han S S 2011 *Appl. Therm. Eng.* **31** 304-10
[9] Zhou Z, Wei Y, Li H, Yuen R and Wang J 2014 *Int. J. Heat Mass Tran.* **70** 578-85
[10] Liu J, He Y, Zhou Z, Yuen R and Wang J 2016 *Appl. Therm. Eng.* **101** 202-16

IOP Conf. Series: Materials Science and Engineering **241** (2017) 012038 doi:10.1088/1757-899X/241/1/012038

The effects of sidewall constraint on temperature distribution of fire-induced thermal flow under an arc-ceiling

Tiannian Zhou[1], Jianzhong Rong[2] and Jian Wang[1,*]

[1]State Key Laboratory of Fire Science, University of Science and Technology of China, Hefei, Anhui, 230026, China
[2]Sichuan Fire Research Institute of Ministry of Public Security, Chengdu, Sichuan, 610036, China

[*]Email: wangj@ustc.edu.cn

Abstract. A numerical study with varying transverse fire locations based on a semicircle tunnel model was carried out and it was used to further compare with the rectangular model in the published literature, aimed to investigate the influence of cross-sectional structure on temperature distribution under the ceiling. Results demonstrate that the reduction in temperature along the tunnel centerline under the flat ceiling is greater than that under the arc-ceiling. In addition, the influence of constraint effect of sidewall under the arc-ceiling is more significant than the flat ceiling because the flame would attach the ceiling when the fire is near the sidewall.

1. Introduction

In recent years, sidewall constraint effects on characteristics of fire-induced thermal flow including temperature distribution, flame shape and length under a tunnel ceiling have received a large amount of attention [1-3]. In a wide multi-lane tunnel, traffic accidents and associated fires may not occur along the centerline of the tunnel but any positions [4]. When the fire approaches the sidewall, in particular, the confined space of tunnels might have an effect on the temperature distribution beneath the ceiling that may lead to the re-layout of hardware, such as heat detectors, smoke detectors and sprinklers corresponding to detection and suppression of tunnel fires [5]. Ji et al. [1] conducted a series of scale model experiments in a tunnel with a rectangular cross section, and found that the dimensionless maximum excess gas temperature under the ceiling can be fitted with the transverse distance of fire to the sidewall by an exponential function. This indicates the maximum ceiling gas temperature would increase significantly once the distance between the fire source and the sidewall decreases to a critical range.

Prior studies in constraint effect of sidewall have focused on temperature distribution under a flat ceiling because the related experiments were carried out in a rectangular model-scale tunnel. In fact, however, the cross section of road tunnels is commonly an arc structure rather than a rectangular structure. In Ref. [5], the longitudinal temperature distributions with two transverse fire locations were tested in a full-scale tunnel with a semicircular cross section. Meanwhile, the simulation technique was also employed to extend the full scale experiments involving more transverse fire locations but the arc-ceiling was replaced by a flat ceiling. Results showed that when the fire located at the centerline of the tunnel, the longitudinal temperature under the flat ceiling was lower than that under the arc-ceiling. In contrast to the arc cross section, the flow stream within the rectangular cross section has a greater perimeter and a shallower smoke layer for the same volumetric smoke flow rate, resulting in a greater heat loss to the surroundings. For the arc-ceiling, furthermore, the flame would attach to the ceiling when the fire is near sidewall. This means the characteristics of temperature distribution under an arc-ceiling may be different from that under a flat ceiling.

In the current work, a numerical simulation tool, Fire Dynamic Simulator (FDS 6.3.2), was employed to build a full scale model with a semicircle cross section based on the tunnel reported in Ref. [5]. The results was used to compare with the previous model with a rectangular cross section, which aimed to investigate the influence of cross-sectional structure on temperature distribution under the ceiling with varying transverse fire locations.

2. Simulation set-up

The total length of the tunnel is 600 m, the maximum width of the tunnel is 14 m and the maximum ceiling height is 7 m. In fact, the tunnel comprises of a 305 m straight section and a 295 m curved section, but it was approximated with a straight tunnel for simplification. The simulation based on this tunnel has been used to study the influence of constraint effect of sidewall on temperature distribution under the ceiling [5]. Wherein, the semicircle cross section was replaced by a rectangle cross section, as shown in figure 1(a).

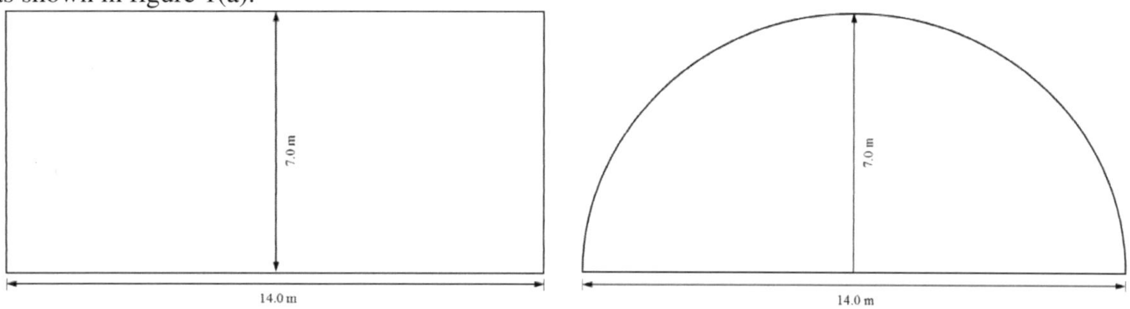

(a) (b)

Figure 1. Cross-sectional dimensions (not in scale). (a) rectangle; (b) semicircle.

In the current study, hence, the main work is to build a more real geometric model with a semicircle cross section, as shown in figure 1(b). Note that only a rectangle grid system is supported in FDS and therefore an insulated obstruction is normally used to produce the inner curved cross section. More detail information can be obtained from Ref. [6].

The longitudinal ventilation was not considered and the openings of the tunnel were set to be "open" boundary conditions. The ambient temperature is 20 ℃. A rectangular fire source of 2 m long, 1 m wide and 0.5 m high with a steady heat release rate of 5 MW was placed at three different transverse locations, namely, d = 7 m (centerline fire), 3.5 m and 0.5 m (wall fire). d is the transverse distance between the center of the fire source and the sidewall. In addition, a multi-grid system with same dimensions from Ref. [5] was employed.

3. Results and discussion

To describe the features of thermal flow under different ceilings, the cross-sectional temperature contours with two kinds of geometric structures: rectangle and semicircle, are shown in figure 2. Wherein, a typical transverse fire locations: 0.5 m, correspond to figure 2 (a) for rectangular cross section and figure 2 (b) for semicircular cross section, respectively. Note that the simulated results of the rectangle tunnel are from the previous literature [5], where the height of fire source is 0 m but the current analysis is not affected significantly by this discrepancy.

For the wall fire, the temperature profile under the arc-ceiling is slender than that under the flat ceiling (figure 3b compared to figure 3a). Nonetheless, the flame under the arc-ceiling directly attaches the ceiling that leads to a greater hazard to the structure of the tunnel. Meanwhile, it should be noted that the flame height of pool fires for the heat release rate of 5 MW is approximately 6.28 m [7], which is less than the maximum ceiling height of 7 m. Therefore, the front of flame could not reach the ceiling when the fire is near the sidewall in the rectangular tunnel. This means the influence of constraint

IOP Conf. Series: Materials Science and Engineering **241** (2017) 012038 doi:10.1088/1757-899X/241/1/012038

effect of sidewall on temperature distribution under the arc-ceiling may be more significant than that under the flat ceiling. To compare the difference between temperature distributions under two ceilings, the quantitative analysis will be used to further discuss this discrepancy.

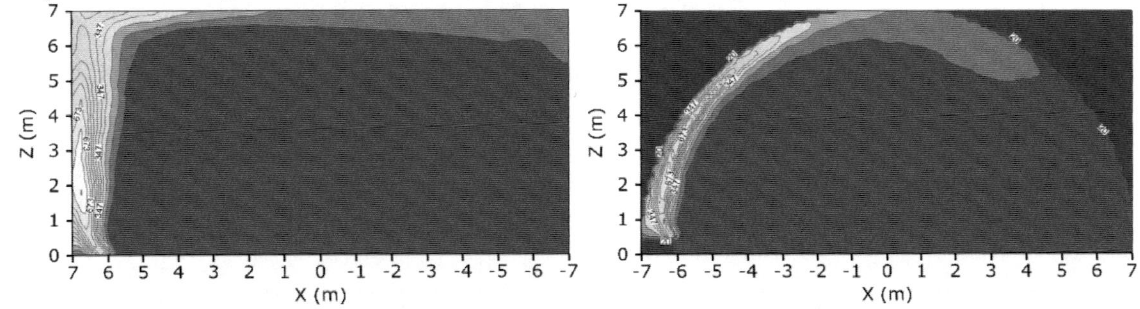

(a) (b)

Figure 2. Cross-sectional temperature contours for $d = 0.5$ m. (a) rectangle; (b) semicircle.

The longitudinal temperature distributions along the centerline of the tunnels are shown in figure 3. It can be seen that the temperature decreases as the longitudinal distance increases. To clarify the trend of the temperature distribution along the ceiling, the decay of temperature is fitted by a power function $\Delta T_{max} \sim r^{a}$. On this basis, an obvious feature can be observed that the maximum temperature rise in the semicircular tunnel is greater than that in the rectangular tunnel, for the reason that it is easier for the buoyancy-driven thermal smoke under the arc-ceiling to flow to the top ceiling and then flow out of the tunnel along the centerline. In contrast, a part of the heat under the flat ceiling lose to the process of smoke movement from the fire to the sidewall, hence greater heat loss to the surroundings in comparison with the arc-ceiling, resulting in greater temperature reduction at the ceiling.

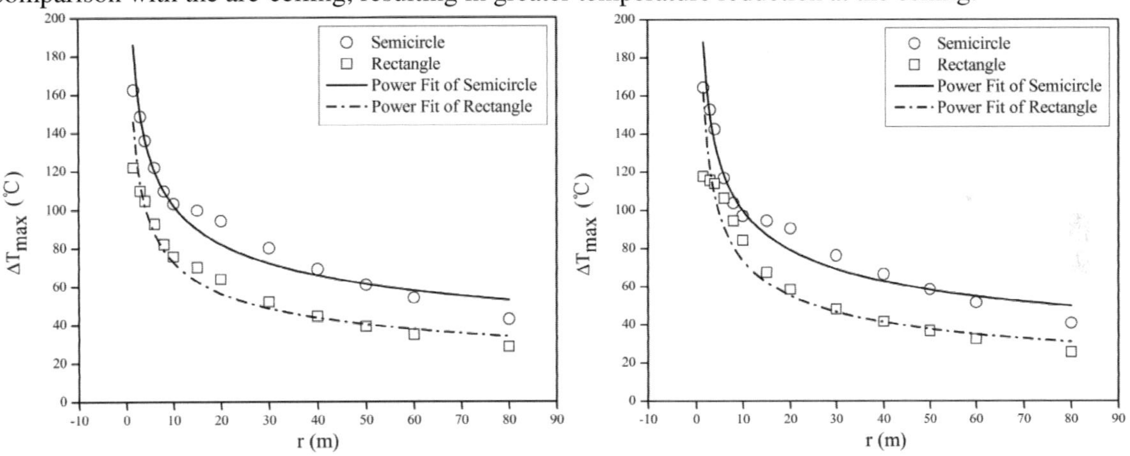

(a) (b)

Figure 3. Longitudinal distribution of maximum smoke temperature at the centerline.
(a) $d = 3.5$ m; (b) $d = 0.5$ m (Wall fire).

In addition, the constraint effect of sidewall under the arc-ceiling has a more significant effect on the maximum ceiling temperature. Figure 4 shows the transverse temperature distribution with varying different fire locations. When the fire is 3.5 m away from the sidewall, the peak of temperature under the arc-ceiling increases significantly, as shown in figure 4(b), compared with the centerline fire ($d = 7$ m) but it remains almost the same under the flat ceiling, as shown in figure 4(a). Although there is also an abrupt change for the flat ceiling when the fire is near the sidewall ($d = 0.5$ m), the maximum ceiling temperature of 450 ℃ is greatly less than 700 ℃ under the arc-ceiling. Thus, the transverse distribution law of maximum temperature under the flat ceiling with varying fire locations in Ref. [5] may be not available under the arc-ceiling and it needs to be further studied in future work.

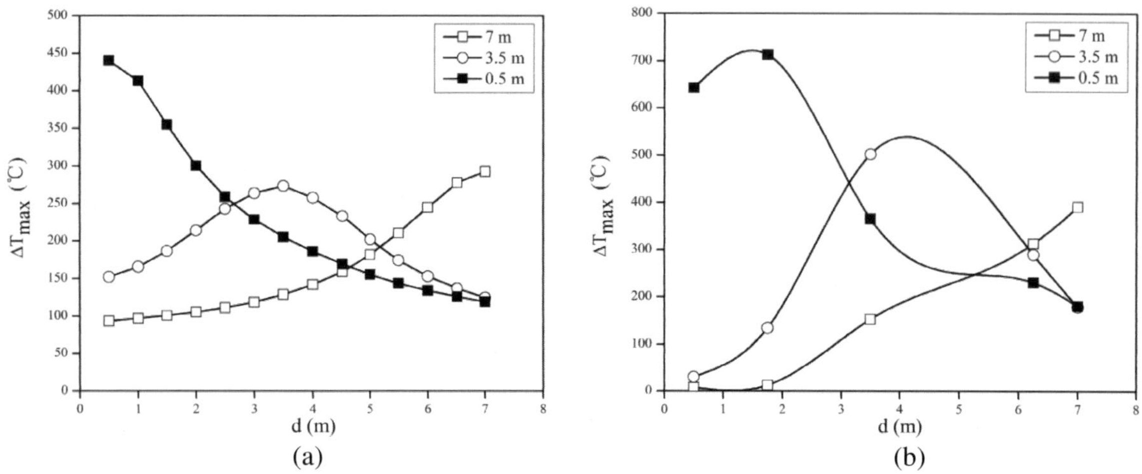

Figure 4. Transverse distribution of maximum smoke temperature at the ceiling. (a) rectangle; (b) semicircle.

4. Conclusions

The flow pattern of fire-induced smoke under the arc-ceiling is different from that under the flat ceiling. For the same fire, the arc cross section of the flow stream has a smaller perimeter and a deeper smoke layer than the rectangular cross section, hence less heat loss to the surroundings, resulting in a greater reduction in temperature along the tunnel centerline under the flat ceiling. In addition, a particular feature for the arc cross section is the flame would attach the ceiling when the fire is near the sidewall, therefore greater ceiling temperature than the rectangular cross section. This indicates the influence of constraint effect of sidewall on temperature distribution under the arc-ceiling is more significant in comparison of the flat ceiling.

Acknowledgements

This work was supported by National Natural Science Foundation of China under Grant No. 51376172. Jianzhong Rong was also financially supported by the Fire Protection and Safety of Illegal Buildings Program under Grant No. 2015XFR11.

References

[1] Ji J, Fan C, Zhong W, Shen X and Sun J 2012 *Int. J. Heat Mass Tran.* **55** 4817-26
[2] Gao Z H, Ji J, Fan C G, Sun J H and Zhu J P 2014 *Energ. Buildings* **84** 13-20
[3] Fan C G, Ji J, Li Y Z, Ingason H and Sun J H 2016 *P. Combust. Inst.* **36** 3121–29
[4] Hu L H, Peng W and Huo R 2008 *J. Hazard. Mater.* **150** 68-75
[5] Zhou T, He Y, Lin X, Wang X and Wang J 2017 *Appl. Therm. Eng.* **112** 932-41
[6] Hu L H, Huo R and Chow W K 2008 *Exp. Therm. Fluid Sci.* **32** 1468-83
[7] Heskestad G 2016 *SFPE Handbook of Fire Protection Engineering* 396-428

IOP Conf. Series: Materials Science and Engineering **241** (2017) 012039 doi:10.1088/1757-899X/241/1/012039

Low air pressure effects on burning characteristics of typical oil with forced irradiance

Li Pan[1,2] , Richard Yuen[2], and Wang Jian[1]

[1]State Key Laboratory of Fire Science, University of Science and Technology of China, Hefei, Anhui, 230026, P. R. China
[2]Department of Civil and Architectural Engineering, City University of Hong Kong, Hong Kong 999077, P. R. China

wangj@ustc.edu.cn

Abstract. In this paper, a report is given on an experimental study of the combustion characteristics of typical oil with forced irradiance under two fixed ambient pressures, which may occurred in real fires. Mass loss and flame axial heat flux distribution were measured to evaluate the hazards. The burning intensity at reduced pressure is relatively lower under the circumstance without incident irradiance because the thin air for reduced pressure may attenuate the combustion. However, the burning intensity at lower pressure is higher due to the lower boiling temperature when the irradiance reaches to 10 kW/m^2.The flame could engulf sufficient air to complete the combustion process for atmospheric pressure condition compared with that under low pressure, and thus resulting in relatively higher flame temperature for a fixed flame height. While in the unified plume region, the weaker air entrainment under lower pressure leads to a poorer cooling effect, i.e. higher plume temperature.

1. Introduction

The combustion characteristics of typical oils with external irradiance have been experimentally studied in the past few decades since the combustible may be radiated by the ceiling or adjacent fire source in real fire. Apte[1] studied the effect of fuel type on combustion characteristics of pool fires using solvents, alcohol and their blends, it is concluded that the smoke extinction area (SEA) increases with the CO and CO_2 yields which in turn increase with the carbon fraction and the aromatic content of the fuel/fuel blend. By testing 14 different crude oils, it is discovered by Iwata [2] that heat release rate (HRR), mass loss rate (MLR), flame radiation, and smoke yield correlates well crude oil density linearly. The ignition and combustion characteristics of three typical oils (diesel, lubricating oil, and aviation kerosene) was studied by Chen[3] and he found that the maximum and average value of HRR, average MLR, and CO/CO_2 ratio seems to be linear with forced irradiance.

In recently year, the pressure effect on the combustion behavior of typical oils has been vastly analyzed since the low ambient pressure environment in plateau districts and cruising airplanes [4-7]. Previous studied and discussion have showed that fire characteristics such as mass loss rate, heat release rate, flame temperature and flame shape is effected by high altitude. However, as far as author's knowledge, no study have been reported on pool fires with external radiant flux under low pressure. Considering the combustion characteristics of typical oils may provide basic knowledge needed for fire protection on Tibet plateau, bench-scale tests of liquid pool fires of typical oil under different external forced irradiance were studied in this study.

2. Experimental Setup

The experimental setup of the chamber fire tests is illustrated in figure 1. Liquid N-Heptane with an industrial purity greater than 99% was selected as testing fuel, whose burning characteristic is similar to diesel. A circular pan with diameter of 10 cm and height of 2 cm was adopted, and the liquid N-heptane was filled to a height of 1.6 cm in each test. Five radiative heat flux, i.e., 0.03, 5, 10 kW/m^2, were selected originating from the conical heater, which is accordance with the heating source of the standard of ISO 5660. An altitude chamber with the internal size of 2×3×2 m was adopted to provide the low air pressure condition. Experiment under two fixed ambient pressure, i.e. 64, and 100 kPa will be conducted with the adjusting accuracy of chamber pressure is ±1 kPa.

An electronic scale with resolution of 0.01 g was used to record the mass loss. A radiative heat flux sensors was located with a horizontal distance of 60 cm from the burner centre. An array of 19 thermocouples spacing 5 cm was located vertically along the axis of the burner. The temperature measurements reported in this study were the direct thermocouple readings without radiation correction which may yield an uncertainty less than 10% [8] depending largely on the level of soot radiation. The sampling rates of electronic scale and thermocouple were both 1 Hz. The fuel was ignited by applying 25 V electric current on 0.3-mm-diameter Ni-Cr wire.

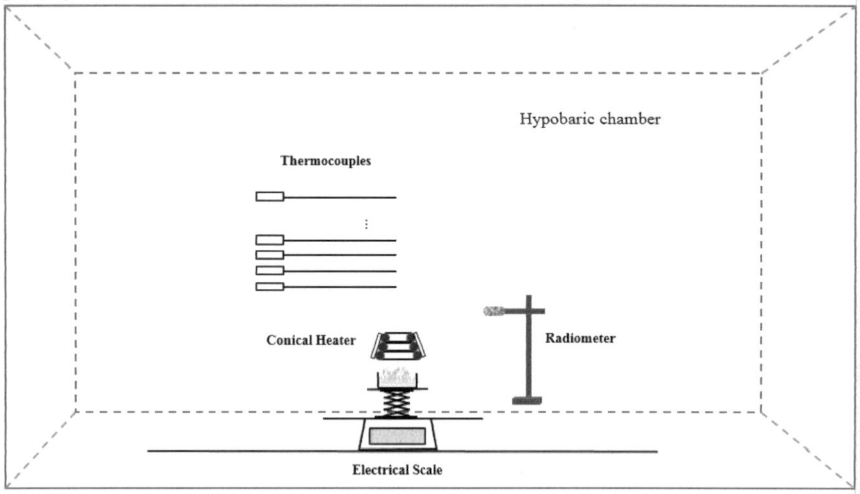

Figure 1. Experimental setup of the chamber fire tests

3. Results and discussion

For precise evaluation on the experimental results, the mass loss rates over the three repeated tests for each configuration are averaged to give the ultimate results. Based on that, the mass loss rates under three irradiances are plotted in figure 2, respectively, where the pressure effect on the results can be clearly observed.

Previous research have indicated that the mass transfer has an exponential relationship with ambient pressure, i.e. $\dot{m} \sim P^{\alpha}$, and thus, the burning intensity at reduced pressure is relatively lower under the circumstance without incident irradiance. For the case of 0.03 kW/m^2, the imposed irradiances are quite small, and play an unobservable role in affecting the heat feedback mechanism compared with that from the flame. Therefore, the combustion behaviours of these two cases seem to follow the basic trend of thin-layer pool fires, where the case under normal pressure underwent four combustion stages, i.e. pre-burning stage, quasi-steady burning stage, boiling stage and decay stage, and for the case under reduced pressure, the mass loss rate curves tend to reach to a plateau after a short pre-burning stage until the commencement of decay stage.

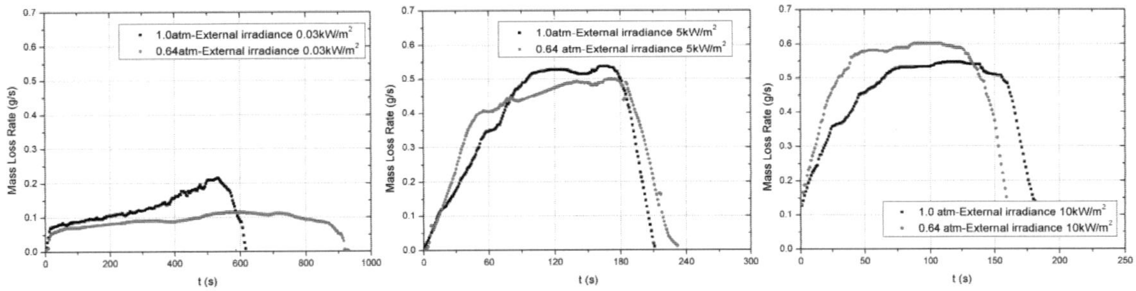

Figure 2 Comparison of mass loss rate under two different pressures

Meanwhile, it is worthwhile to note that for reduced pressure condition, the maximum burning rate at steady stage gradually approaches to the value at normal pressure, and even exceeds it at 10 kW/m^2. Correspondingly, the duration of combustion is gradually shortened with increasing irradiance, and ultimately shorter than the case of 1 atm. In effect, under these two irradiances, the burning intensities are extremely violent, showing significantly higher mass loss rates than the cases of 0.03 kW/m^2.The fact that the differences in mass loss rate between two pressures appears to be vague for irradiances larger than 5 kW/m^2 leads to an inference that a transition region between 0.03~5 kW/m^2 may exist, where the incident radiative heat flux together with the heat feedback from the flame rapidly heat up the liquid fuel to the boiling point, resulting in the drastic, but steady combustion process.

In general, the ambient pressure will influence the boiling point of the fuel. For n-heptane in current study, its boiling temperatures are 98.5 ^{o}C and 89 ^{o}C in Hefei and Lhasa [9~10], respectively, which corresponds to the tested pressures in the altitude chamber. Though the thin air for reduced pressure may attenuate the combustion, the mass loss rate may be dominated by the vaporization rate caused by the elevated incident heat flux. As shown in figure 2, as the irradiance reaches to 10 kW/m^2, the burning intensity at lower pressure is higher due to the lower boiling temperature.

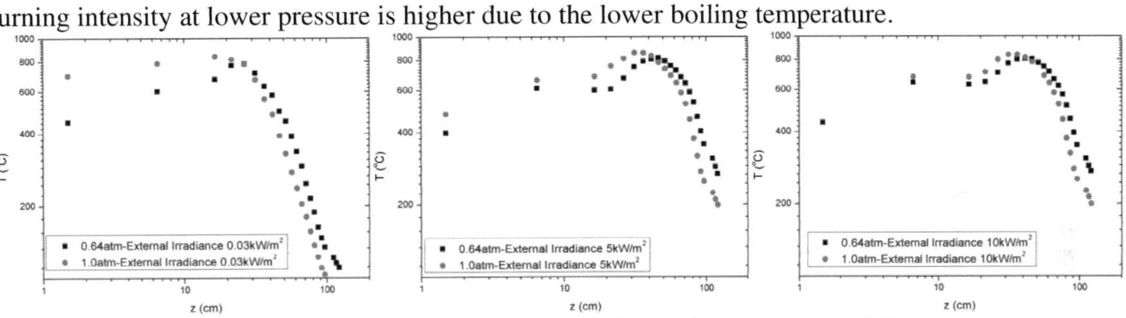

Figure 3 Comparison of centerline temperature distribution under two different pressures

Temperature is the vital parameter in fire analysis, which has been measured along the axis of the burner and plotted in figure 3. The maximum flame temperatures are about 800 ^{o}C. It is clear that there is a consensus among the three figures, i.e. under low pressure condition, the temperature is smaller at lower vertical height, and subsequently exceeds the temperature at normal pressure. The transition point lies between the measurement points of 30~45 cm.

Here, we simply partition the axial centreline temperature into two regions according to the correlation developed by Heskestad [11]: 1) flame region, where the temperatures are larger than 500 ^{o}C; 2) the oscillating flame zone and none-combusting plume regions are roughly marked as the unified plume region. As seen in figure 3, the flame temperatures at reduced pressure are lower than that for atmospheric pressure. This may be because in the flame region, the flame could engulf sufficient air to complete the combustion process for atmospheric pressure condition compared with that under low pressure, and thus resulting in relatively higher flame temperature for a fixed flame height. While in the unified plume region, the temperature will gradually decrease due to the cooling effect caused by the air entrainment. Through the experimental investigation on temperature profile of buoyant spill plume in a reduce pressure atmosphere at high altitude, Tang et al., found that the air entrainment of the buoyant spill plume is weaker in the reduced pressure atmospheric, being about 0.8 time of that in the normal pressure condition. Consequently, the weaker air entrainment under lower pressure leads to a poorer cooling effect, i.e. higher plume temperature.

4. Conclusion

In this study, an experimental study of the combustion characteristics of typical oil with forced irradiance were conducted to reveal the pressure effects on mass loss rate and flame axial temperature distribution. The burning intensity at reduced pressure is relatively lower under the circumstance without incident irradiance because the thin air for reduced pressure may attenuate the combustion. However, the burning intensity at lower pressure is higher due to the lower boiling temperature when the irradiance reaches to 10 kW/m^2.The flame could engulf sufficient air to complete the combustion process for atmospheric pressure condition compared with that under low pressure, and thus resulting in relatively higher flame temperature for a fixed flame height. While in the unified plume region, the weaker air entrainment under lower pressure leads to a poorer cooling effect, i.e. higher plume temperature.

Acknowledgments

This research was supported by the National Natural Science Foundation of China (No. 51376172) and the grant from the Research Grant Council of the Hong Kong Special Administrative Region, China (contract grant number CityU 11301015).

References

[1] Apte, V., 1998. Fire Saf. J., **31**(4) 283-298.
[2] Iwata, Y., Koseki, H., Janssens, M. L., & Takahashi, T. 2001. Fire Mater., **25**(1), 1-7.
[3] Chen, X., Lu, S., Li, C., Zhang, J., & Liew, K. M. 2014. Fire Mater., **38**(3), 409-417.
[4] Li, Z. H., He, Y., Zhang, H., & Wang, J. 2009. Proc. Comb. Inst., **32**(2), 2481-2488.
[5] Hu, X., He, Y., Li, Z., & Wang, J. 2011. Proc. Comb. Inst., **33**(2), 2607-2615.
[6] Hu, L., Tang, F., Wang, Q., & Qiu, Z. 2013. Fuel, **111**, 298-304.
[7] Tang, F., Hu, L., Zhang, X., Zhang, X., & Dong, M. 2015. Fuel, **139**, 18-25.
[8] Luo, M., He, Y., & Beck, V. 1997. Fire Saf. J., **29**(1), 1-25.
[9] Zhou, Z., Wei, Y., Li, H., Yuen, R., & Jian, W. 2014. Int. J. Heat Mass Tran., **70**, 578-585.
[10] Fang, J., Tu, R., Guan, J. F., Wang, J. J., & Zhang, Y. M. 2011. Fuel, **90**(8), 2760-2766.
[11] Heskestad, G. 1983. Fire Saf. J., **5**(2), 109-114.

Evaluation of Preduster in Cement Industry Based on Computational Fluid Dynamic

E L Septiani[1,a], W Widiyastuti[2,b], A Djafaar[3], I Ghozali[3], H M Pribadi[3]

[1] Department of Chemical Engineering, Universitas Internasional Semen Indonesia (UISI), Kompleks PT. Semen Indonesia (Persero) Tbk., Gresik 61122 Indonesia
[2] Department of Chemical Engineering, Industrial Technology Faculty, Institut Teknologi Sepuluh Nopember, Surabaya 60111, Indonesia
[3] Directorate of Production, PT. Semen Gresik (Persero) Tbk., Ds SumberArum, Kerek, Tuban 62356 Indonesia

[a]eka.septiani@uisi.ac.id, [b] widi@chem-eng.its.ac.id

Abstract. Ash-laden hot air from clinker in cement industry is being used to reduce water contain in coal, however it may contain large amount of ash even though it was treated by a preduster. This study investigated preduster performance as a cyclone separator in the cement industry by Computational Fluid Dynamic method. In general, the best performance of cyclone is it have relatively high efficiency with the low pressure drop. The most accurate and simple turbulence model, Reynold Average Navier Stokes (RANS), standard k-ε, and combination with Lagrangian model as particles tracking model were used to solve the problem. The measurement in simulation result are flow pattern in the cyclone, pressure outlet and collection efficiency of preduster. The applied model well predicted by comparing with the most accurate empirical model and pressure outlet in experimental measurement.

1. Introduction

The highest heat loss as hot air in cement industries obtained from kiln have been used as heater media in calciner and multi-stage preheater with countercurrent flow. After heating raw materials from preheater, hot air still used to reduce water content of coal, fuel of calciner and kiln, in the coal mill. However, this hot air is existing ash content therefor separating process is needed first before sent to coal mill. Preduster, cyclone separator, has been used for removing large particles in many industries because it has low capital cost, very high duct loading, ability to operate at high temperature, and low maintenance requirement but it have low efficiency especially for small particles [1,2]. Therefore, evaluation of preduster efficiency is needed. Many researchers have been looking for cyclone efficiency both in operating cost and collection efficiency by experiment or theoretic. Chuah et. al. [3] compared experimental and theoretical pressure drop in cyclone by Shepherd and Lapple (1939), Casal and Martinez (1984), Dirgo (1985) and Coker (1993) model. Their study shows Coker model predicted cyclone pressure drop, which is related to operating cost and collection efficiency. Bogodage and Leung [4] estimated overall collection efficiency of cyclone by experimental method, then improve the cyclone separator performance by variation of down-comer tubes. Nevertheless, these method are not feasible to predict cyclone efficiency in industrial scale due to manufacturing scale-up parameter and operation cost. Now a adays, theoretical cyclone efficiency widely evaluated by Computational Fluid Dynamics (CFD) method after the first CFD investigation presented by Boyson et. al.. They reported pressure drop that is predicted by CFD has a great agreement with measured data.

CFD offered momentum, heat, and mass transfer data to simulate cyclone operation which is based on combination of mass rotational forces, drag forces, and other forces [5]. Brennan et. al [6] validated performance of industrial cyclone separators by CFD modelling. Their study compared Differential Reynold Stress (DRSM), Large Eddy Simulation (LES), and Laminar flow model. They reported that DRSM and LES model predicted well. Kepa [7] also reported that CFD modelling allows the analysis of large-size cyclones that are very difficult to analyze in experimental studies by RSM turbulence model. Since LES and RSM method burdened with a lot of computational effort [8-10] although let the result that closed to experimental measurement, Reynold Average Navier Stokes (RANS), the simpler model gave accurate estimation [11], may be considered to provide this present study. For particles tacking is defined by Lagrangian calculation that have been commonly used. Since collection efficiency is affected by velocity inlet [12] and the rate of hot air is can't be changed, here the influence of inlet dimension in Le/D ratio to pressure drop and collection efficiency of the preduster was examined.

2. Methodology

Preduster performance in this study was calculated by commercial CFD, ANSYS FLUENT 16.2. The geometry was defined in 3-D calculation domain. Figure 1(a) shows the sketch of preduster dimension in millimeter size.

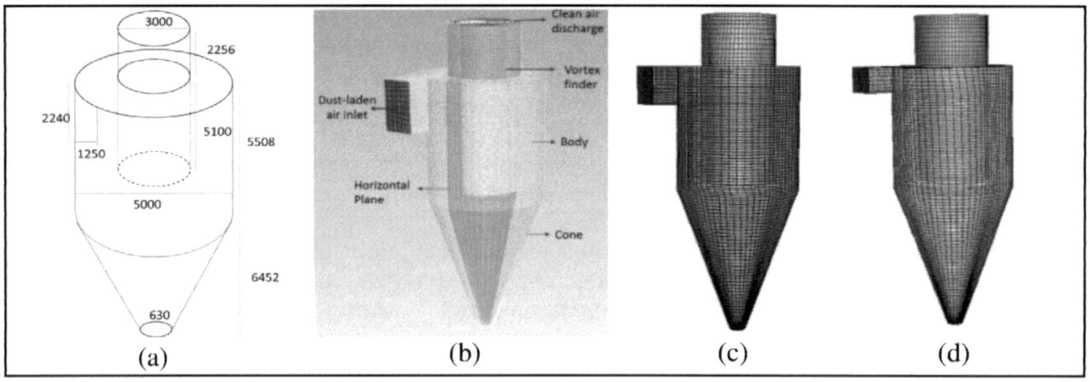

Figure 1. (a) Preduster geometry, (b) Name section and Preduster grid of Le/D = 0.5 (current preduster), (c) Peduster grid of Le/D = 0.3, (d) Preduster grid of Le/D = 0.2.

The geometry was defined by Workbench Design Modeller software then should be devided by Meshing software into several grids. Here, the grids of current preduster were generated in 84,148 grids with skewness 0.153 as can be seen at Fig. 1(b). While for Le/D=0.3 was divided by 85,734 meshes with skewness 0.159 and 84,914 grids with skewness 0.158 for Le/D=0.2. Its mean the calculated domain has relatively high quality. Reynold Average Navier Stokes model is used to describe turbulence model by assuming fully turbulence flow. This model solved two equations [5,13], turbulence kinetic energy (k) and dissipation rate (ε), here

$$\frac{\partial}{\partial t}(\rho k)+\frac{\partial}{\partial x_i}(\rho k u_i)=\frac{\partial}{\partial x_j}\left[\left(\mu+\frac{\mu_t}{\sigma_k}\right)\frac{\partial k}{\partial x_j}\right]+G_k+G_b-\rho\varepsilon-Y_M+S_k \quad (1)$$

$$\frac{\partial}{\partial t}(\rho \varepsilon)+\frac{\partial}{\partial x_i}(\rho \varepsilon u_i)=\frac{\partial}{\partial x_j}\left[\left(\mu+\frac{\mu_t}{\sigma_k}\right)\frac{\partial \varepsilon}{\partial x_j}\right]+C_{1\varepsilon}\frac{\varepsilon}{k}(G_k+C_{3\varepsilon}G_b)-C_{2\varepsilon}\rho\frac{\varepsilon^2}{k}+S_\varepsilon \quad (2)$$

Where $G_k=-\rho\overline{u_i'u_j'}\frac{\partial u_j}{\partial x_i}$, eddy viscosity calculated by $\mu_t=\rho C_\mu\frac{k^2}{\varepsilon}\mu_t$. C_μ is constant 0.09. $C_{1\varepsilon}, C_{2\varepsilon}$ are 1.44; 1.92, $\sigma_k, \sigma_\varepsilon$ are Prantdl Number 1, 1.3. S_k, S_ε are user defined source. Meanwhile, the particles movement was defined by Lagrangian model, Discrete Phase Model (DPM), which solved that force balance of particle is

$$\frac{du_p}{dt}=F_D(\vec{u}-\vec{u_p})+\frac{\vec{g}(\rho_p-\rho)}{\rho_p}+\vec{F} \quad (3)$$

$\vec{F}$ is additional acceleration, $F_D(\vec{u}-\vec{u_p})$ is drag force each particle mass, and F_D presented by $F_D = \dfrac{18\mu}{\rho_p d_p^2}\dfrac{C_D \text{Re}}{24}$. While $\vec{u}$ and $\vec{u}_p$ are fluid velocity and particle velocity [5]. Contour of pressure in the preduster was observed from horizontal plane that shown at Fig. 1(b). Overall collection efficiency was calculated by

$$\eta_o = \sum \eta_j m_j \tag{4}$$

$$\eta_j = \frac{1}{1+(\frac{d_{pc}}{d_{pj}})^2} \tag{5}$$

$$d_{pc} = \left[\frac{9\mu W}{2\pi N e V_i(\rho_p - \rho_g}\right]^2 \tag{6}$$

m_j is mass fraction of particles in the jth particle size range. d_{pj} is characteristic diameter of the jth particle size range (in μm). Ne is number of effective turns and Vi is velocity inlet.

3. Result and Discussion

The result of this work indicates well prediction. It can be seen in Fig. 2(a), the air and particles through inlet of preduster entered from cyclone top. This dust-laden air flow in a downward spiral against the cyclone body's outer wall, that is called outer vortex while particles were separated by centrifugal force. The outer vortex gets smaller as the air flows downward along the cone. Furthermore, the vortex hopped up in smaller spiral flow, it is called inner vortex, leaves the preduster to outlet in the top by passing the vortex finder.

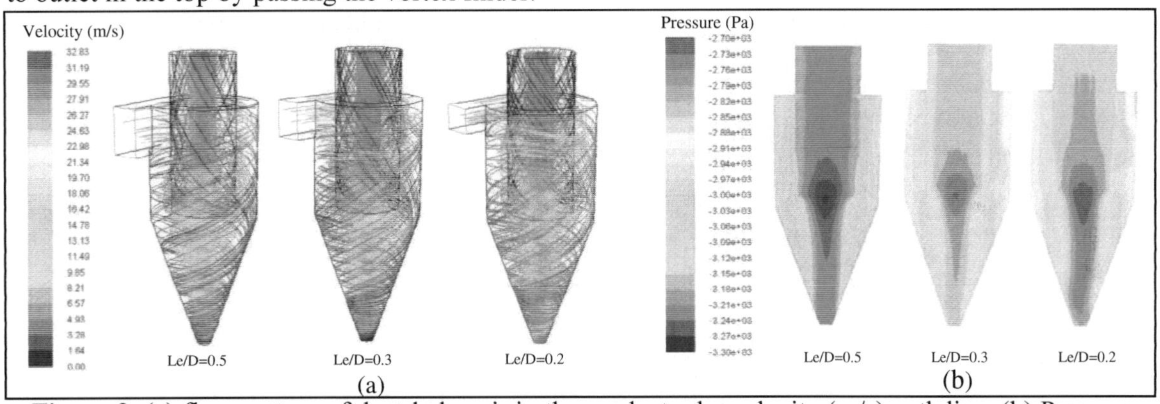

Figure 2. (a) flow pattern of dust-laden air in the preduster by velocity (m/s) path line, (b) Pressure contour in Horizontal plane (Pa)

Figure 2(b) shows similarly approximated pressure outlet to measured pressure in plant. CFD method predicted pressure outlet in current preduster -3205 Pa, while the actual pressure inlet of hot air inlet in the coal mill is -2546.5 Pa. this difference may caused by the number of geometry mesh, besides the friction losses by elbow and pipe which have length about over 10 meters. The pressure drop in CFD method give well comparison with the most accurate empirical model, Shepherd and Lapple method, as can be seen in Fig. 3(a). Increasing velocity inlet due to reducing Le/D ratio will be rise the pressure drop in the preduster.

Ash particles content in hot air were being approached by Rosin-Rammler distribution in particle range between 0.5 μm and 25 μm with 10% of hot air flow rate. particles passed both in clean air discharge and collected dust discharge was calculated in steady state condition. From this, the over-all collection efficiency calculation for current preduster using eq. (4) is 47.59%. This relatively low efficiency may be affected by the dimension of preduster so that the particles also swirl in the inner vortex from the bottom to the gas outlet and the largest number of particles loading in which smaller than 5 μm. Whereas decreasing Le/D ratio to 0.3 increase the overall efficiency until 66.95%, it can be caused by some particles go down directly without flowing in the inner vortex from the bottom as seen as Fig. 3(b). however, reducing Le/D ratio to 0.2 did not increase overall collection efficiency because the particles being trapped in the swirling hot air flow in the preduster.

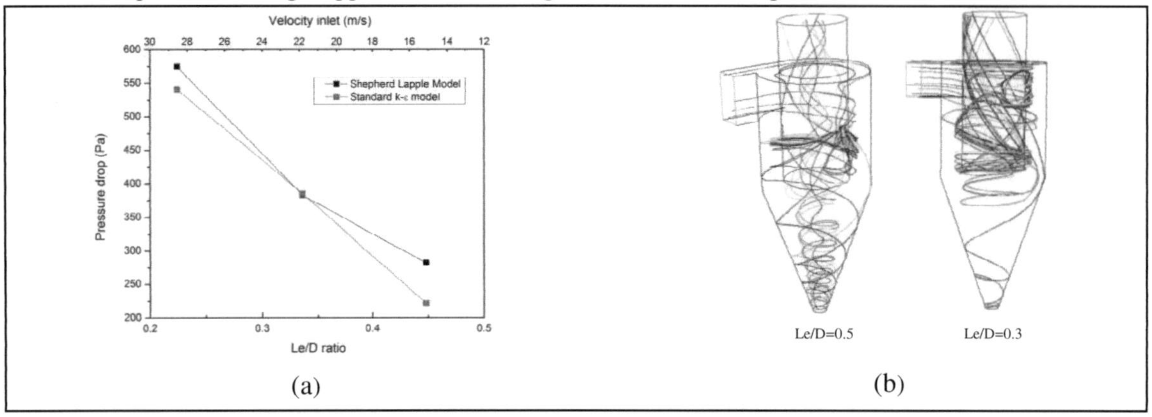

Figure 3. (a) Pressure drop comparison between Shepherd and Lapple and Standard k-ε method, (b) Particle tracking in the preduster

4. Summary

This present study proved that CFD can predicted the performance of cyclone well that is shown by flow pattern of air containing dust, pressure outlet approximation in current preduster and good comparison pressure drop to Shepherd and Lapple model. The overall collection efficiency was relatively low for the current preduster. Increasing velocity due to Le/D ratio of 0.3 rise the overall collection efficiency, however reducing Le/D to 0.2 drop the overall collection efficiency.

Acknowledgement

Research grant allowed by LPPM-UISI is gratefully acknowledged. PT Semen Indonesia (Persero) Tbk. Is also acknowledged for providing necessary data.

References

[1] Y. Kosaki, T. Hirai, Y. Yamanaka, K. Takeshima. 2015 *J. Powder Technology* 277 p. 22–35
[2] F. Ficici, V. Ari, M. Kapsiz.2010 *International Journal of the Physical Sciences* Vol. 5 (6), p. 804-813
[3] T. G. Chuah, J. Gimbun, T. S. Y. Choong. A. Fakhrulrazi.2003 *J. Chem. Eng. and Environmental* Vol. 2, No.2, p 67 – 71
[4] S. G. Bogodage and A. Y. T. Liung 2016 *J. Hazardous Material* 311 p. 100 – 114.
[5] Anonim, 2011 *ANSYS FLUENT 14.0 Teory Guide* (Canonsburg: ANSYS Inc,)
[6] M. Brennan, P. Holtam, M. Narasimha. 2014 *International journal of Minerals Process* 133 p.1-12
[7] Kepa. 2013 *J. Separation and Purification Technology* 118 p. 105–111
[8] G. Wan, G. Sun, X. Xue, M. Shi. 2008 *J. Powder Technol.* 183, 94.
[9] J.J. Derksen, H.E.A. van den Akker, S. Sundaresan.2008 *AIChE J.* 54 (4) 872.
[10] F.J. Souza, R.V. Salvo, D.A.M. Martins. 2012 *J. Separat. Purificat. Technol.* 94, 61.
[11] E. L. Septiani, W. Widiyastuti, S. Winardi, S. Machmudah, T. Nurtono, and Kusdianto. Effect of turbulence modelling to predict combustion and nanoparticle production in the flame assisted spray dryer based on computational fluid dynamics.2016 *Padjadjaran International Physics Symposium 2016. AIP Conference Proceedings*, 1712, 040002
[12] Sgrott Jr., O. L., Noriler, D., Wiggers, V. R., Meier, H. F. 2015 *J.Powder technology* 2015.02.039
[13] Versteg, H. K. and Malalasekera W. 2007 *An introduction to computational fluid dynamics* (England: Pearson Education Limited)

IOP Conf. Series: Materials Science and Engineering **241** (2017) 012041 doi:10.1088/1757-899X/241/1/012041

Feasibility Study of Seawater Electrolysis for Photovoltaic/Fuel Cell Hybrid Power System for the Coastal Areas in Thailand

A Srisiriwat[1] and W Pirom[2]

[1]Department of Advanced Manufacturing Technology, Faculty of Engineering, Pathumwan Institute of Technology, Bangkok 10330, Thailand
[2]Department of Information Technology, Faculty of Science and Technology, Pathumwan Institute of Technology, Bangkok 10330, Thailand

srisiriwat@gmail.com

Abstract. Solar photovoltaic cell and fuel cell are the practicable options to realize as a possible hybrid power system because the power of the sun cannot be utilized at night or cloudy days but hydrogen has been found as an ideal energy carrier for being transportable, storable, and converting energy though fuel cell. Hydrogen storage is chosen for its ability to obtain a clean energy option. Electrolysis, which is the simplest process to produce hydrogen, can be powered by the dc voltage from the photovoltaic cell instead of using the battery as power supply. This paper concentrates on a feasibility study of seawater electrolysis using photovoltaic power integrated fuel cell system for the coastal cities in Thailand. The proposed system composed of photovoltaic arrays, seawater electrolyzer and fuel cell is presented when the 10-kW of fuel cell electrical power is considered. The feasibility study of hydrogen production and energy analysis of this proposed system is also evaluated.

1. Introduction

As the development of renewable energy in Thailand increased gradually and the perspective of Ministry of Energy of Thailand in energy security, economy and ecology to increase domestic renewable energy production has been focused, the Alternative Energy Development Plan (AEDP2015) was developed to set a target of renewable energy sources to replace 30 percent of final energy consumption (in form of electricity, heat and Bio-fuel) by the year 2036 [1]. The target of electricity generation from solar energy is approximately 6,000 MW. In order to reach the goal for developing renewable energy sources, a hybrid power system have been adopted to increase the system reliability and security. Moreover, one of the most promising hybridization approaches is the combination of photovoltaic (PV) panels to other resources [2]. In this paper, we have applied the solar power system via PV to generate hydrogen for energy storage purposes of the hybrid power system. Hydrogen can be produced by a water electrolysis process powered by the electrical energy and thereafter it can be used to produce electricity through fuel cell (FC). The coastal regions, especially in the remote area power supply, are focused to develop the PV/FC hybrid power system using seawater electrolysis. When we consider water as the raw material for hydrogen production, it would be a noticeable advantage to consume these abundant salt water resources for electrolysis, rather than using fresh water. Moreover, the high amount of hydrogen produced from electrolysis of seawater [3-5] or electrolyte added with sodium chloride (NaCl) [6] was observed. Slama [6] found that the addition of NaCl in various types of electrolyte can increase the hydrogen production although it also increases the consumed electrical power because of higher electrolyte conductivity. Normally, seawater in the world's oceans has about 3.5% of dissolved salts predominantly NaCl.

Many literatures have investigated the hydrogen production via seawater electrolysis integrated solar system, for example to examine the variation of the electric tension and energy consumption of the natural seawater electrolysis comparing with the electrolysis of 15%NaOH solution [4] and to predict the hydrogen production rate in terms of panel temperature, solar irradiation, and current [5]. Abdel-Aal et al. [7] estimated the economic feasibility of a solar-energy stand-alone system including PV modules, water electrolyzer and FC. In addition, it is important to investigate the optimal design of a PV/FC hybrid power system focused on the economical performance [2] and the effects of current, voltage, electrical power on the performance of electrolyzer for hydrogen production [8]. However, a study on the hybrid PV/FC power system integrated seawater electrolysis is limited. As the Phi Suea House project in Chiang Mai, Thailand was introduced as the world's first 24-hour solar-powered hydrogen storage in a multi-house compound [9], it is a motivation of this work to apply the PV/FC hybrid system by using the electrolysis from seawater as a stand-alone hybrid power system for the coastal areas in Thailand.

In this work, the seawater electrolysis for PV/FC hybrid power system is preliminary investigated as the stand-alone hybrid power system for the coastal areas in Thailand. This paper presents a feasibility study of the PV/FC hybrid power system when seawater electrolyzer is used for hydrogen production when the 10 kW of electrical power of the whole FC stack produced from this hybrid power system is set as the target.

2. System Description

A schematic of seawater electrolysis PV/FC hybrid system is demonstrated in figure 1. The main components are PV arrays, seawater electrolyzer, hydrogen storage and FC. Electricity is firstly generated by PV cells for day time at the appropriate solar radiation. Seawater electrolyzer is then powered by electricity generated from PV to produce hydrogen. The electrode processes at seawater electrolysis [4] are:

- Hydrogen production at cathode, conform to the reaction:

$$2H_2O + 2e^- \rightarrow H_2 + 2OH^- \tag{1}$$

- Oxygen and/or chlorine production at anode, conform to the reactions:

$$6H_2O \rightarrow O_2 + 4H_3O^+ + 4e^- \tag{2}$$

$$2Cl^- \rightarrow Cl_2 + 2e^- \tag{3}$$

The produced hydrogen is compressed in the hydrogen storage tank and it could be noted that the high amount of power is utilized for storing hydrogen at high pressures. However, compressor costs are based on the amount of work done by the compressor, which depends on the inlet pressure, outlet pressure, and flow rate. For night time, electricity is produced by FCs when hydrogen is utilized as fuel. FCs also produce fresh water as a by-product. Moreover, oxygen produced in electrolysis process can be utilized, instead of air, in the chemical reaction taking place in the FC. Inside the PEMFC, the reactions that take place at anode and cathode to generate both of the electricity and water can be described as

$$\text{Anode:} \quad H_2 \rightarrow 2H^+ + 2e^- \tag{4}$$

$$\text{Cathode:} \quad O_2 + 4H^+ + 4e^- \rightarrow 2H_2O \tag{5}$$

$$\text{Overall:} \quad H_2 + \tfrac{1}{2}O_2 \rightarrow H_2O + \text{electricity} \tag{6}$$

The merits of the hybrid power system are summarized as follows:

- It can supply electricity for day and night use.
- It can utilize availability of hydrogen gas for energy generation using FCs.
- It can supply fresh water as a by-product from the FC.
- It can supply heat source as a by-product from the FC (depending on types of FC).

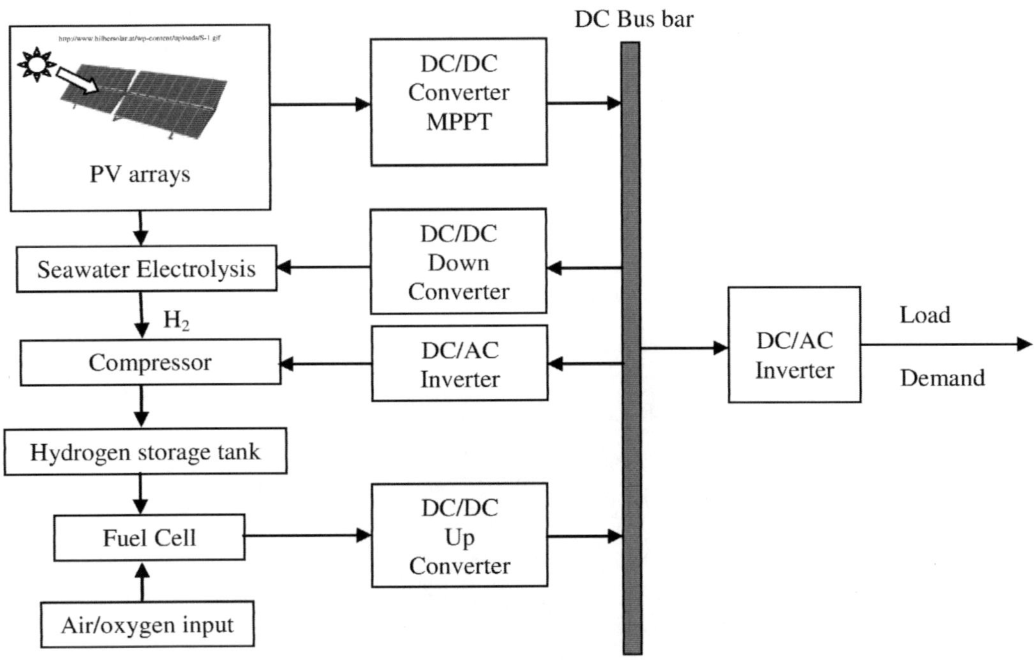

Figure 1. Schematic of the seawater electrolysis of PV/FC hybrid power system (adapted from [2]).

3. Fuel Cell

Firstly, we considered the electric power production from FC when the produced hydrogen from electrolysis was utilized. In this work, the target of 10-kW electrical power produced from FC for day time or cloudy time was investigated to design the size of electrolysis and numbers of PV panels using in this process. The rate of usage of hydrogen is derived from the basis operation of the FC that there are two electrons from each mole of hydrogen as given in equation (7) and equation (8)

$$H_2\ usage = \frac{I}{2F}\ moles^{-1} \tag{7}$$

and

$$H_2\ usage = 1.05 \times 10^{-8} \times \frac{P_e}{V_c}\ kg\,s^{-1} \tag{8}$$

Where I is the DC current, F the Faraday's constant (96,484 C/mol), P_e the electrical power of the whole FC stack, and V_c is the average voltage of each cell in the stack. In this paper, the using V_c = 0.65 V will give a good approximation [11]. For the 10-kW FC operating for 1 h, at the cell voltage of 0.65 V, the H_2 usage is 1.62 x 10^{-4} kg s^{-1} (108.12 liters h^{-1}); the O_2 usage is 1.28 x 10^{-3} kg s^{-1} and the rate of water production is 1.44 x 10^{-3} kg s^{-1}. This corresponds to an efficiency of 43.9% (using HHV).

4. Hydrogen Production using Seawater Electrolysis

Electrical energy from the PV system is used to split water into hydrogen and oxygen through the process of seawater electrolysis. Hydrogen is an energy carrier and it is stored until the FC utilizes it. The FC will convert the hydrogen back into electrical energy to supply the loads during periods when the PV system is not producing any power.

For the preliminary test for seawater electrolysis, the experiments were set up as illustrated in figure 2 and the seawater electrolysis testing parameters and the hydrogen production were given in table 1. In theoretically, electrolysis at ambient temperature and pressure requires a minimum voltage of 1.481 V and therefore, a minimum energy of 39.4 kWh kg^{-1}hydrogen [12] whereas this experiment gave the

IOP Conf. Series: Materials Science and Engineering **241** (2017) 012041 doi:10.1088/1757-899X/241/1/012041

energy of 190.4 kWh kg^{-1}hydrogen and efficiency of 41.7%. According to the designed 10-kW FC power output, for about 18 h during periods when the PV system is not producing any power, the hydrogen rate 1.62 x 10^{-4} kg s^{-1} (108.12 liters h^{-1}) was fed to FC to produce electricity and for the total time of FC operation, the hydrogen of about 2,000 liters was utilized; consequently the hydrogen storage should be sufficiently prepared for support the PV/FC hybrid system. However, the detail of hydrogen storage design and energy consumed for compressing hydrogen will be considered and proposed in the future work because there are many types of hydrogen storage tanks. In the hydrogen production via seawater electrolysis of this work, hydrogen was produced 1 liter h^{-1} with a power demand of 9.6 W whereas the results from the Phi Suea House project was found that 1 liter h^{-1}of hydrogen should be produced with a power demand of 4.2 W.

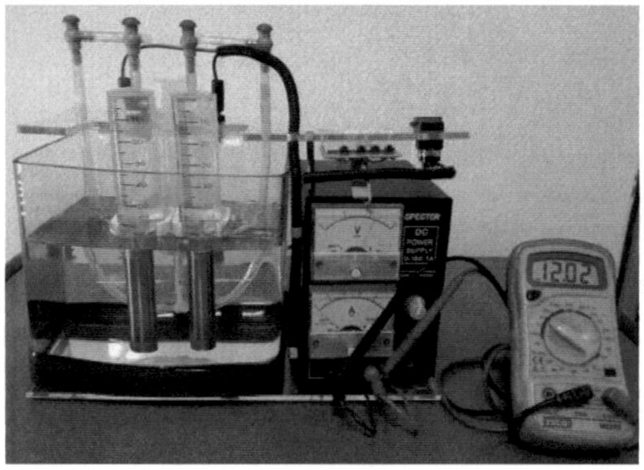

Figure 2. Seawater electrolyzer used in experiments.

Table 1. Seawater electrolyzer parameters.

Seawater electrolyzer parameters	Value
Temperature	26 °C
Cathode material/area	316 SS / 125.6 cm^2
Anode material	316 SS / 125.6 cm^2
Electrolyte	Seawater at Samut Sakhon Province, Thailand
Input voltage	3 V
Electrolysis current density	3.974 mA/cm^2
Hydrogen production rate	155.65 ml/h
Energy requirement	190.4 kWh kg^{-1} hydrogen
Efficiency	41.7%

5. Photovoltaic Cell

There are three basic types of solar PV cells, monocrystalline silicon, polycrystalline silicon and amorphous silicon. The efficiency of different types of silicon based PV cell is shown in table 2. Based on the Phi Suea House project [9], the polycrystalline PV cell is also chosen for this feasibility study. Therefore, the selected PV cell has an efficiency of approximately 15%. According to the initial experiment the power requirement from PV arrays is 12.3 kW. However, PV array site is determined from the availability of space, typically 1 kWp requiring 8 m^2 of roof/space, that faces south with slope of around 30 – 40 [10]. Figure 3 shows the solar radiation in different areas in Thailand (figure 3(a)) and the monthly average daily solar radiation in Thailand (figure 3(b)). This can be noted that the coastal areas of Thailand are appropriate for using seawater electrolysis integrated PV/FC hybrid system as a stand-alone hybrid power system.

Table 2. Efficiency percent of different types of silicon based PV cells [10].

Type/Material of PV cell	Level of efficiency in Laboratory (%)	Level of efficiency in Production (%)
Monocrystalline Silicon based	24 (Approximate)	14-17
Polycrystalline Silicon based	18 (Approximate)	13-15
Amorphous Silicon based	13 (Approximate)	5-7

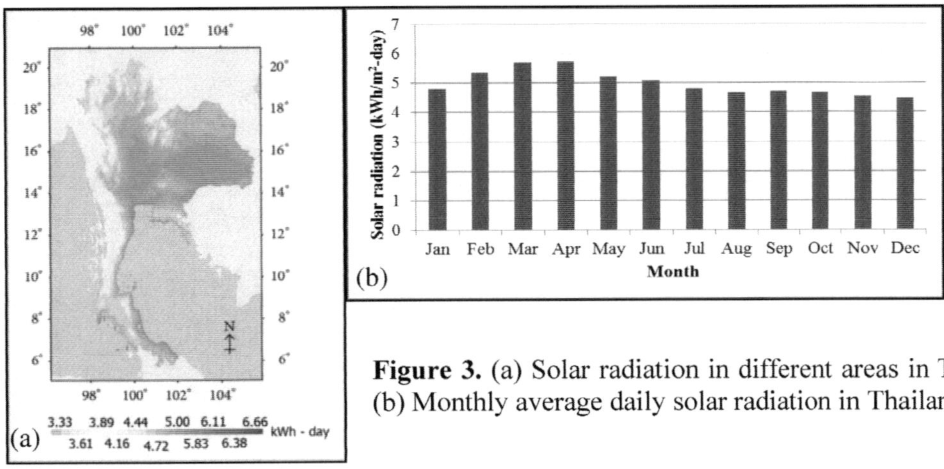

Figure 3. (a) Solar radiation in different areas in Thailand [13] and (b) Monthly average daily solar radiation in Thailand [14].

6. Summary

The feasibility study of seawater electrolysis for PV/FC hybrid system was investigated in order to investigate the system parameters when the target of 10-kW of FC electrical power generation was evaluated. The appropriate amount of hydrogen produced by seawater electrolysis was calculated for using in FC to produce electricity at night time. In this system, the PV arrays supply the electricity for electrolyzer to produce hydrogen at day time.

However, the technical challenge is real time production, the safe and convenient storage and the efficient of the seawater electrolysis for PV/FC hybrid system. The optimal design by using Hybrid Optimization Model for Electric Renewable (HOMER) and the operational control are considered to perform in the future work. Moreover, the economic challenge of this hybrid power system is the reasonable cost of hydrogen production, hydrogen storage and FC system is also considered in the future work as well.

Acknowledgements

Authors gratefully acknowledge Department of Chemical Engineering, Pathumwan Institute of Technology for equipment support.

References

[1] Information on http://www.eppo.go.th/images/POLICY/ENG/AEDP2015ENG.pdf
[2] Mohammed O H et al. 2014 *Proc. 2014 First International Conference on Green Energy* pp 119-123
[3] Abdel-Aal H K, Zohdy K M and Kareem M A 2010 *The Open Fuel Cells Journal* **3** 1
[4] Badea G E et al. 2007 *ANNALS of the ORADEA UNIVERSITY Fascicle of Management and Technological Engineering* **VI** pp 224-249
[5] Saeed F et al. 2013 *Int. J. Electrochem. Sci.* **8** 6311
[6] Slama R B 2013 *Computational Water, Energy, and Environmental Engineering* **2** 54
[7] Abdel-Aal H et al. 2012 *Smart Grid and Renewable Energy* **3** 204
[8] Lopes Francisco da Costa and Watanabe E H 2009 *Proc. 2009 Brazilian Power Electronics Conference* pp 775-782
[9] Information on http://www.phisueahouse.com/downloads/User%20Report.pdf
[10] Banerjee S et al. 2015 *J Fundam Renewable Energy Appl* **5** doi:10.4172/20904541.1000193
[11] Larminie J and Dicks A 2003 *Fuel Cell Systems and Explained*, 2nd edition (John Wiley & Sons Ltd, England)
[12] Züttel A 2008 *in: Hydrogen as a Future Energy Carrier*, ed A Züttel, A Brogschulte and L Schlapbach (Wiley-VCH Verlag GmbH&Co. KGaA, Weinheim)
[13] Information on https://solarsmileknowledge.com
[14] Information on http://www.dede.go.th

IOP Conf. Series: Materials Science and Engineering **241** (2017) 012042 doi:10.1088/1757-899X/241/1/012042

Characterization of patterns of Localized Doping Using Stamping technique for Selective n-Emitter Solar Cell Structure

A. Mangkornkaew[1] and T. Fangsuwannarak[1]

1 School of Electrical Engineering, Institute of Engineering, Suranaree University of Technology, Nakhon Ratchasima, Thailand 30000

E-mail: thipwan@g.sut.ac.th

Abstract. In the present, a novel cost-effective process scheme for single step selective emitter diffusion was implemented. It is based on the fabrication of acid-resist pattern using a stamping technique with collaboration of a spin on dopant (SOD) and chemical etched-back emitter methods. The SOD diffusion process provided heavily doping n-emitter. Acid-resist pattern without exploitation of a complex method as a photolithography, was stamped as a metal contact pattern for prevention of a localized heavy-dope region from etching back. Phosphorus doping profiles were controlled by etching back time to provide the formation of n-type selective emitter. Sheet resistance is tunable from 10 to 180 Ohm/Sq on localized n-layer. After removal of the patterned acid-resist, the selective n-emitter solar cell structure was obtained under one-step diffusion to achieve a better blue-light response and low contact resistance.

1. Introduction

Today, photovoltaic (PV) technology is not only important to require renewable energy but also has more concern to clean environment and sustainable energy sources. One of the challenges for commercial PV cell manufacturers adapted is to further reduce the cost and to simultaneously improve the efficiency of PV cells. One key limitation of present silicon industrial solar cells is the production based on homogeneous emitter region with a low sheet resistance of about 50-70 Ω/Sq. Meanwhile, a junction depth of about 0.5-1 μm is required for a compromise between a sufficient contact quality of silver grid metallization and a low recombination rate at the front surface. It causes a problem of shorter life time of generated charge carriers because a low sheet resistance of homogeneous emitter from high dopant concentration provides high density of carrier recombination centres [1, 2]. Thus, conventional PV cell efficiency is usually below 17%.

The increase in sheet resistance of n-emitter enables to reduce the Auger recombination, resulting in a better blue-light response and thus a higher I_{sc}. Moreover, it is also able to provide a better front surface passivation, leading to a higher V_{oc} [3]. Nonetheless, another effect considered due to low dopant concentration is a severe contact quality between metal and silicon emitter, resulting in a lower fill factor and solar cell efficiency decreased [4].

In order to improve solar cells efficiencies, optimization of emitter to require higher blue response and better contact quality has been investigated. One of the most potential method is selective emitters (SE) that feature a heavily-doped contact area underneath the metallized region and a lightly-doped emitter area between front fingers. Crystal silicon SE solar cell is able to provide its efficiency improvement with a competitive production cost [5, 6].

In the present, commercially available researches involve dopant ink through an additional screen printing step and laser-based SE process, which are available for forming different regions including heavily-doped emitter region with about below 50 Ω /Sq and lightly-doped emitter region with about 90-110 Ω /Sq. Si ink based SE technology has achieved an efficiency of up to 19.2% on screen-printed c-Si solar cells [7,8]. However, both of the SE processes have relied on more masking steps, laser ablate openings and scribe grooves. These additional processes affect the rising cost, the complex laser processing, and the use of relatively expensive materials and equipment, which cause the main reasons for the industry's reluctance to adopt these technologies. Therefore, simple SE processing to replicate that based on low cost materials is required.

A stamp-patterning fabrication using a cheap flash foam is an alternative patterning technique to rapidly transfer acid-resist pattern in both micro-scale and nano-scale structures that is applicable to devices on planar, curved, flexible, or soft substrates, especially when low cost processing is required [8]. Although the stamping and etching steps are added in the process, the using repeat of inexpensive flash foam and etching solution is a highlight advantage. This approach is related to a pattern fabrication with stamping process to produce acid-resist area localized underneath the metallized region. A selective emitter structure with one diffusion step to form n-p shadow junction with highly doping concentration. This paper highlights the ability to achieve a wide range of dual doping concentrations using the flash form stamp and additional etch-back processing step. The study of variation in sheet resistance via etched back processing times is demonstrated. The localized emitter feature which is patterned by using flash foam stamp, is characterized.

2. Experimental Details

Sol-gel consists mainly of phosphorus solution as a SOD source, Tetraethylorthosilicate (TEOS, 98% Fluka), Ethanol absolute (EtOH, 99% BDH) and H_2O with a volume ratio of 5:10:1.5. Subsequently, the mixed solution was stirred for 5 minutes at 70°C to ensure homogeneity. Phosphoric acid (H_3PO_4, 85%Ajax) was then added by a volume ratio of TEOS: H_3PO_4 at 3:1 and stirred for 40 minutes. The obtained mixture solution was cooled down to room temperature for 30 minutes.

P-type poly-crystalline silicon wafers (~1 Ω·cm) as substrates were textured by HF chemical etching and then were cleaned by standard Radio Corporation of America (RCA) method and subsequently were treated in dilute hydrofluoric acid (HF) for 1 minute. The mixture of SOD solution was spun on the Si substrates at 2-speed steps at 1000 rpm/min for 10 seconds and 3000 rpm/min for 30 seconds. Subsequently, the wafers coated by SOD film were sintered at 1000°C for 60 minutes. The high heat-treatment acts as a thermal diffusion to drive the phosphorus atoms through the silicon wafer for formation of the shallow n^{++}/p junction. The low sheet resistance of n^{++}emitter layer was measured by a 4-point probe technique.

Flash foam stamp (FFS) acts as photosensitive material, is a type of micro porous polymer. FFS surface exposed by flash light will be burnt and sealed solution. A grid mask on top of the flash foam is exposed transferring the grid pattern to FFS. Surface morphologies of FFS comprising 2 regions: a liquid sealed part and an unsealed part, are created on a FFS. Acid resist solution was stored in the FFS and then stamped onto poly-Si substrate to flow through the unsealed pattern area. The acid resist features the grid metallization pattern stamped on n^{++} emitter surface, thus the heavily doping area was kept fixed by the acid resist. After drying at 120°C for 30 minutes, to minimize the dopant concentration the etching step as a etched-back emitter method was processed in $HF:HNO_3$ solution with the volume ratio of 200:1 in times variation. The influence of etching time on a decrease in sheet resistance of emitter layer was investigated.

3. Results and Discussions

FFS feature was changed when the FFS below the designed grid mask was exposed by light flashing. Fig.1 (a)-(b) shows FFS pattern composed of 2 different surface morphology areas: porous area (unsealed area) and sealed solution area. As can be seen the scanning electron microscope (SEM) image in Fig.1(c), the locally unsealed region shown as porous openings has the size less than 20 μm, which allows the solution flowing through. The size of FFS pattern obtained is consistent with the

IOP Conf. Series: Materials Science and Engineering **241** (2017) 012042 doi:10.1088/1757-899X/241/1/012042

mask designed in a feature of PV grid contact pattern. Acid resist solution was stamped onto n^{++} Si surface by flowing through the patterned FFS.

(a) (b) (c)

Fig. 1 Pattern images of FFS after transferring grid layout to feature in (a), magnified image in (b) and unsealed surface SEM image in (c).

After etching back n^{++} Si emitter, different values of emitter sheet resistance were obtained by varying the etching times as shown in Fig. 2. The sheet resistance of the processed emitters on poly-Si wafers was measured with a four-point probes technique. Each wafer was measured on five positions. Etching time extension from 0 - 45 second is able to provide the sharply change of emitter sheet resistance from 10 Ω/Sq to 180 Ω/Sq in average. It is found that sheet resistivity of n^{++} emitter is adjustable from varying etching times. The higher resistivity range is usually required between 90-110 Ω/Sq that is form as a lightly doped emitter region for PV efficiency improvement.

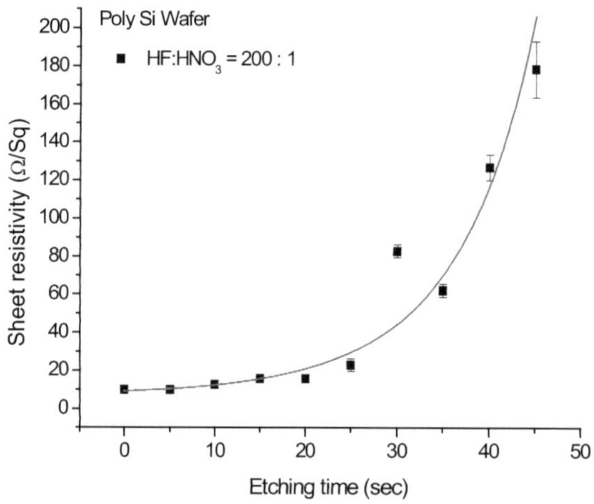

Fig. 2 Sheet resistance at different emitter etching times.

As shown in Fig. 3 the dark color pattern on poly-Si wafer showed the acid resist region which is able to prevent chemical etch in heavily doped area. When the n^{++} Si surface was patterned by FFS stamp to transfer acid resist as a grid pattern on the surface, the n^{++} emitter region was changed into lightly doped region during etched-back process. After removing acid-resist, the feature in dark color of localized n-Si emitter after etched-back process is shown in Fig.4 when we used the emitter etching time about 37 sec. Meanwhile, n^{++} Si region featured light color, thus selective emitter performing the both of lightly doped region (110 Ω/Sq) and heavily doped region (< 20 Ω/Sq) are clarified as shown in light color pattern and dark color pattern, respectively. It is noted that different phosphorus concentration regions is fabricated by simple stamping technique without complex process as a photolithography.

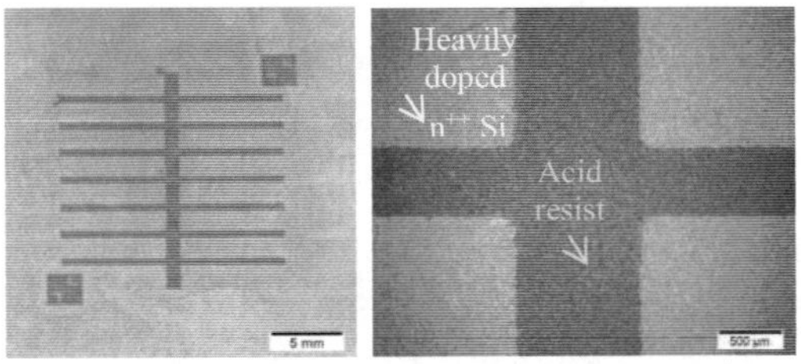

Fig. 3 Grid pattern images of acid resist pattern in dark color to be transferred from designed FFS.

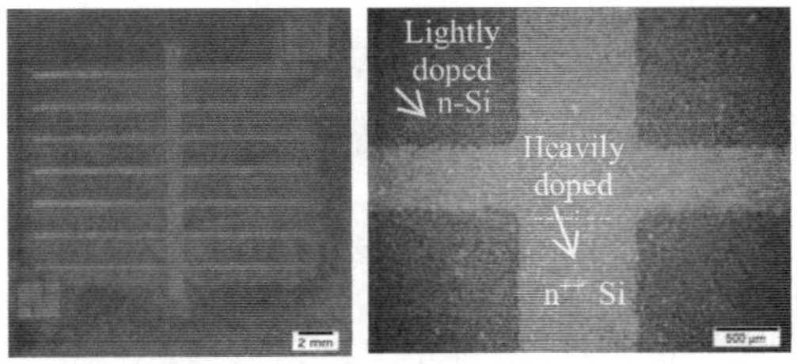

Fig. 4 Grid pattern feature of SE surface after etched-back process: lightly doped region with dark color and heavily doped region with light

4. Conclusion

The results of this study were revealed that the simple stamp coating is a potential method to produce a selective emitter in only one thermal diffusion step with one additional etching back step. By transferring grid pattern to n^{++} Si emitter region, locally lightly dopant regions were performed at 37 seconds for etching time without employing a complex procedure of photolithography. The n^{++} Si region coverage with patterning acid resist was provided with less than 20 Ohm/Sq and there was 110-120 Ohm/Sq at lightly doped region.

The SE pattern was controlled by designing the feature of FFS. The proposed simple process employing stamping technique also shows clearly the potential to become more cost-effective than existing selective emitter processes relying on other additional complex procedure and expensive tools such as laser doping and photolithography steps.

Acknowledgments

This study was supported by Suranaree University of Technology and research funding from National Research Council of Thailand (NRCT 2015). The author would like to gratitude to Solartron Public Company Limited Thailand and Synchrotron Light Research Institute, Thailand for equipment support.

References

[1] Ulrich, Mahir O, et.al, Selective emitter by laser doping from phosphosilicate glass 2009 24th European PV Solar Energy Conf. and Exhibition 21–25.

[2] Ruby D S, Albuquerque N M, et.al, Recent progress on the self-aligned selective-emitter silicon solar cell, 1997 26th IEEE PVSC 39–42.

[3] Bentzen A, Holt A, et.al, Stokkan Gettering of transition metal impurities during phosphorus emitter diffusion in multicrystalline silicon solar cell processing 2006 J. Appl. Phys. 99 093509.

[4] Wagner H, Shirazi A D, et.al, Optimizing phosphorus diffusion for photovoltaic 2016 J. Appl. Phys., 119 185704.

[5] Kim M, Kim D, et.al, Laser etch back process to fabricate highly efficient selective emitter 2014 Solar Energy 109 105–110.

[6] Horzel J, Szlufcik J, et.al, A simple processing sequence for selective emitters 1997 26th PVSC 139–140.

[7] Scardera G, Meisel A, et.al, High efficiency phosphorus emitters for industrial solar cells: comparing advanced homogeneous emitter cells and selective emitters using silicon ink technology 2012 27th EU PVSEC 923-5.

[8] Sastrawan R, Aigner F, et.al, Combing Laser Doping and Wet Chemical Etch Back for Industrial Selective Emitter Solar Cells 2012 27th EU PVSEC 1563 - 15659 [9] Qin D, Xia Y, et.al, Soft lithography for micro-and nanoscale patterning 2010 Nat. Protoc. 5 491–502.

IOP Conf. Series: Materials Science and Engineering **241** (2017) 012043 doi:10.1088/1757-899X/241/1/012043

Biomass Briquette Investigation from *Pterocarpus Indicus* Leaves Waste as an Alternative Renewable Energy

Willyanto Anggono[1], Sutrisno[2], Fandi D. Suprianto[3], Jovian Evander[4]

[1,2,3,4]Mechanical Engineering Department, Petra Christian University, Surabaya, Indonesia

[1,2,3,4]Centre for Sustainable Energy Studies, Petra Christian University, Surabaya, Indonesia

E-mail: willy@petra.ac.id

Abstract. Indonesia is a tropical country located in Southeast Asia. Indonesia has a lot of variety of plant species which are very useful for life. *Pterocarpus indicus* are commonly used as greening and easily found everywhere in Surabaya city because of its characteristics that they have dense leaves and rapid growth. *Pterocarpus indicus* leaves waste would be a problem for residents of Surabaya and disturbing the cleanliness of the Surabaya city. Therefore, the *Pterocarpus indicus* leaves waste would be used as biomass briquettes. This research investigated the calorific value of biomass briquettes from the *Pterocarpus indicus* leaves waste, the effect of tapioca as an adhesive material to the calorific value of biomass briquettes from the *Pterocarpus indicus* leaves waste, the optimum composition for *Pterocarpus indicus* leaves waste biomass briquette as an alternative renewable fuel and the property of the optimum resulted biomass briquette using ultimate analysis and proximate analysis based on the ASTM standard. The calorific value biomass briquettes from the *Pterocarpus indicus* leaves waste were performed using an oxygen bomb calorimeter at various composition of *Pterocarpus indicus* from 50% to 90% rising by 10% for each experiment. The experimental results showed that the 90% raw materials (*Pterocarpus indicus* leaves waste)-10% adhesive materials (tapioca) mixtures is the optimum composition for biomass briquette *Pterocarpus indicus* leaves waste. The lower the percentage of the mass of tapioca in the biomass briquettes, the higher calorific value generated.

1. Introduction

Energy crisis is a concern because of the diminishing natural resources. Diminishing energy sources influence to the fuel prices. The fuel price will be increased due to the needs of a huge market but limited resources. Therefore, the renewable energy needed to meet the energy needs. Biomass is an alternative renewable energy. Biomass can be produced from wild plants, forest plants, plant waste, garbage, grass, etc. [1].

Biomass itself is much in demand by many people all over the world. It has been observed that the biomass is an alternative energy that can replace fossil fuels in the future. Indonesia is a tropical country with various types of plants. The plants in Indonesia have the potential to be used as biomass briquettes. *Pterocarpus indicus* is a plant road side which grow tall and large in Indonesia as shown at figure 1. The leaves of this plant often fall to the ground and the highway when the rain and heat as shown in figure 2 [2-5].

Figure 1. *Pterocarpus indicus* tree in Surabaya

Figure 2. *Pterocarpus indicus* leaves waste in Surabaya

Pterocarpus indicus plant is grown in tropical areas, particularly in parts of Southeast Asia. Surabaya is one of the city in Southeast Asia with a tropical climate, therefore *Pterocarpus indicus* trees can grow in the area of Surabaya, Indonesia. The *Pterocarpus indicus* tree can grow up to 15 meters' height and the rejected material of municipal waste can actually be used as biomass briquettes [6-9]. The *Pterocarpus indicus* leaves waste can be used as an alternative energy through the right process.

Table 1. Calorific values comparison of various wood [10]

Wood Type	Moisture Content (%)	Calorific Value (kcal/kg)
Leuchaena leucocephala	10.13	4197
Samanea saman	10.36	3926
Sesbandia grandiflora	6.83	3965
Glirisidia maculate	23.97	4168
Enterolobium cyclocarpum	14.21	3948
Hibiscus arborea	10.33	4266
Gmelina arborea	9.24	4282

From the Table 1, it appears that *Gmelina arborea* (9.24 % moisture contents) has the highest calorific value of wood (4282 kcal/ kg), while *Samanea saman* (10.36% moisture contents) has lowest calorific value of wood (3926 kcal/ kg). The coal with better quality has the highest calorific value (6300 kcal/ kg at 2.1% moisture content) compare to all various wood [10].

2. Experimental Method

This study used *Pterocarpus indicus* leaves waste that they have already fallen from the plant. After collected, the leaves waste must be sun dried for three days. Biomass briquettes were made by crushing dried *Pterocarpus indicus* leaves waste into the desired particle size (60 Mesh), mixing them with tapioca flour as a binder material, and compacting the mixtures under pressure (2 MPa). The calorific value biomass briquettes from the *Pterocarpus indicus* leaves waste were performed using a 1341 Plain Jacket oxygen bomb calorimeter Parr Instrument at various composition of *Pterocarpus indicus* leaves waste. The experiments in this paper were performed at various compositions of *Pterocarpus indicus* leaves waste biomass briquettes from 50% *Pterocarpus indicus* leaves waste (using 50% tapioca as a binder material) to 90% *Pterocarpus indicus* leaves waste (using 10% tapioca as a binder material) rising by 10% for each experiment. After that it will be tested heat values by using a bomb calorimeter.

3. Result and Discussion

Based on the experimental investigation using an oxygen bomb calorimeter, the calorific value of 100% *Pterocarpus indicus* leaves waste was 4909.89 kcal/ kg and the calorific value of 100% tapioca as a binder material was 3574.47 kcal/kg. The calorific value of biomass briquette from leaves waste *Pterocarpus indicus* at various composition mixtures are shown in Table 2 and the summary of the results from calorific value or calorific value of *Pterocarpus indicus*–tapioca mixtures at various composition are shown Figure 3.

The greater amount of tapioca the lower the calorific value of the biomass briquette *Pterocarpus indicus* leaves waste. The highest calorific value of the biomass briquette *Pterocarpus indicus* leaves waste was the biomass briquette using 90% *Pterocarpus indicus* leaves waste-10% tapioca mixtures as shown in figure 3. The 90% *Pterocarpus indicus* leaves waste-10% tapioca mixtures successfully creates the biomass briquette model as shown in figure 4. The composition of 90% *Pterocarpus indicus* leaves waste-10% tapioca mixtures using 60 mesh particle size was used as the best biomass briquette model.

Table 2. *Pterocarpus indicus* leaves waste biomass briquette calorific value of at various composition

Biomas Briquette Composition	Calorific Value (kcal/kg)
50% *Pterocarpus indicus*-50% tapioca mixtures	3913.16
60% *Pterocarpus indicus*-40% tapioca mixtures	4031.77
70% *Pterocarpus indicus*-30% tapioca mixtures	4155.85
80% *Pterocarpus indicus*-20% tapioca mixtures	4366.76
90% *Pterocarpus indicus*-10% tapioca mixtures	4648.15

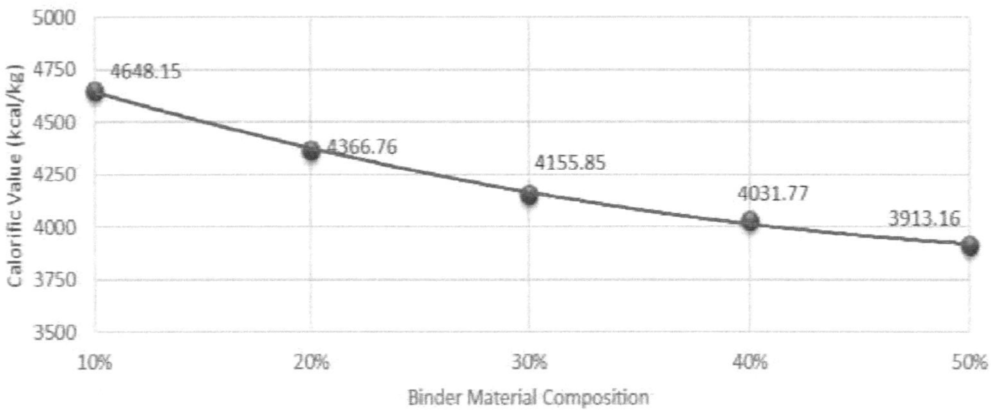

Figure 3. Effect of binder material to the calorific value *Pterocarpus indicus* leaves waste briquette

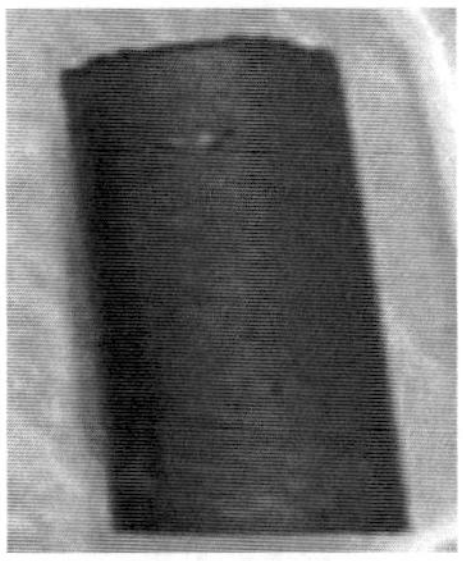

Figure 4. *Pterocarpus indicus* leaves waste biomass briquette

Tapioca as a binder material reduces the calorific value of the biomass briquette from leaves waste *Pterocarpus indicus*. The greater amount of tapioca, the lower the calorific value of the biomass briquette from *Pterocarpus indicus* leaves waste. In term of cost, the effective composition also 90%:10% because *Pterocarpus indicus* leave waste is cost free. The higher the precentage of tapioca in the biomass briquette from leaves waste *Pterocarpus indicus* the higher the cost of the biomass briquettes.

The proximate analysis and ultimate analysis also have been examined through a laboratory test. The test was carried using some ASTM standardization. The moisture content using ASTM D 2961-11. Fixed carbon calculation was examined using ASTM D 3172-13. The ash content was examined using ASTM D 3174-12. The volatile matter was examined using ASTM D 3175-11. The total sulfur was examined using ASTM D 4239-14E1. The gross calorific value was examined using ASTM D 5865-13. The proximate analysis result of *Pterocarpus indicus* leaves waste biomass briquettes (90% *Pterocarpus indicus* leaves waste and 10% tapioca) shows total moisture, fixed carbon, ash content, volatile matter, total sulfur and gross galorific value as shown in Table 3.

Table 3. Proximate analysis result of *Pterocarpus indicus* leaves waste biomass briquette

Test Method	Parameters	Unit	Value
ASTM D 2961-11	Total Moisture	%wt	5.2
ASTM D 3172-13	Fixed Carbon	%wt	15.3
ASTM D 3174-12	Ash Content	%wt	6.2
ASTM D 3175-11	Volatile Matter	%wt	73.3
ASTM D 4239-14E1	Total Sulfur	%wt	0.25
ASTM D 5865-13	Gross Calorific Value	kcal/kg	4648.15

Table 4. Ultimate analysis result of *Pterocarpus indicus* leaves waste biomass briquette

Test Method	Parameters	Unit	Value
ASTM D 5373-14	Carbon	%wt	49.12
ASTM D 5373-14	Hydrogen	%wt	6.67
ASTM D 5373-14	Nitrogen	%wt	0.19
ASTM D 4239-14E1	Sulfur	%wt	0.25
ASTM D 5373-15	Oxygen	%wt	33.40

The ultimate analysis result of *Pterocarpus indicus* leaves waste biomass briquettes (90% *Pterocarpus indicus* leaves waste and 10% tapioca) used to investigate the carbon, hydrogen, oxygen, nitrogen and sulfur concentration on the *Pterocarpus indicus* leaves waste biomass briquettes. The examination of the proximate analysis using ASTM D 5373-14, ASTM D 5373-14E1 and ASTM D 5373-15. The result of ultimate analysis of *Pterocarpus indicus* leaves waste biomass briquettes shown in Table 4. The *Pterocarpus indicus* leaves waste biomass briquette has the highest calorific value compared to *Cerbera manghas* leaves waste biomass briquette (4164.00 kcal/kg), sawdust biomass briquette (4161.08 kcal/kg), sugarcane biomass briquette (3902.96 kcal/kg), rice straw biomass briquette (3926.86 kcal/kg) and coconut coir biomass briquette (4146 kcal/kg) [1,2].

4. Conclusion

Biomass briquette from *Pterocarpus indicus* leaves waste is an alternative renewable and sustainable energy and tapioca as a binder material play a role to the calorific value on the biomass briquette from *Pterocarpus indicus* leaves waste. The lower the amount of tapioca as a binder material, the greater the calorific value of the biomass briquette from *Pterocarpus indicus* leaves waste and the cheaper the cost of the biomass briquette from *Pterocarpus indicus* leaves waste. The biomass briquettes have been developed as energy source from *Pterocarpus indicus* leaves waste using 90% *Pterocaraus indicus* leaves waste and 10% tapioca as the optimum composition.

Acknowledgments

Thanks to Petra Christian University Indonesia and Direktorat Jendral Pendidikan Tinggi Kementerian Riset Teknologi dan Pendidikan Tinggi Republik Indonesia (Hibah Penelitian Produk Terapan 2016-2017) for their support during this research.

References

[1] Sutrisno, Anggono W, Suprianto F D, Kasrun A W and Siahaan I H 2017 *ARPN Journal of Engineering and Applied Sciences* **12** 931
[2] Anggono W, Suprianto F D, Sutrisno and Kasrun A W 2016 *International Journal of Industrial Research and Applied Engineering* **1** 1
[3] Sellin N, Oliveira B G, Marangoni C, Souza O, Oliveira A P N and Oliveira T M N 2013 *Chemical Engineering Transactions* **32** 349
[4] Rosua J M and Pasadas M 2012 *Renewable and Sustainable Energy Reviews* **16** 4190
[5] Raju C A I, Jyothi K R, Satya M and Praveena U 2014 *International Journal of Research in Engineering and Technology* **3** 54
[6] Prasityousila J and Muenjina A 2013 *Procedia Environmental Sciences* **17** 603
[7] Yank A, Ngadi M and Kok R 2016 *Biomass and Bioenergy* **84** 22
[8] Jittabut P 2015 *Energy Procedia* **79** 2
[9] Po Z X, Hu P Y, Sheng L M, Li K S and Qing Z J 2010 *Asian Pacific Journal of Tropical Medicine* 109
[10] Cahyono T D, Coto Z, dan Febrianto F 2008 *Forum Pasca Sarjana* **31** 105

IOP Conf. Series: Materials Science and Engineering **241** (2017) 012044 doi:10.1088/1757-899X/241/1/012044

A method of estimating the contents of components, structural and physical-mechanical properties of rocks

A A Kurmankozhayeva and A S Azhibekova

Department of Mathematics, Institute of Information and Telecommunication Technologies, Kazakh National Research Technical University named after K.I. Satpayev, 22a Satpayev street, Almaty, Kazakhstan

aliya.azhibek@gmail.com

Abstract. The method for estimation of the mean value and the mode of the property becomes important with increasing the depth growth of mining operations and the complexity degree of excavation sites, the growth of number of cases, when geological information volume representing the level of values of structural and physical-mechanical properties is insufficient. The reliability of estimates is achieved through the involvement of different characteristics of the studied property. The developed method is based on calculation of prediction estimates of the mean value and the mode, which allow providing the accuracy of the estimates with incomplete information, and when the type of the studied property distribution is unknown.

1. Introduction

The efficiency of preliminary surveying of deposit reserves, drilling and blasting, loading and unloading operations, ore extraction and separation depends on certainty and accuracy of methods for assessment, determination and forecasting of not only contents of components, but also values of structural, physical and mechanical properties of rocks. Yield of boulders and fine grinded blasted mined rocks leads to the increase of expenses for extraction, losses and clogging of ore, and to the decrease of the production capacity [1, 2]. Well-known traditional arithmetic mean and weighted mean methods for determination of mean metal content values, structural, physical and mechanical properties of ores are carried out without considering peculiar properties of formation of geological, statistical conditions and patterns of distribution of properties. The knowledge about the distribution of the property of mineral deposits is highly important in solving geological and geometrical, mining and technological problems [3, 4]. The estimation of average values of structural and mechanical properties of real values of the main component content is often followed by systematic errors due to weak correlations of such interrelations.

The abovementioned aspects induce to substantiate new methods for assessing the distribution of structural and mechanical properties of mineral deposits, the use of which would allow taking into account diversity, interconnectivity and features of geological and statistical conditions of their distributions in solid. The developed method is based on calculation of prediction estimates of mean value and the mode, which allow providing the accuracy of the estimates with incomplete information, and when the type of the studied property distribution is unknown.

2. Estimating technique

To study empirical distributions the set of structural properties have been used, including the power of ores and inter-mineral inclusions, bedding angles and fracturing rates: the set of physical and mechanical properties comprises density values, volume weight, angles of internal friction, porosity, the Poisson's ratio, static and dynamic modules of elasticity, compressive and tensile resistance, specific cohesion, deformation modulus, rock failure angles [5].

The estimation procedure of the mean value of structural and mechanical properties includes two methods, corresponding to two geological and static conditions of the exploration degree of deposit property:

a) cases, where the mode of the deposit property is known, and available information allows to determine it;

b) cases, where the mode of the property is unknown, as well as its distribution function is unknown.

The first method for estimation of the mean value of the property derives from the direct approach principle and it is based on the use of structural and regression equations. In this case the estimation of the mean value of the property is performed using the derived structural and regression equations presented in table 1.

The statistical analysis has been carried out with the major structural properties and physical and mechanical properties. Further, the following distribution parameters have been selected as initial basic parameters: the mean (X_{av}), the mode of (X_0) the property and of the property distribution characteristics: the mean square deviation (σ), the variation amplitudes (d), the degree of uniformity of the property distribution in the solid ore (H).

At the same time one or two-parameter equations are appropriate to use for prediction estimate when there are no other initial parameters:

$$X_{av} = A_1 \exp(X_0),\ \ X_{av} = A_2 F(X_0,\sigma),\ \ X_{av} = A_3 F(X_0,d),\ \ X_{av} = A_4 F(X_0,H). \qquad (1)$$

It is recommended to use three or four-parametric structural regression equations to estimate the average value of the property:

$$X_{av} = A_5 F(X_0,\sigma,d),\ \ X_{av} = A_6 F(X_0,\sigma,H)\,. \qquad (2)$$

Structural and regression equations have inherent high values of the correlation ratio and the coefficient of determination (R = 0.94). It proves their suitability for estimation of the mean values of the property. The sequence of calculating the mean value of the property by this method is simple, but it is necessary to determine the type of distribution. The density function of the determined type of the property distribution can also be used for estimation of the mean value.

The stage of a direct approach, which does not require obligation of a preliminary determination of the type of the considered property distribution, has been set on the second method. This makes this estimation method highly important. The method is based on the use of derived structural and regression equations which are composed of parameters: amplitude, the mean square deviation and the uniformity coefficient of the property distribution. In this case, the type of the property distribution and its mode are omitted.

There are used one-parameter structural and regression equations for the prediction estimate of the mean value:

$$X_{av} = B_1 F(k_1,\sigma),\ \ X_{av} = B_2 \cdot F(k_2,d)\,. \qquad (3)$$

It is recommended to use two or three-parameter equations for estimation of the mean value of the property:

$$X_{av} = C_1 F(\sigma,d),\ \ X_{av} = C_2 F(\sigma,H),\ \ X_{av} = C_3 F(d,H),\ \ X_{av} = C_4 F(\sigma,d,H)\,. \qquad (4)$$

Exploration data can be used to determine values of the variation amplitude, standard deviation and indicators of uniformity property of the property distribution, when their values cannot be directly calculated. It is not always possible to reliably determine the type of the distribution function for each studied property due to insufficient information, and also because of the complexity of finding a suitable distribution type. The developed rational estimation procedure of the property mode allows eliminating these deficiencies and it provides for the accuracy of determination of the property mean value.

Table 1. Structural relationships between the distribution parameters and the property distribution characteristics.

Structural and regression equations	Statistical parameters	
	Correlation coefficient	Number of observations
I. Single-parameter models		
$X_{av} = 0.86 + 1.15X_0$	0.73	104
$X_{av} = 2.69\sigma^{0.59}$	0.94	107
$X_{av} = 5.62 - 0.42H$	-0.21	107
$X_{av} = 0.63d^{0.75}$	0.82	105
$X_0 = 0.62d^{0.601}$	0.61	94
$X_0 = 13.8H^{-4}$	-0.12	113
$\sigma = 0.51d^{0.71}$	0.89	111
$\sigma = 0.30 + 0.2d$	0.94	113
$\sigma = 6.01H^{-0.47}$	-0.16	118
$d = 19.0H^{-0.47}$	-0.15	113
II. Two-parameter models		
$X_{av} = 0.13X_0 + 0.70\sigma + 0.09$	0.94	104
$X_{av} = 0.73X_0 - 0.14H - 0.16$	0.73	103
$X_{av} = 0.80\sigma - 0.35H - 0.45$	0.94	104
$X_{av} = 0.78d - 0.03H - 0.50$	0.82	104
$X_{av} = 0.94X_0 + 0.57d - 0.03$	0.87	104
$X_{av} = 0.93\sigma + 0.04d + 0.15$	0.94	104
$X_0 = 1.23\sigma + 0.10d + 1.11$	0.70	103
$X_0 = 1.23d - 0.03H - 0.81$	0.61	103
$X_0 = 0.72\sigma - 0.07H - 0.85$	0.70	103
$X_0 = 0.43X_{av} + 0.38\sigma + 0.16$	0.73	104
$X_0 = 2.44X_{av} + 0.81d + 1.57$	0.73	104
$X_0 = 0.73X_{av} - 0.06H - 1.06$	0.73	104
III. Three-parameter models		
$X_{av} = 0.12X_0 + 0.82\sigma + 0.03d - 0.54$	0.94	104
$X_{av} = 0.91\sigma + 0.55d - 0.08H + 0.46$	0.70	103
$X_0 = 0.81\sigma + 0.04d - 0.02H - 0.78$	0.74	103
IV. Four-parameter models		
$X_{av} = 0.15X_0 + 0.82\sigma - 0.04d - 0.07H + 0.21$	0.94	103
$X_0 = 0.92X_{av} + 0.05\sigma + 0.55d - 0.07H - 0.35$	0.74	104

The estimation procedure of the mode of a structural and mechanical property is based on deducting the structural and regression ratio between distribution parameters and property distribution characteristics. The mode as the most probable property value is the one of its important numerical characteristics. It is less affected by errors as well as the median. In comparison with the median and other static parameters, the mode is characterized by relatively high information content, a high degree

of influence on the level of the mean value and the efficiency of estimation, prediction and stabilization of the ore quality, when solving problems on estimation of the opening strength.
The prediction estimation of the mode of the property by applying the developed method is carried out with one or two-parameter structural and regression equations:

$$X_0 = B_{01} \cdot F(\varphi_1, \sigma),\; X_0 = B_{02} \cdot F(\varphi_2, d),\; X_0 = B_{03} \cdot F(\varphi_3, H). \quad (5)$$

To calculate the mode of the property two and three-parameter equations are used:

$$X_0 = C_{01} \cdot F(\sigma, d),\; X_0 = C_{02} \cdot F(d, H),\; X_0 = C_{03} \cdot F(\sigma, H),\; X_0 = C_{04} \cdot F(\sigma, d, H). \quad (6)$$

The analysis of the results of correlation analysis gave us:

- a close linear relationship in the dependence $X_{av} = F(X_0)$ and there is a less inversely proportional relationship in the dependences: $X_{av} = F(H)$, $X_0 = F(H)$, $\sigma = F(H)$;
- a significant nonlinear directly proportional relationship in the dependence: $X_0 = F(d)$;
- a close nonlinear directly proportional relationship in the dependences: $\sigma = F(d)$, $X_{av} = F(\sigma)$, $X_{av} = F(d)$;
- a close directly proportional relationship in the dependences: $X_{av} = F(\sigma, H)$, $X_{av} = F(X_0, \sigma)$, $X_{av} = F(X_0, d)$, $X_0 = F(X_{av}, \sigma)$, $X_0 = F(X_{av}, d)$, $X_{av} = F(X_0, d, \sigma)$, $X_{av} = F(X_0, \sigma, d, H)$;
- a significant directly proportional relationship in the dependences: $X_{av} = F(X_0, H)$, $X_{av} = F(d, H)$, $X_0 = F(X_{av}, H)$, $X_0 = F(X_{av}, \sigma, d, H)$, $X_0 = F(\sigma, d, H)$;
- an insignificant direct linear relationship in the dependences: $X_0 = F(\sigma, d)$, $X_0 = F(d, H)$, $X_0 = F(\sigma, H)$.

Based on the analysis of the results of the correlation analysis, it is found that the strength of a relationship between the mode and property distribution characteristics is slightly less (R = 0.67-0.82) whereas the degree of dependence between the mean value of the property and its distribution characteristics is high (R=0.90-0.94). The presence of a significant relationship between the distribution parameters and property distribution characteristics refers to the genesis of statistical structures and the geometry of deposit property distributions. The results of a comparative evaluation of the empirical distributions of structural and physical-mechanical properties characteristics according to the proposed estimation procedure has allowed to conclude the following:

- the distributions of values of ore and rock volume weights are best described by the normal distribution and the Weibull distribution. They are characterized by symmetrical curves and small values of asymmetry and kurtosis;
- the distributions of values of the density of ores and rocks are described by the lognormal distribution and the Weibull distribution; They have forms of curves that are close to symmetric, but related to the right symmetric distribution;
- the distributions of values of the angle of internal friction and the specific cohesion of rocks are described by the normal distribution and the Weibull distribution;
- the distributions of the values of porosity, modulus of deformation, static elasticity modulus of rocks, resistivity and fracture, the Poisson's ratio are described by the Weibull distribution. They have slightly right-handed forms of curves, for which the asymmetry values are above zero;
- the distributions of the values of compressive and tensile resistance are described by the exponential distribution and the Weibull distribution. They are characterized by moderately right-symmetric radial curves and significant values of asymmetry and kurtosis;
- the distributions of structural properties are characterized by symmetrical curves. The distributions of physical and mechanical properties are mostly mixed: symmetric, right-symmetric radial curves.

The dependences for the mean value and the mode of the property are carried out in the form of multifactorial equations. The results of the correlation-regression analysis of the relationship between the distribution parameters and property distribution characteristics are presented in table 2.

Table 2. The relationships between the distribution parameters and property distribution characteristics.

#	Types of dependence	Correlation coefficient	Number of observations	Equations of the relationship	
				Nonlinear	Linear
	I. Dependences between the mean value and other characteristics				
1	$X_{av} = F(X_0)$	0.73	104	$X_{av} = 1.44\exp(0.93X_0)$	$X_{av} = 0.86 + 1.15X_0$
2	$X_{av} = F(\sigma)$	0.94	107	$X_{av} = 2.69\exp(0.59\sigma)$	$X_{av} = 0.50 + 1.310\sigma$
3	$X_{av} = F(d)$	0.80	107	$X_{av} = 0.6\exp(0.75d)$	$X_{av} = 0.62 + 0.28d$
	II. Dependences between the mode value and other characteristics				
1	$X_0 = F(\sigma)$	0.70	103	$X_0 = 1.482\exp(0.638\sigma)$	$X_0 = 0.77 + 0.63\sigma$
2	$X_0 = F(d)$	0.61	115	$X_0 = 0.622\exp(0.61d)$	$X_0 = 0.69 + 0.151d$
3	$X_0 = F(H)$	-0.12	113	$X_0 = 13.H^{-1.01}$	$X_0 = 4.94 - 0.37H$
	III. Dependences between property distribution characteristics				
1	$\sigma = F(d)$	0.89	108	$\sigma = 0.51\exp(0.71d)$	$\sigma = 0.30 + 0.21d$
2	$\sigma = F(H)$	-0.16	118	$\sigma = 6.01H^{-0.52}$	$\sigma = 4.22 - 0.301H$
3	$d = F(H)$	-0.50	103	$d = 19.95H^{-0.47}$	$\sigma = 15.17 - 1.04H$

3. Conclusions

The method for estimation of the mean value and the mode of the property becomes important with increasing the depth growth of mining operations and the complexity degree of excavation sites, the growth of number of cases, when geological information volume representing the level of values of structural and physical-mechanical properties is insufficient. The reliability of estimates is achieved through the involvement of different characteristics of the studied property. The efficiency of the procedure is associated with possibility to determine the mean value and the mode regardless of its distribution function and when the volume of information is insufficient. The developed procedure can be used in making short- and long-term plans, technical and economic and process design solutions, in optimizing ore excavation conditions and mining operations, in substantiating the density of an exploration grid. Factual data from geological and surveying materials from Don, South Kempirsai chromite and Krasnooktyabrskiy bauxite mines have been used as experimental data.

References

[1] Yershov V V 1982 *Geological-Surveying Fundamentals of Ore Quality Management in the Underground Mining of Non-Ferrous Metals Deposits* (Moscow)

[2] Timofeyenko Ye P 1987 *Scientific Fundamentals of Modeling and Forecasting Properties of Ore Deposits* (Moscow)

[3] Smolich S V 1980 *Substantiating the Rational Density of the Exploration for the Planning of Mining Operations in Fields with Intermittent Mineralization* (Moscow)

[4] Kuz'min V I 1977 *Estimation of Parameters of geological and mine survey information for the geometrization and calculation of the reserves of deposits* (Moscow)

[5] Kurmankozhayeva A A 1993 *Correlation-Structural Estimates of the Determination of the Mean Value and the Mode of Deposit Properties* (Alma-Ata)

IOP Conf. Series: Materials Science and Engineering **241** (2017) 012045 doi:10.1088/1757-899X/241/1/012045

Research on Thermal Stability of Cellulose-based Lithium Battery Separator Paper

Lizhen Zhang[1,*], Chuanshan Zhao[2], and Wenjia Han[3], Yifei Jiang[4]

[1] Lizhen Zhang, Qilu University of Technology,Changqing,Jinan,Shandong Province, P. R. China
[2] Chuanshan Zhao, Qilu University of Technology,Changqing,Jinan,Shandong Province, P. R. China

18363000439@163.com

Abstract. With the application of battery becoming more and more widely in our life, the importance of battery production with high quality has been increasingly emphasized. Battery separator is one of the important factors that affect the performance of the battery, so the high functions separator production is particularly important. This article preliminary investigates the effect of pulp types, composition and beating degree on the performances of separator. Main research work:Beating softwood pulp, hardwood pulp and dissolving pulp for different degree by PFI, and papermaking, then detecting the physical properties of paper sheets. Through comprehensive analysis conclude that softwood pulp, hardwood pulp and dissolving pulp have excellent physical properties in the 40° SR, and thermal stability up to 200℃in the process of heat treatment. The paper sheets are made using different ratio of softwood pulp, hardwood pulp and dissolving pulp, then test the physical properties. According to the comprehensive analysis of experimental datum obtained that the sheets made using mixed pulp (40° SR dissolving pulp: 40° SR softwood =5:5) have excellent physical performance, and sheet strength index decreased slightly at 200℃,compared to other ratios.

1. Introduction

In many batteries, lithium battery has been widely used in the field of mobile phones, portable computers and other digital products and pharmaceutical industry, electric vehicles and space, because of it has the advantages, such as high output voltage, self discharge rate, low specific energy, good cycle performance, and no memory effect, on the environment pollution-free. Separator as part of the lithium battery in the battery, plays a barrier between the positive and negative pole electronic conductance, allows for electrolyte ions through freely so as to realize the important role of ionic conduction, so the separator is an important determinant of the battery capacity and cycle ability and safety performance. Development of new membrane materials to balance and even improve the performance and safety of the separator is a new demand for power lithium battery[1].

According to the structure and composition, the separators are mainly divided into five categories: (1) the traditional microporous polymer membrane (2) modified microporous membrane (3) non-woven membrane (4)composite membrane (5) electrolyte membrane. Separator is mainly produced by dry method, wet method and polymer extrusion into the net [2].

PE and PP microporous membrane is the first commercially used lithium-ion battery separator, and is also the most common lithium ion battery separator. Polyolefin separator has the characteristics of high strength, good chemical stability, high thermal stability and low price. But the polyolefin microporous membranes have poor wettability with liquid electrolytes, low temperature resistance, which makes the development of lithium battery with technical barrier. The use of other raw materials which have better liquid properties, higher temperature performance to produce the new membranes owning more superior performance is an important direction of the current development and research[3]. Recently, cellulose has drawn considerable attention as a very promising material for a variety of applications, due to its excellent properties such as high dielectric constant, good chemical stability, superior thermal stability (initial decomposition temperature 270℃), biocompatibility, and environmental benignancy. Given that cellulose is abundant renewable biomaterial with superior thermal stability, several studies have been carried out to explore the potential applications as high-performance LIB separators. But the cellulose separator is too thick, and their poor mechanical strength as well as big hole ,limiting its aaplication[4, 5].

In this paper, we preliminary investigates the sheet thermal stability of different pulps at different beating degree and after treatment at different temperatures for a while; at the same time, the effects on the pages thermal stability of different pulps in different proportion of processing temperature; the pages thermal stability of dissolving pulp, mixed pulp at different ratios between dissolving pulp and common pulp.

2. Experimental

2.1. Materials

Softwood pulp and hardwood pulp(By Laiwu City, Shandong province Vinda Paper Co., Ltd.), dissolving pulp (laboratory-made).

2.2. Preparation and characterization of LIB separators

(1)The first we use PFI refiner treat softwood pulp, hardwood pulp, dissolving pulp until the beating degree reach the 20°SR and 40°SR, 60°SR, 80°SR. Then we prepare cellulose membrane with German sheet mechanism by wet papermaking process. By thermogravimetric and the tensile strength, tear strength, permeability and thickness test of separator to determine the best pulping degree .(2) Under the best beating degree we made of softwood pulp, hardwood pulp, dissolving pulp diaphragm, after treatment for 60 min at 160℃, 180℃, 200℃, 220℃and 240℃, to determine the maximum heat temperature of the separator by tensile strength test. (3) Under the best pulp beating degree, we determine the optimum ratio of paper pulps by the tensile strength and tear strength, permeability and thickness test.

3. Results and discussion

Table 1. The physical performance test of different beating degree softwood

Pulp	Tearing index ($mN.m^2/g$)	Air permeability (μm/(pa.s)(mL/min))	Thickness (mm)	Absorbabil-ity (cm)	Tensile index (N.m/g)
Softwood-20 °SR	6.0	44.8	0.13	3.3	0.98
Softwood-40 °SR	8.8	5.4	0.12	2.0	1.15
Softwood-60 °SR	7.8	1.7	0.11	1.5	1.26
Softwood-80 °SR	6.2	—	0.11	1.2	1.24

Table 2. The physical performance test of different beating degree hardwood

Pulp	Tearing index (mN.m²/g)	Air permeability (μm/(pa.s)(mL/min))	Thickness (mm)	Absorba-bility (cm)	Tensile index (N.m/g)
Hardwood -20 °SR	5.1	60.91	0.15	7.8	0.52
Hardwood -40 °SR	7.4	16.55	0.13	3.4	1.03
Hardwood -60 °SR	6.2	6.36	0.12	2.3	1.01
Hardwood -80 °SR	5.9	3.59	0.12	1.8	0.99

Table 3. The physical performance test of different beating degree dissolving pulp

Pulp	Tearing index (mN.m²/g)	Air permeability (μm/(pa.s)(mL/min))	Thickness (mm)	Absorba-bility (cm)	Tensile index (N.m/g)
Dissolving pulp - 20 °SR	4.6	69.32	0.17	10.3	0.43
Dissolving pulp - 40 °SR	6.0	48.61	0.15	7.7	0.89
Dissolving pulp - 60 °SR	5.2	31.14	0.14	6.8	0.75
Dissolving pulp - 80 °SR	4.8	19.10	0.11	5.2	0.64

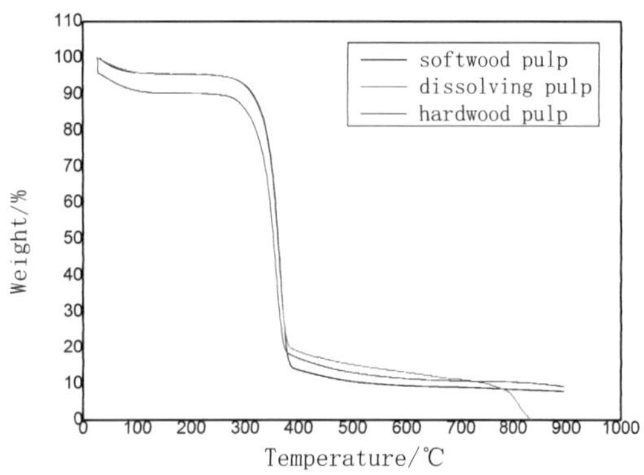

Figure 1. The thermogravimetric analysis chart of different pulp

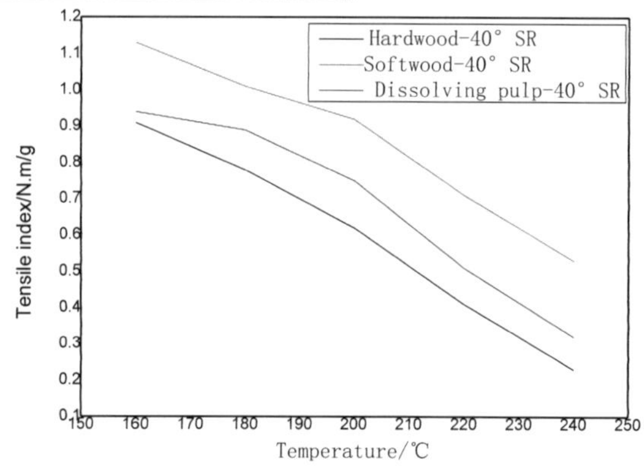

Figure2. The strength changes of papers after treated 60min at different temperature

According to the above table1, table 2 and table 3, we can draw conclusions. Softwood, hardwood and dissolving pulp have similar trends in some physical properties, as the beating degree increasing, tearing index and tensile index increased first and then decreased. This is because the fiber swell and fibrillate faster at the early stage of the beating, make the fiber specific surface area increases, more free hydroxyl, promote hydrogen bonding between the fibers and make the adhesion strength of the fibers increased, and cause paper strength increases. The main factors influencing the intensity of this stage is fiber bonding force. When the strength reaches the maximum, continue beating, although fiber bonding force continues to improve, but with the decrease of the average length of fiber, strength began to fall. In the last phase of the beating, the main factors influencing the strength is the average length of the fiber. The strength index of softwood is the largest, followed by hardwood, and the minimum dissolving pulp. The permeability, absorption, thickness reduced with the beating degree increased; mainly because with the improvement of beating degree, the hydrogen bonding and fiber bonding increase, and reduce the size and number of holes in paper, make the absorbability and the air permeability decreased. Compared with hardwood and softwood, dissolving pulp has good air permeability, absorption and high whiteness, but low strength. However, the lithium battery separator paper not only needs high strength but also needs good air permeability and absorbability, thus softwood and dissolving pulp are good raw materials preparation of lithium battery separator paper. Various physical properties,we determine the softwood and dissolving pulp best beating are 40°SR.

From the figure 1 we can be observe the change trend of each kind of pulp three curves are almost the same. The three curves are gradually decreased with the increase of temperature, the change trend is more stable before 300°C, the main reason is the decrease of the moisture content of paper page. The turning point is about 300°C, the pace of declines increased significantly in the 300-400°C, after 400°C the change trend is smaller, this is demonstrated a sharp reduction in the quality at 300°C-400°C. It can be concluded that the maximum bearing temperature of plant cellulose fibers is about 300°C, and it has good thermal stability properties before 300°C.

Figure2 shows the strength changes of 40°SR softwood, hardwoods and dissolving pulp fiber after treatment 60min at different temperature, it can be seen from the figure, the three kinds of fibers sheet strength decreased with the temperature increasing, compared to the three kinds of fiber strength, softwood largest, followed by dissolving pulp and hardwood minimum. With the increase of treatment temperature, the main component of paper (pulp cellulose ingredient), under the action of heat, amorphous region of cellulose react thermal oxidative degradation, the degree of polymerization of the cellulose molecules is reduced, which leads to decrease of adhesion between the fibers, reducing the strength of the fiber itself, the tensile strength of the paper decreased.

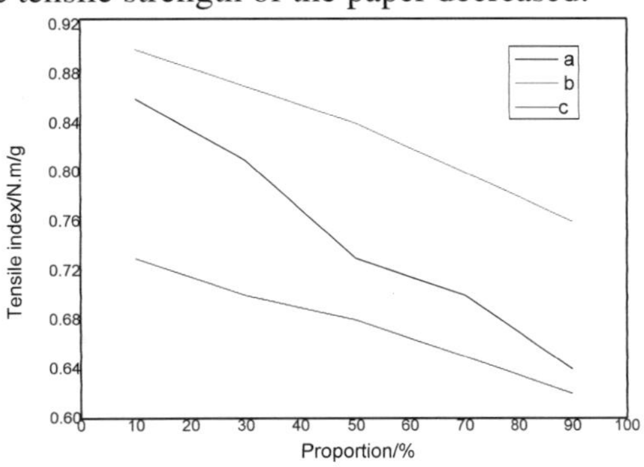

Figure3. The thermal stability of plant fibers
(a)softwood: hardwood =9:1, 7:3, 5:5, 3:7, 1:9;
(b)softwood: dissolving pulp =9:1, 7:3, 5:5, 3:7, 1:9;
(c)dissolving pulp: hardwood =9:1, 7:3, 5:5, 3:7, 1:9

Figure3 shows the sheet tensile index of different plant fiber at different ratio after 60 min treatment at 200°C, it can be seen from the figure, softwood and dissolved pulp sheet has the largest tensile index after treated at high temperature, the more softwood the more tensile index.

Table 4 shows the physical properties of softwood and dissolving pulp paper at different ratio. From the table, softwood pulp and dissolving pulp paper has the high strength of the softwood , and good permeability, the excellent absorption of dissolving pulp, prepared successfully lithium battery separator paper. with good thermal stability and permeability ,as well as good absorption. Integrated all kinds of physical properties, when the softwood： dissolving pulp =5:5, the separator has comprehensively excellent performance .

Table 4 The physical performance test of different proportion of softwood and dissolving pulp

Proportion	Tearing index （mN.m²/g）	Air permeability （μm/(pa.s)(mL/min)）	Thickness （mm）	Absorba-bility （cm）	Tensile index （N.m/g）
Softwood 90%	8.6	13.54	0.13	1.2	6.4
Softwood 70%	8.2	20.76	0.12	1.5	5.2
Softwood 50%	7.8	28.17	0.11	2.0	4.7
Softwood 30%	7.3	30.09	0.11	3.3	3.1

4. Summary

This experiments with ordinary softwood and hardwood as well as dissolving pulp as raw material, through the use of different beating degree of plant fibers to make papers, and then the sheets were treated at different temperatures for some time to detect the thermal stability of the common pulps; then we test the thermal stability of the paper that under the best beating degree at different proportions of plant fibers after treatment at a high temperature.Through analysis data we can draw the following conclusions: Softwood pulp, hardwood pulp and dissolving pulp have excellent physical properties in the 40°SR, and thermal stability up to 200°C in the process of heat treatment. The paper sheets are made using different ratio of softwood pulp, hardwood pulp and dissolving pulp, then test the physical properties. According to the comprehensive analysis of experimental datum obtained that the sheets made using mixed pulp (40°SR dissolving pulp: 40°SR softwood =5:5) have excellent physical performance, and sheet strength index decreased slightly at 200°C,compared to other ratios.

References

[1] T.H. Cho, et al.Battery performances and thermal stability of polyacrylonitrile nano-fiber-based nonwoven separators for Li-ion battery[J].Journal of Power Sources, 155-160(2008)
[2] X.x.h.g.c. Huang.Performance evaluation of a non-woven lithium ion battery separator prepared through a paper-making process[J].Journal of Power Sources, 96-101(2014)
[3] F. Jiang, et al.Bacterial cellulose nanofibrous membrane as thermal stable separator for lithium-ion batteries[J].Journal of Power Sources, 21-27(2015)
[4] J.H.Kim,etal.Functionalized Nanocellulose-Integrated Heterolayered Nanomats toward Smart Battery Separators[J].
[5] G.E. Blomgren.The Development and Future of Lithium Ion Batteries[J]. 5019-5025(2017).

Institute of Physics
Dirac House, Temple Back
Bristol BS1 6BE UK

ISSN: 1757-8981
ISBN 978-1-5108-5198-6

CURRAN ASSOCIATES INC.
proceedings.com